Automedica, 1989, Vol. 11, pp. 1–3
Reprints available directly from the publisher only
Photocopying permitted by license only

EDITORIAL

"MODIFY THE DISABILITY—NOT JUST THE ENVIRONMENT"

CHANDLER A. PHILLIPS

I have recently returned from California where I attended the "Disabilities Expo 88" at the Los Angeles Convention Center. Among the various marvels of technology for the wheel-chair disabled were stair-climbing wheelchairs, self-raising and lowering kitchen cabinetry, and even a completely accessible "dude ranch" experience. At the same time, as a guest of Dr. Paul Berns of the National Spinal Cord Injury Association, I was part of a small booth (among the over two hundred exhibitors) in which we had spinal cord injured people up and walking with an orthotic (bracing) system developed by Dr. Roy Douglas and used at the Peers Spinal Injury Program in Los Angeles. I had a young man, a C6/7 level quadriplegic, walking with electrical muscle stimulation and bracing. The system is described in an article which appears on pages 247–261 of this issue.

As these "disabled" persons walked erect and upright among their wheelchair bound colleagues and took long, confident strides past exhibits extolling the latest technological virtues of yet another "new" wheelchair, I was struck by the paradox of it all.

What a majority of these paralyzed people were really looking for was an alteration of their disability so that they could more normally function (in an unaltered environment). What over 95% of the exhibitors were offering was an alteration of the environment so that they could more normally function (with an unaltered disability).

One association of spinal cord individuals has tersely condensed these observations to simply "cure not care". But they also miss the mark!

For in their zeal to modify their disability, they go to the extreme of complete removal of their disability. The quest for "neural regeneration" of the injured spinal cord is lauditory. But it won't happen tomorrow, and more important, it isn't here today. Not, at least, at the human species level. For these spinal cord injured individuals to continue to wait expectantly for their "cure" (and ignore modern prosthetic advances!) is just as foolish as the amputee who eschews the artificial arm, hand, leg, foot or other body part while waiting for "limb regeneration".

What is here today are modern prosthetic devices ranging from internal devices (ligaments, bones, knees, hips, etc.) to external devices (the Seattle foot, the Utah arm, etc.) and most recently, neuroprostheses (cochlear implants, functional electrical stimulation, etc.). The decade of the 80s has seen tremendous development in these areas, and in one area especially—extremity prostheses.

The neuromuscularly disabled as well as the amputee are benefiting from modern technology which is restoring lost extremity function and lost extremities through a variety of modalities. But what makes these upper and lower extremity prostheses "effective"? What separates them from their older counterparts such as "peg legs" and "electrical splinting"? This is the question that we shall address in this issue.

In editing this issue of *Automedica*, I have somewhat arbitrarily divided the topic of "Effective Upper and Lower Extremity Prostheses" into four areas. These areas are further subdivided as follows (with the page numbers of their respective papers in this issue given in parentheses):

Although the material is somewhat diversified, I feel that this particular order represents the general thematic progression. I am indebted to each of the participating authors for their excellent contributions.

These articles summarize and commemorate much of our technological progress during the decade of the 80s. More important, they point the way to our future directions during the decade of the 90s.

I hope that the following articles will stimulate you to appreciate how much has already been accomplished toward "Effective Upper and Lower Extremity Prostheses". More than that, I hope that they arouse the "Divine Discontent" within you—so much more needs to be done!

Chandler A. Phillips
July 21, 1988
Dayton, Ohio, USA

REFERENCES

1. *Electrical Current Distribution Analysis for FES of Paraplegics* by N. Debnath, S. Anand and S. K. Guha.
2. *Magnetic Neuromuscular Stimulation in Humans* by W. V. Ellis.

3. *Transcutaneous Transducer Garments: An Advancement in Transcutaneous Delivery and End User Compliance* by H. Granek and M. Granek.
4. *Electrocutaneous Sensations Elicited Using Subdermally Located Electrodes* by R. Riso, A. R. Ignagni and M. W. Keith.
5. *Characterization of the Lower Limb of Quadriplegics for the Design of a Controller for Functional Electrical Stimulation* by B. N. Ezenwa, D. B. Reynolds, B. A. Rowley and P. T. Danset.
6. *A Flexor Tendon Model of the Hand* by D. J. Giurintano and D. E. Thompson.
7. *A Preliminary Study of Rhythmic Movement* by H. Hemami, J. S. Bay and J. B. Evans.
8. *Dynamics and Control of A Muscle-Load System During Electrical Nerve Stimulation* by P. H. Veltink and J. A. van Alste.
9. *Simulation and Modeling of a Microcomputer Controlled Above Elbow Prosthesis* by J. S. Philippe-Auguste, D. T. Gibbons and M. D. O'Riain.
10. *A Mathematical Model to Analyze Control of Upright Posture and Balance* by J. R. Buhrman and C. A. Phillips.
11. *Control Strategies with Application to Limb Prostheses* by D. W. Repperger, J. W. Frazier and C. Goodyear.
12. *Human Performance with Prosthetic Devices and Surgically Modified Skeletal Elements* by R. Seliktar, J. Mizrahi, T. Vachranukunkiet, M. Besser and D. Kuenzig.
13. *CNS Strategies in Human Gait: Implications for FES Control* by D. A. Winter.
14. *Rule-Based Control of a Hybrid FES Orthosis for Assisting Paraplegic Locomotion* by B. J. Andrews, R. W. Barnett, G. F. Phillips, C. A. Kirkwood, N. Donaldson, D. N. Rushton and J. A. Perkins.
15. *Foot Controlled Artificial Arm with Complimentary Activation by Body Power* by S. K. Guha, S. Anand and J. Sharma.
16. *A Master-Slave Type Multi-Channel Functional Electrical Stimulation (FES) System for the Control of the Paralyzed Upper Extremities* by N. Hoshimiya and Y. Handa.
17. *A Portable Multi-Channel FES System for Restoration of Motor Function of the Paralyzed Extremities* by Y. Handa, K. Ohkubo and N. Hoshimiya.
18. *A Practical Approach to Multifunctional Above Knee Prosthesis* by S. K. Guha and K. K. Chaudhry.
19. *An Interactive System of Electronic Stimulators and Gait Orthosis for Walking in the Spinal Cord Injured* by C. A. Phillips.
20. *Locomotion of Paraplegic Patients by Functional Neuro-Stimulation* by J. Holle, H. Thoma, M. Frey, H. Kern, W. Mayr, G. Schwanda and H. Stoehr.

Automedica, 1989, Vol. 11, pp. 5–13
Reprints available directly from the publisher only
Photocopying permitted by license only

ELECTRICAL CURRENT DISTRIBUTION ANALYSIS FOR FUNCTIONAL ELECTRICAL STIMULATION (FES) OF PARAPLEGICS

N. DEBNATH, SNEH ANAND and SUJOY K. GUHA

Centre for Biomedical Engineering, Indian Institute of Technology and All India Institute of Medical Sciences, New Delhi 110016, India

Using surface electrodes for functional electrical stimulation of paraplegics, the importance of the current density distribution in the three layer thigh model is highlighted. The electrode placement for optimal selection of a group of muscles in respect of restoring movement is described. Two, three or four electrode configurations have been suggested for obtaining knee flexion, extension, adduction and abduction independently.

Keywords: Functional electrical stimulation, current density distribution, relative intensity, muscle group, movement.

1 INTRODUCTION

Functional electrical stimulation (FES) of limbs to obtain movement in cases of neurological paralysis is a rapidly developing field of research. The feasibility of obtaining muscle contractions with electrical current stimulation at a level which can be tolerated has already been demonstrated. The task ahead is to obtain selective activation of different muscle groups in the sequence desired. Since stimulation with surface electrodes is the method of choice, difficulties arise in estimating the current passing through specific zones deep within the limb. So far a more or less empirical approach has been taken [1–3]. Because of the electrically nonhomogeneous and also nonisotropic character of the body, an intuitive estimation of current distribution generally gives inadequate results. A systematic study is required. Probably owing to the inadequate quantification of the current distribution generally the simplest excitation mode, that is the bipolar electrode system, has been used. Experience shows that adequately selective stimulation of muscle groups is not obtained. Therefore, multielectrode configurations need to be considered and the current distribution patterns for such systems are to be determined to arrive at optimal excitation parameters.

It is possible to take a totally experimental approach in the study. In a subject, probes can be inserted into the muscles at different locations to measure potential gradients on the application of the external electrical field and from this data infer the current densities based upon an assumption regarding the tissue resistivity. In humans and even in animals such a measurement can be performed at very few points and so the perspective of the current distribution cannot be obtained. Also the assumption regarding the resistivity precisely at the probe tip may not apply. On account of these limitations of the experimental approach a simulation study is required to obtain an understanding about the interrelationships of the variables involved.

Stimulation current density at a particular region depends upon the location of electrodes; shape of the electrodes; orientation of the muscles in the thigh; thickness and electrical conductivities of the contents in the region besides the frequency and strength of stimulus. For the complex geometry of the human limbs it is not practicable to deal with all these factors with classical analytical mathematics. Numerical methods give more flexibility. Hence, all the variables have been incorporated in a simulation model based upon the Laplace field distribution equation and solved by a numerical technique. With the model then various combinations of excitation modes can be studied.

2 THEORETICAL MODEL

The Laplace equation has already been applied to obtain the current distribution in brain tissue with the source being internal [4]. Also, a similar analysis has been carried out for the thorax with the electrical source being external [5]. The same basic direction is being taken in the present analysis with the difference that multiple sources are considered and the body segment analysed is the thigh. While earlier investigations had taken into account skin, muscle and bone, the anatomy of the thigh makes it mandatory to include fatty tissue in the analysis since, in the thigh, especially in women, there is enough fat layers to become an important determinant of the current distribution.

The problem requires that the Laplace equation for the potential V in the conduction current field, that is

$$\text{del squared } v = 0 \tag{1}$$

be solved for the region of the thigh. The geometry is too complex to be considered in all its detail. As an approximation a representative section of the thigh [6] was approximated by straight line boundaries and some of the regions of similar tissue properties have been lumped together (Figure 1). To solve the Laplace equation by the finite difference method the region has to be divided into meshes around nodes. A large mesh size leads to loss in detail while a very small mesh size increases computation time enormously. Muscle, bone and fat layers are quite thick and can be handled with a coarse mesh. Skin is relatively thinner. Setting the mesh size to the skin thickness dimension makes the total number of nodes extremely large. For this first estimate study the skin layer has not been considered and a mesh size of 10×10 mm was adopted. The thigh in the mid region above the insertions of the adductor magnus muscles is fairly uniform in structure along the axial direction. As an approximation two-dimensional analysis would serve to determine the current distribution pattern so long as the excitation electrodes cover the entire segment uniformly in the axial direction. Adding additional layers to the analysis will not alter the relative current densities in a plane as long as the above mentioned constraints on the electrode position are maintained. On the other hand, the absolute magnitudes of the current density will be affected by additional planes since the uppermost and lowermost planes will have air as the media on one side. Quantitatively, the effect on the midplane current density on account of the air boundaries will be a maximum when there are three layers in consideration and the deviation will decrease as the number of layers in the analysis is increased. In order to obtain an estimate of this extreme case in the present study three planes have been taken into account. Mesh interval axially which is the z direction was taken as 10 mm to simplify the formulation of the numerical form of the Laplace equation. In this way there are about 185 nodes in each plane and a total of 555 nodes in the system.

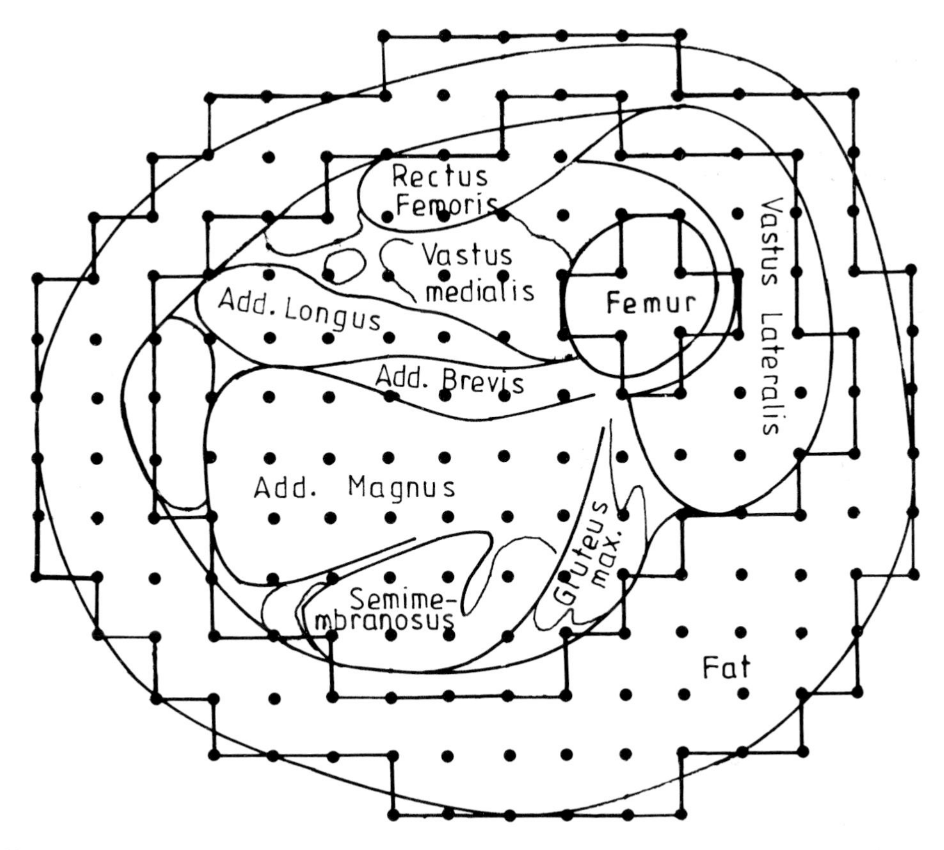

Figure 1 Thigh section approximated by steps and straight lines at the outer boundaries and at the interfaces.

Labelling the nodes as $0(i, j, k)$ where i, j, k are the node positions in the x, y and z directions respectively (Figure 2) the numerical equivalent of the Laplace equation is [7],

$$\begin{aligned} V(i, j, k) = [&V(i+1, j, k)(P+Q+U+V) + V(i, j+1, K)(R+Q+U+V) \\ &+ V(i-1, j, k)(S+R+X+W) + V(i, j-1, k)(S+P+U+X) \\ &+ V(i, j, k+1)(P+Q+R+S) + V(i, j, k-1)(W+U+V+X)] \\ &/[6.0(P+Q+R+S+W+V+U+X)], \end{aligned} \tag{2}$$

where P, Q, R, S, W, X, U, V represent the conductivities of the different regions. The regions have fat, muscle and bone with conductivities in the ratio 0.4, 1.7, 0.5 respectively. On the upper side of the uppermost layer and the lower side of the lowermost layer there is air with zero conductivity. Beyond the outer margin of the model in the plane of the paper the medium is air. At each node the eight conductivity values of the cubes around the nodes have to be appropriately taken. The other important boundary condition is the potential on account of the excitation inputs and these are indicated in the individual diagrams of current density patterns.

The set of simultaneous equations, one for each node, was solved by the Gauss-Siedel iteration method with Young's relaxation [8]. A relaxation factor of 1.65 gives rapid convergence. Computation was carried out on an ICL 2900 computer and the computer time for each analysis was about two minutes. From the potential difference values the current densities were calculated by multiplying the potential gradient by the conductivity.

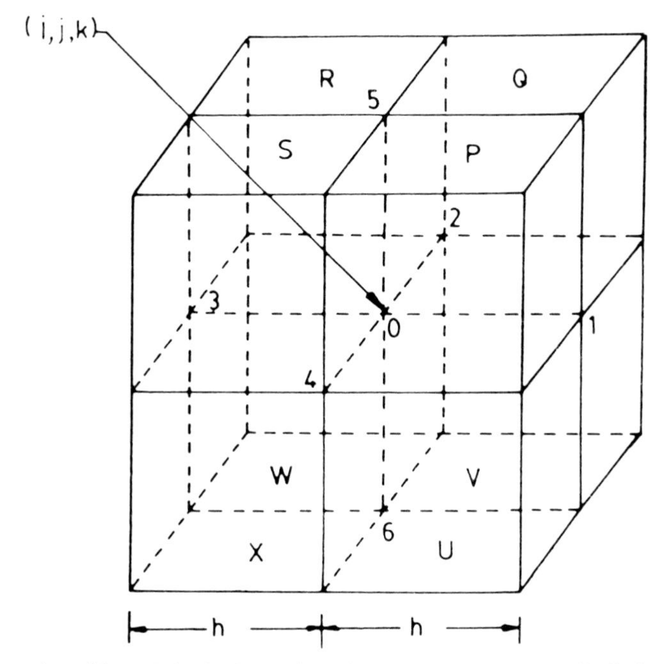

Figure 2 Location of boundaries in three dimensions. All eight quadrants *P, Q, R, S, U, V, W* and *X* have different conductivity and *h* represents the mesh interval.

3 OPTIMIZATION OF ELECTRODE CONFIGURATIONS

In functional electrical stimulation of paraplegics for locomotion the movements of importance are extension, flexion and adduction. Here, only the extension and adductor groups are being considered and the muscles are:

—Quadriceps (Rectus femoris, vastus lateralis, vastus intermedius and vastus medialis)

—Semitendinosus

—Adductor magnus, longus and brevis

The objective is to stimulate these groups selectively as best possible.

In the first instance two-electrode stimulation was given. On the lateral side near the region of the vastus lateralis a potential of zero volts is assigned and almost diametrically opposite on the medial side a potential of 10 V was assigned. The absolute value of these potentials are not important since relative current densities are to be determined. The current densities have been grouped into six levels: below 0.4, 0.4–0.59, 0.6–0.79, 0.8–0.99, 1.0–1.2 and above 1.2. Once again it is the relative magnitudes which are relevant and so no units are ascribed. Figure 3 shows the current density distribution. There is distinct asymmetry even in the regions near the two electrodes. Current density immediately in the vicinity of the electrodes is not the highest because this layer is predominantly fat which has a low conductivity. The relative distance of the bone from the surface on the lateral and the medial sides is quite different and this factor too contributes to the inequality on the two sides. The most significant feature is that the current density in the region of the rectus femoris is very low. Even though the vastus lateralis portion does carry some current concentration with the low current in the rectus femoris area the extension effect will be low. So the flexor group also has low density. This electrode placement is not appropriate for obtaining the extension and flexion movements.

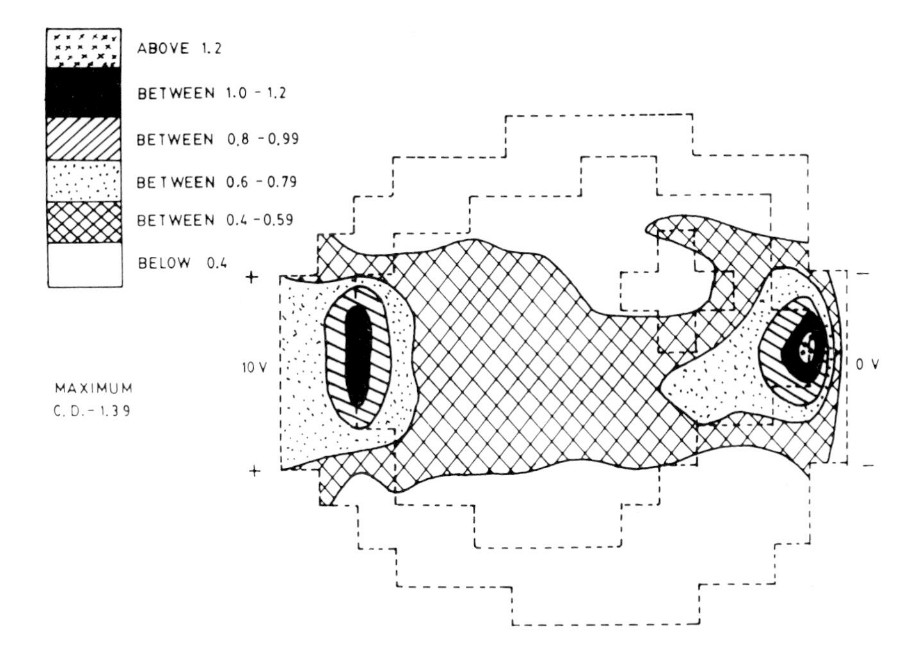

Figure 3 Showing the relative current density (CD) distribution in the thigh. Excitation electrodes are placed across the thigh and all muscle groups are simultaneously stimulated.

Next, potentials on the medial and posterior sides were ascribed as indicated in Figure 4. In this case the adductor magnus and the semitendinosus area, adductor magnus and gracilis areas receive current while the extensor area is not activated. This placement is then proper to selectively cut out extensor action of the knee but flexion of the knee, adduction and some degree of medial rotation can be obtained.

Figure 5 indicates placement of two sets of electrodes. The sources for the two are electrically isolated and so there is no electrical shorting link modelled between the two zero volt electrodes and the two 10 V electrodes. In this case the rectus femoris area is strongly excited and the semitendinosus area also receives substantial current. The tendency will be to stiffen the leg with possible abduction. Another dual excitation shown in Figure 6 gives high densities in all quadrants.

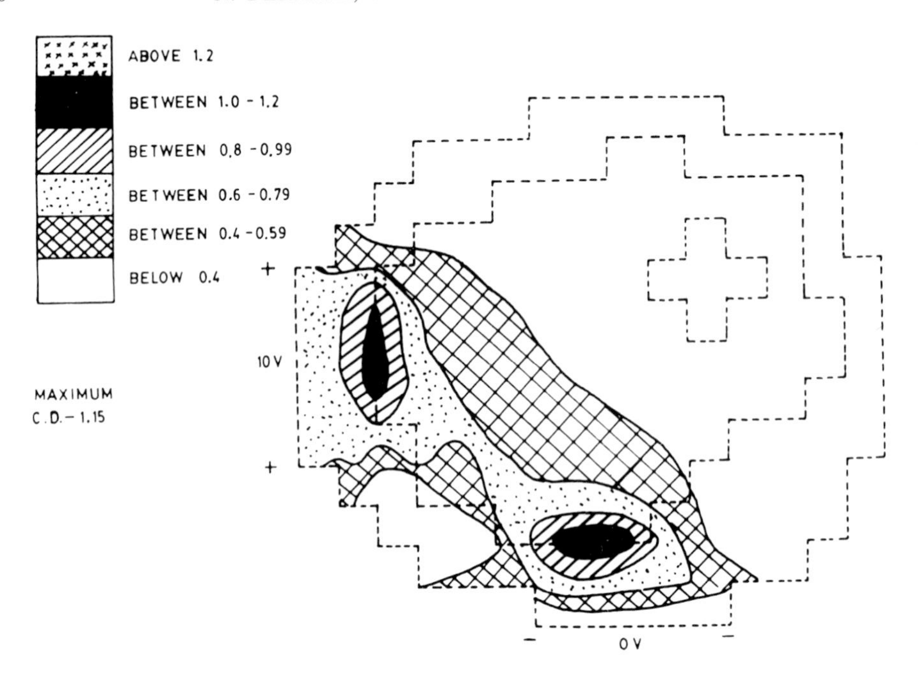

Figure 4 Relative current density levels on the medial and posterior sides of the thigh selectively stimulating the abductus magnus and semitendoinous area. This cofiguration is suitable for knee flexion adduction and medial rotation.

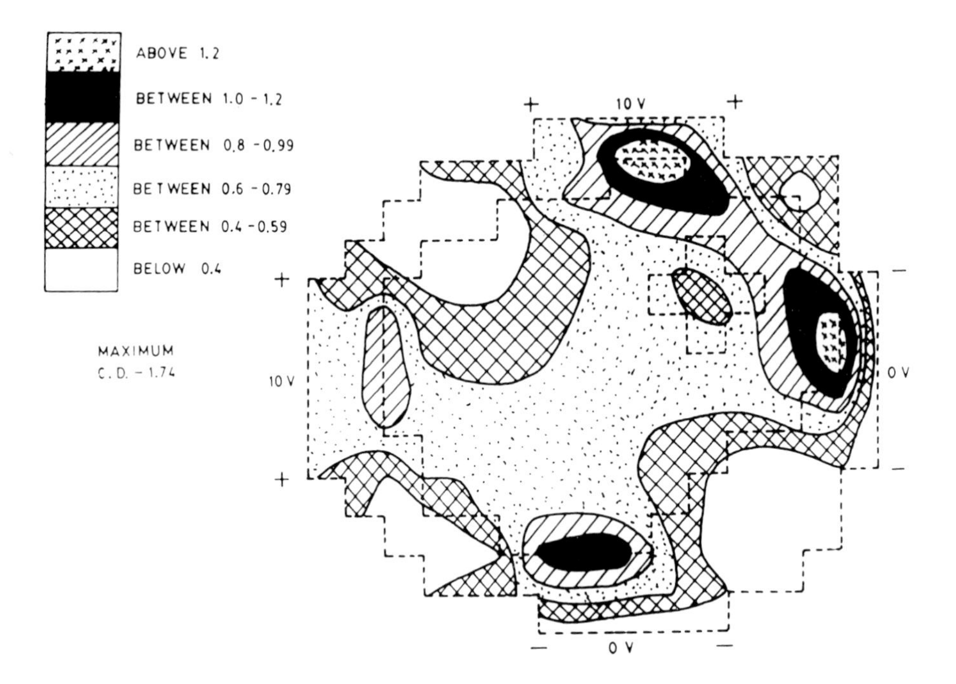

Figure 5 Indicates the placement of two sets of electrodes for specific stimulation of rectus femoris strongly and semitendinosus partially causing the stiffening of the leg with possible abduction.

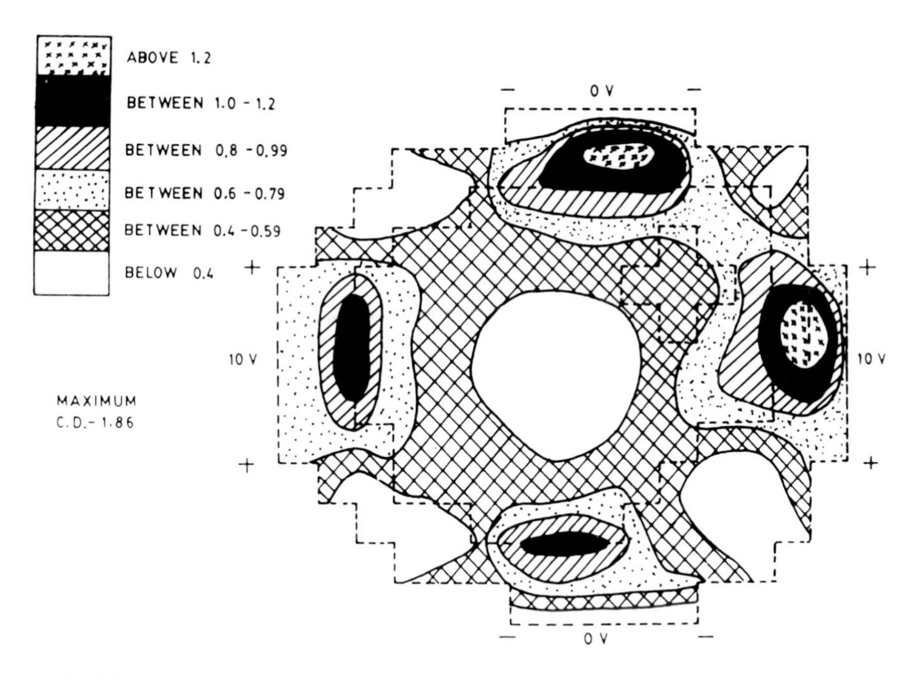

Figure 6 Showing the placement of two sets of electrodes for high current densities in all quadrants, appropriate to fix the leg in extension for a standing posture.

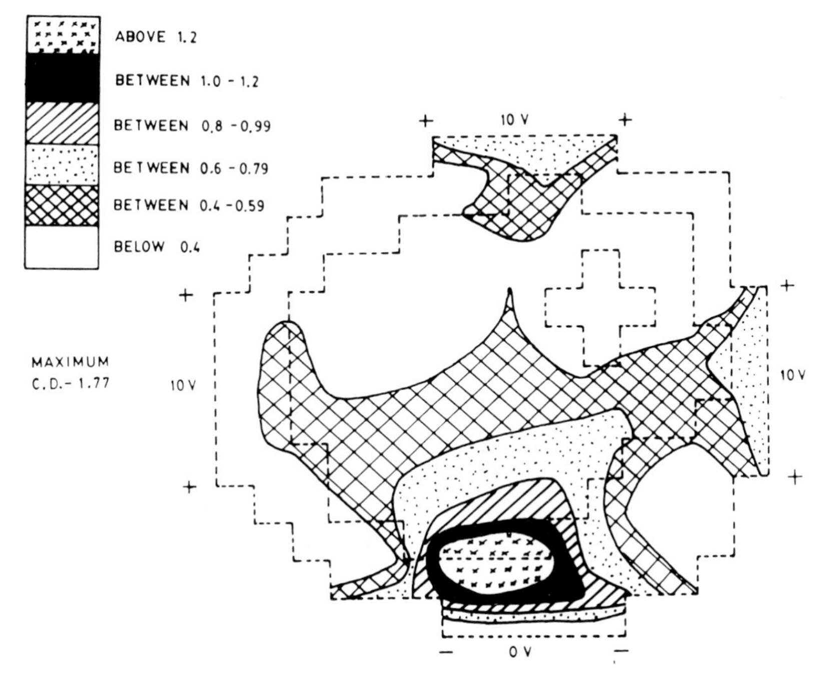

Figure 7 Indicating the placement of two sets of electrodes for low stimulus in the entire anterior compartment for knee flexion.

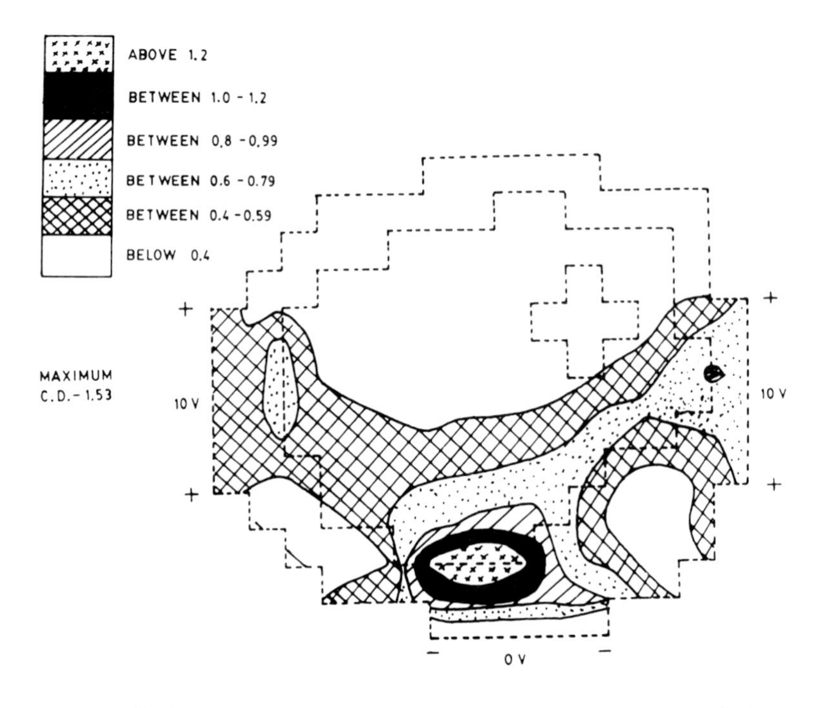

Figure 8 Showing the placement of three electrodes suggested for knee flexion.

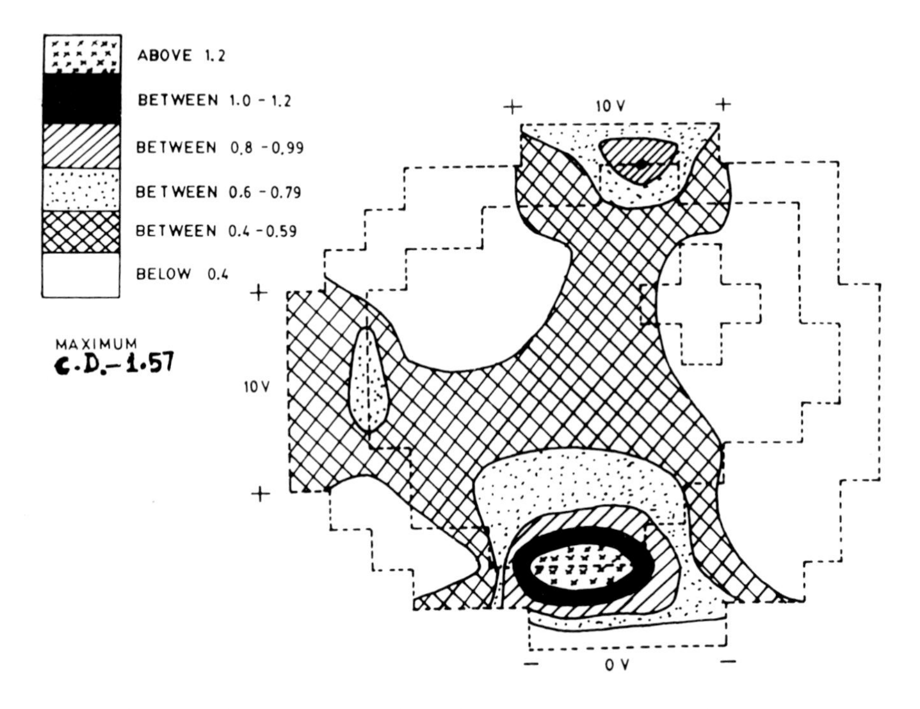

Figure 9 Three electrodes configuration suitable for adduction of the leg.

Such an excitation will then be appropriate to fix the leg in extension as is required to maintain a standing posture. A change in placement (Figure 7) alters the picture markedly with the entire anterior compartment receiving low stimulus. This mode can be of value in obtaining knee flexion. More specificity for flexion is obtained with three electrode mode (Figure 8) where the rectus femoris receives very low current and even a large part of the vastus lateralis is not stimulated. In this mode the zero potential points of the two sources are joined but not the 10 V points. Finally the configuration of Figure 9 with three electrodes was assessed. This placement is more specific to adduction but not very much so.

4 CONCLUSIONS

In the present study, the effect of a single source as well as two isolated sources have been studied. The assumption is that the two sources are in phase. It is clear from the data that marked changes in the current density pattern can be obtained by changing the electrode placements. Specificity to extension and flexion can be realised. Adduction activity cannot be achieved in an isolated manner. It has, however, to be noted that whereas in respect of flexion and extension the reference is to the knee joint, in respect of flexion it is to the hip joint. Hence, in practical terms there is an isolation. Nevertheless, fairly simple combinations have been investigated. By introducing phase shifts between the sources and a greater number of sources with smaller electrode size it may be possible to effectively obtain sequential activation of muscle groups in a more desired manner.

Analysis also shows that the current density in the central plane is lower than in the two bounding planes bounded by air medium. In actual practice the difference will not be as much as is indicated by the theoretical model since although there will be tissue above and below the electrode, it is still important that electrodes should not be placed just above or below bony areas such as the hip and knee joints since bone has a low conductivity. The presence of high conductivity synovial fluid in the joint will not compensate for this effect.

A number of assumptions have been made in the formulation of the model. Most of these approximations will not materially alter the relative current density pattern even if they were not made. Ignoring the skin layer, however, does introduce errors since the skin helps to distribute the current. A more detailed study is therefore required to assess this factor.

REFERENCES

1. P. E. Crago, P. H. Peckham, J. T. Mortimer and J. P. Vander, The Choice of pulse duration for chronic electrical stimulation via surface, nerve and intramuscular electrodes, *Ann. Biomed. Eng.*, **2,** 252 (1974).
2. M. Milner, A. O. Quanbury and J. V. Basmajian, Surface electric stimulation of lower limb, *Arch. Phys. Rehabil.*, **51,** 540 (1970).
3. H. E. Nelson, M. B. Smith, R. L. Waters and B. Browman, Electrode Comparison in FES, *Ibid*, **58,** 519 (1977).
4. J. G. Witwer, G. J. Trezekk and D. L. Jewett, The effects of media inhomogeneities upon intracranial electrical fields, *IEEE Trans. Biomed. Eng.*, **19,** 352 (1972).
5. S. K. Guha, M. R. Khan and S. N. Tandon, Electrical field distribution in human body, *Physics Med. Biol.*, **18(5),** 712 (1973).
6. J. E. Anderson, *Grant's Atlas of Anatomy*, Williams and Wilkins, Baltimore/London (1984).
7. D. Vitkovitch, *Field Analysis Experimental and Computational Methods*, D. Van Nostrand Co. Ltd., London (1965).
8. D. Young, Iterative methods for solving partial differential equations of elliptic type, *Trans. Amer. Math. Soc.*, **92.** 72 (1954).

Automedica, 1989, Vol. 11, pp. 15–18
Reprints available directly from the publisher only
Photocopying permitted by license only

MAGNETIC NEUROMUSCULAR STIMULATION IN HUMANS

WLADISLAW V. ELLIS

2445 Carleton, Berkeley, CA 94704, U.S.A.

(Received September 1988)

Rapidly varying, high-intensity magnetic fields have been shown to depolarize neural tissues in man, as well as in other mammals. Clinical magnetic stimulation evolved during the past decade in England and has aroused interest in the United States during the last several years. A historical summary, current research, and potential applications will be the objective of this article.

Keywords: Magnetic stimulation, neuromuscular activation.

The stimulation of nerves and muscles has been of paramount importance in furthering our understanding of physiology. Long-term human stimulation either has required surgical procedures with their attendant risks or has proven too painful to be practicable. My own work in Functional Electrical Stimulation (FES) has made me acutely aware of this last problem, and I was pleased to discover that magnetic stimulation of neuromuscular preparations has been successfully accomplished by a number of groups in England fairly recently. The attractive feature of this new modality is that it evokes no painful sensations. Additionally, it has the ability to stimulate at depth, as well as transcranially, opening up the possibility of direct, non-invasive modulation of the central nervous system (CNS).

HISTORY

The clinical application of magnetism has a long medical history dating back, anecdotally, probably to the discovery of the first loadstone and of action at a distance by an invisible force. The magical antecedents can be best seen in the circumstances surrounding Mesmer's coining the word "animal magnetism" as a description for the psychological phenomena he observed.

Popular interest in the clinical use of magnetism has waxed and waned rather regularly for the last several centuries. Currently, there has been a minor resurgence in the use of small, constant-field magnets in the treatment of a variety of disorders, primarily in traditional Oriental medicine.

The first rigorously described effects of magnetic fields occurred at the turn of the century with d'Arsonval in 1896 and Thompson in 1910 [1, 2]. Thompson described a flickering nimbus that consistently appeared when he, or a number of other volunteers, stuck their heads between the poles of a large, rapidly oscillating magnet. This was later thought to be due to retinal excitation.

Since then, interest in the direct effects of magnetic fields has been desultory until the last decade. This renewed interest has split roughly into two major groupings: those involved with low-intensity and those involved with high-intensity magnetic fields. The former group is perhaps best represented by Becker and

Delgado, who demonstrated the biological effects of pulsed, weak magnetic fields in radically different experimental systems.

Bassett, capitalizing on earlier observations using direct, extremely low-current electrical fields in regeneration, showed conclusively that weakly pulsed magnetic fields promote healing in bony non-fusions [3]. The presumed mechanism is the direct induction of pico amp. current flow in the area of injury. This technique has been available commercially for some time and has shown consistently productive results in select cases [4].

Delgado, using magnetic fields in the micro-Tesla range, showed that they produced teratogenic effects in chick embryos [5, 6, 7]

High-intensity magnetic stimulation has been developed in England with Bickford and Fremming reporting peripheral nerve excitation in 1965 [8]. It was not until 1982, however, that Polson *et al.* first reported motor action potentials as a result of magnetically induced peripheral nerve stimulation [9]. Barker subsequently cooperated with Merton and obtained clear-cut motor responses to cerebral transcranial brain stimulation in May, 1985 at the National Hospital, Queens Square, England. This technique has since been found to be valuable in the diagnosis of long motor tract disturbances with a number of studies corroborating its effectiveness [10].

American interest arose subsequently with a number of laboratories using magnetic stimulation clinically. As well, a major manufacturer of this device has been based in the state of Washington. The University of New Jersey sponsored the first International Symposium on Magnetic Stimulation in Clinical Neurophysiology on 20 November 1987 [11].

Other areas of interest involving magnetic fields would include magnetic field-induced drug permeability, magnetoencephalography (especially in epilepsy), electromagnetic stabilization of fluids, and the study of animal magnetoreception [12–15].

CURRENT RESEARCH

Evidenced best perhaps by the contents of the above-mentioned symposium, current efforts have been focused on determining baselines of information regarding long tract conduction times in a variety of diseases, on exploring the safety of the method, and on elucidating the mechanisms of activation. The conclusion emphasized the usefulness of this methodology in evaluating deep motor nerves, such as the femoral, sciatic, cervical and lumbosacral roots, as well as in assessing central motor conduction time.

Work from a number of groups has substantiated the following observations: Sensory and muscular responses to magnetic stimulation are largely a matter of the intensity of the magnetic flux; muscular responses are primarily a function of the pulse magnitude; and the appropriate geometry is necessary for the event to take place [16, 17]. All investigators describe minimal sensation with the complete absence of pain. The actual conduction time is longer than with electrical stimulation, but, interestingly, willful recruitment of the target muscle reinforces the magnetic stimulation [18]. Clinical studies have documented abnormalities in cervical myelopathy, radiation myelopathy, hereditary spastic paraparesis, as well as in multiple sclerosis. Responses to magnetic stimulation have been of a large amplitude, longer latency, and a shorter duration, suggesting more homogenous excitation than with direct electrical stimulation [10].

The limitations of current technology accentuate the uncertainty about the actual geometry of the magnetic lines of force as they course through the tissues, which results in a very vague sense of the actual localizability of the stimulation. Repetition rates of less than 3 Hz preclude neuromuscular summation.

FUTURE DIRECTIONS

Given the fairly recent emergence of this technology, current efforts, as best I know, are aimed at extending baselines of applicability. Its current use lies primarily in investigating conduction pathways and long motor tracts.

My own goal is to develop the methodology so that one will be able to selectively and non-invasively stimulate the CNS as well as peripheral nerve without undue pain. The major immediate limitation is current technology, which uses capacitative discharge via high-power thyristors to discharge into a variety of coil shapes [19]. This perforce limits the stimulation to very low frequencies and, although useful in neurodiagnostics, does not have the capacity either to ensure muscular contractions against resistance or to stimulate deep-seated CNS structures. I am currently working on technology that would allow for a higher magnetic flux, as well as for repetition rates up to 200 Hz. I envision that, with the use of multiple programmed electrodes, one can achieve satisfactory stimulation of deep-seated structures, such as the autonomic ganglia, as well as core CNS formations. I also feel that a word of caution is necessary as we are dealing with largely unknown quantities, which have been shown to have an effect on calcium efflux at certain frequency "windows" [20, 21].

In summary, high-intensity magnetic stimulation has been found useful for the evaluation of central motor conduction time, the stimulation of deep-seated peripheral nerves, and holds out the promise of focusable and selective non-invasive, as well as non-painful, stimulation of the nervous system generally.

REFERENCES

1. d'Arsonval, Dispositifs pour la mesure des courants alternatifs de toutes frequencies, *C.R. Soc. Biol. Paris*, **2,** 450–457 (1896).
2. S. P. Thompson, A physiological effect of an alternating magnetic field, *Proc. Roy. Soc. Lond. (Biol.)*, 396–398 (1910).
3. C. A. L. Bassett, R. J. Pawluk and A. A. Pilla, Acceleration of fracture repair by electromagnetic fields, a surgically non-invasive method, *Ann. N.Y. Acad. Sci.*, **238,** 242–261 (1974).
4. R. K. Aaron and D. Ciombor, Treatment of osteonecrosis of the femoral head with pulsed external magnetic fields. The first colloquium in biologic science, *N.Y. Acad. Sci.* (1985).
5. J. R. Delgado *et al.*, Embryological changes induced by extremely weak, extremely low-frequency electro-magnetic fields, *J. Anat.*, **134,** 533–551 (1982).
6. A. Ubeda *et al.*, Pulse shape of magnetic fields influences chick embryogenesis, *J. Anat.*, **137,** 513–536 (1983).
7. T. C. Rozzell, Extremely low-frequency magnetic fields effect chick embryos, *E.S.N.*, **38**(6), 297–301 (1984).
8. R. G. Bickford and B. P. Fleming, Neuronal stimulation by pulsed magnetic fields in animals and man, *Digest of the Sixth International Conference on Medical Electronics and Biological Engineering*, Tokyo (1965).
9. M. J. R. Polson, A. T. Barker and I. L. Freeston, Stimulation of nerve trunks with time-varying magnetic fields, *Med. Biol. Eng. Comput.*, **20,** 243–244 (1982).
10. K. R. Mills *et al.*, Magnetic and electrical transcranial brain stimulation: physiological mechanisms and clinical applications, *Neurosurg.*, **20,** 164–168 (1987).

11. *First International Symposium on Magnetic Stimulation and Clinical Neurophysiology*, given by the University of Medicine and Dentistry of New Jersey, Department of Neurology, New Brunswick, New Jersey, November 20, 1987.
12. R. P. Liburdy and T. S. Tenforde, Magnetic-field-induced drug permeability in liposome vessicles, *Rad. Res.*, **108,** 001–010 (1986).
13. D. F. Rose *et al.*, Magnetoencephalography and epilepsy research, *Sci.*, **238,** 329–335 (1987).
14. C. F. Ivory *et al.*, Electromagnetic stabilization of weakly-conducting fluids, *Sci.*, **238,** 58–65 (1987).
15. J. B. Phillips, Two magnetoreception pathways in migratory salamander, *Sci.*, **233,** 765–767 (1986).
16. D. McRobbie and M. A. Foster, Thresholds for biological effects of time-varying magnetic fields, *Clin. Phys. Physiol. Meas.*, **5,** 67–78 (1984).
17. M. J. R. Polson, A. T. Barker and S. Gardiner, The effect of rapid rise-time magnetic fields on the ECG of the rat, *Clin. Phys. Physiol. Meas.*, **3,** 231–234 (1982).
18. R. Q. Cracco, Evaluation of conduction in central motor pathways: techniques, pathophysiology, and clinical interpretation, *Neurosurg.*, **20,** 199–203 (1987).
19. D. McRobbie, Design and instrumentation of a magnetic nerve stimulator, *J. Physics E: Scientific Instruments*, **18,** 74–78 (1985).
20. C. V. Byus, S. E. Pieper and W. R. Adey, The effects of low energy 60 Hz environmental electromagnetic fields upon the growth-related enzyme, Ornithine Decarboxylase, *Carcinogenesis*, **8,** 1385–1389 (1987).
21. W. R. Adey, Effects of microwaves on cells and molecules, *Nature*, 332–401 (1988).

Automedica, 1989, Vol. 11, pp. 19–23
Reprints available directly from the publisher only
Photocopying permitted by license only

TRANSCUTANEOUS TRANSDUCER GARMENTS®: AN ADVANCEMENT IN TRANSCUTANEOUS DELIVERY AND END USER COMPLIANCE

HERB GRANEK and MURRY GRANEK

Bio-Stimu Trend Corp.™, 14851 N.W.27 Av., Opa-locka, FL 33054, USA

(Received September 1988)

The broad spectrum of electric modalities is reliant upon transference of current through the epidermis via electrodes transcutaneously. Drawbacks associated with conventional electrode systems (e.g. carbon silicon, karaya, polymers, metal discs) are myriad. Technically most electrode systems are prone to intermittent contact, disconnections with normal activity, distortions of transmitted signals and relatively short optimum life spans which can give rise to skewed results, punctuate burns and erythema. The patient may also need "Herculean" dexterity and skill to apply conventional electrode systems properly without assistance especially if they are disabled. Transcutaneous Transducer Garments® are a lightweight non-bulky stretchable prosthesis (garment) incorporating a permanent array of highly conductive durable long lasting electrodes. The electrodes can be made in various configurations with practically no limit to the amount of sites to be stimulated. TTG® are a practical and effective mode of stimulating or monitoring biomedical signals individually or simultaneously.

Keywords: Multiple site stimulation, electrode, TTG®, FES, NMES.

INTRODUCTION

Transcutaneous electro modalities are playing a larger role in progressive medicine. Among the types of electromedical devices are electrocardiogram (EKG), electroencephalogram (EEG), Holter Monitoring, electromyogram (EMG), Transcutaneous Bone Growth, Functional Electrical Stimulation (FES), Transcutaneous Electrical Nerve Stimulation (TENS) and Neuromuscular Electric Stimulation (NMES). The aforementioned devices are reliant on electrodes for conveyance of the electric signal through the skin.

Skin, which is the body's largest organ is composed of two principal layers: (1) the epidermis, which includes the stratum corneum, stratum granulosum, stratum germinatium, and (2) the dermis, which covers the subcutaneous tissue. FES, TENS and NMES devices convey the electrical impulse to the electrodes which then pass the signal through the skin via sense, innervating spinal and cervical nerves. Adverse skin reactions associated with electrodes can be classified into four broad categories:

1) Mechanical.
2) Electrical.
3) Chemical.
4) Allergenic.

Bio-Stimu Trend Corp.™ TTG® are a practical and simple method of alleviating common skin problems arising from transmission of pulsatile signals from FES, TENS, or NMES devices and the application of electrodes. The garments (TTG®)

are constructed of lightweight lycra-spandex fabric with interstices which mold to the body part to be stimulated. Affixed above the fabric are Selectrodes® which is a soft electrode consisting of an impermeable cover, external gel orifice, gel cap, flexible fabric conductor and wicking member. Activation of a Selectrode® site is accomplished by introducing Selectrode® Gel into the Selectrode® gel orifice which can expand to accommodate a reservoir of electrolyte enabling long term transmission. The interconnection between the selectrode and the pulsatile device consists of a conductive rail which is a stretchable non-dangling conductor affixed in a parallel series to the TTG®. This unique feature protects against accidental disconnection to the individual Selectrodes® and allows an unlimited amount of areas to be stimulated from one channel of a device or asynchronous stimulation.

ELECTRODE CHARACTERIZATION AND EFFECTS

It is generally recognized that abrasion of the skin (shaving, adhesives, depilatories) makes the skin more susceptible to irritants. Electrical stimulation can create and exacerbate adverse skin conditions and stimulation should be discontinued until there is healing.

Tam and Webster [1], without using electrical stimulation, compared various conventional electrodes, adhesives, electrolytes and their effects on the forearm over seven days. The electrodes were removed daily, cleansed and reapplied with fresh accessories to avoid probable irritation due to elevated bacterial population. Irritation and reddening appeared on subjects. When the skin was mechanically abraded twenty times and the tests repeated, significantly increased levels of skin irritation manifested themselves including immediate swelling and pain upon application requiring the test to be discontinued. Utilizing custom formulated electrolytes most of the discomfort and negative effects continued and could be attributed to the adhesive carrier. It was deduced that abrasion exposed more nerve endings causing a stronger adverse reaction.

Adhesive tape reactions are attributed to mechanical irritation [2]. The application of adhesive tape or electrodes with binding agents is in itself enough to cause shearing stress. Irritation from adhesives include pruritus due to restricted perspiration, puckering and pain when the adhesive carrier is removed. Stretching of the skin changes its capacitance and its perception of a pulsatile signal. The introduction of a pulsatile signal over an irregularly abraded surface is a main reason patients complain of a stinging sensation from stimulation [3]. This mechanical sheer stress factor is complicated by body movement.

Micropunctuate burns occur when current density is concentrated under a small part of the electrode. This can be attributed to improper electrolyte application, electrolyte evaporation, electrode deterioration, oxidation, or poor interface of electrode and skin because of body contours and movement. Chemical reactions with electrical stimulation and electrodes occurs when there is an imbalance of ion displacement within the tissues [4]. This problem is usually associated with monophasic current of long duration.

Allergic reactions associated with electrodes are most commonly attributed to propylene glycol in the electrolyte [5, 6]. Adhesives used to maintain contact between the electrode and skin are also a significant cause of allergic reactions. An additional complication utilizing adhesive agents directly upon the skin is the abrasion of the epidermal layers leaving the user prone to adverse skin reactions from any electrolyte formula.

Current density patterns of various electrodes have been established [3, 8]. Metal electrodes have higher concentration of current at the periphery. Rubber, carbon silica and disposable electrodes have higher concentration at their core. The consensus holds that an easily applied electrode with more uniform current dispersement will have a wide application in electro-medicine. Proliferate electrode research in regard to function and design has spawned design modification enhancing electrode performance [3, 8, 9–12]. It has been found that electrodes in a convex shape will allow the electrolyte to vary its depth and resistance [13]. Incorporating plastic between an electrode's connector and skin increases capacitance at the core encouraging a more even current dispersement [1]. It has also been found that segmenting of electrodes with external series resistors or multiple connected electrodes in parallel series is a safer dispersion method even with the use of RF current in electro surgers [9, 10].

In a study comparing surface electrodes with electrical stimulation (NMES) the primary goal was to find an electrode that was "commercially available, reasonably inexpensive and easy for the patient to use" [11]. Previously the clinicians were using a hand fabricated electrode consisting of a 3M Myocare™ electrode sandwiched between a sponge and naugahyde cover that was not durable enough. Six neurologically intact volunteers tested the following electrodes:

1) 3M Myocare™ = water activated adhesive surface.
2) Unipatch® Encore # 697 = hypo-allergenic self adhering gel.
3) Hand fabricated electrode = water activated.
4) Medtronic™carbon silica = gel activated.
5) Uni-patch® # 555 = gel activated.
6) Bio-Stimu Trend Corp.™ TTG® = gel activated.

The conventional electrodes were placed at approximately 5.5 and 20.5 cm above the superior border of the patella. The TTG® quadricep garment was configured in the same manner and worn over the leg. Peak force, peak voltage (to determine impedance), subjective discomfort and skin coloration were recorded in response to one of three stimuli at 40 and 60 mA as tolerated. The non-gelled based electrodes (1, 2, 3) produced the highest discomfort levels with impedance as high as 1162 ohms. The water and gel activated conventional electrodes (1, 3, 4, 5) produced slight skin reddening. Reddening was never seen while using the TTG® (6). The TTG® also produced the highest average peak force as measured on a Gould polygraph.

CONCLUSIONS

It has been documented that problems associated with the use of conventional electrodes and their accessories are:

a) Abrading of the epidermal tissue upon application and removal.
b) Sheer stress.
c) Micropunctuate burns, erythema and electrical burns.
d) Allergic reactions.
e) Non-uniform current density distribution across the electrode surface.
f) Inadaptability for asynchronous stimulation.

g) Inability in conforming to anatomy.
h) Inability for self or correct placement.

Transcutaneous Transducer Garments ® have addressed and overcome many of the preceding shortcomings associated with conventional electrodes. Their unique construction and self-incorporated electrode rail apparatus feature a more uniform capacitance and resistive properties previously cited. The external gel orifice permits re-gelling while the garment is being worn even while under a cast without disrupting stimulation. When the Selectrode® reservoir is completely filled, the convex shaped Selectrode® will provide for long term transmission of the pulsatile signal. By nature of its unique construction TTG® increases the rate of peripheral contact impedance to central contact impedance. The multiple connected Selectrodes® in series is an additional safeguard for TENS usage. Elimination of adhesives and direct contact of the garments conductors from the skin eliminates the primary causes of skin irritation associated with electro medicine (erythema, allergenic, micropunctuate burns). It is obvious that TTG® ease of use, application and cost effectiveness regardless of areas to be stimulated will enable more end users the benefits of electrical medical devices. Transcutaneous Transducer Garments® represent a biomedical engineering advancement by virtue of their unique construction, versatility and systems approach in the conveyance of pulsatile signals.

ADDENDUM

Bio-Stimu Trend Corp™, TTG®, Selectrode®, Selectrode System®, are registered trademarks. Bio-Stimu Trend Corp™ TTG® are covered by one or more of the following patents:

US Patent 4,580,572
US Patent 4,583,547
US Patent 4,729,377
UK Patent 2,143,135
Patents Pending Worldwide.

REFERENCES

1. H. W. Tam and J. G. Webster, Minimizing electrode motion artifact by skin abrasion, *IEEE Trans. Biomed. Eng.*, **24,** 31–39 (1977).
2. Johnson and Johnson, Professional use of adhesive tape, Third-Edition (1972), pp. 13–20.
3. C. M. Readdy and J. G. Webster, Uniform current density electrodes for transcutaneous electrical nerve stimulation, *IEEE Frontier of Engineering and Computing in Health Care,* 187–190 (1984).
4. J. A. Orpin, Unexpected burns under skin electrodes, *CMA Journal,* **127,** 1106.
5. C. Zugerman, Dermatitis from transcutaneous electrical nerve stimulation, *J. Amer. Acad. Dermatol,* **6,** 936–939 (1982).
6. A. A. Fisher *et al.*, Allergic contact dermatitis due to ingredients of vehicles, *Arch. Dermatol,* **104,** 286–290 (1971).
7. A. A. Fisher, Dermatitis associated with trans. elec. nerve stimulation, *Cutis,* **21,** 24–27 (1978).
8. J. G. Webster, Minimizing cutaneous pain during electrical stimulation, *Ninth Annual Con. of the IEEE Eng. in Medicine and Biology Society* (1987) (in press).
9. Wiley and Webster, Analysis and control of current dist. under circular dispersive electrodes, *IEEE Trans. Biomed. Eng.,* **29,** 381–389 (1982).
10. Wiley and Wester: Distributed equiv.-circuit models for circular dispersive electrodes, *IEEE Trans. Biomed. Eng.,* **29,** 385–389 (1982).

11. C. J. Robinson, J. M. Bolam and N. A. Kett, Comparing surface electrodes electrical muscle stimulation: resistance, comfort, force produced and other effects, American Paraplegic Society, Las Vegas (1988) (in press).
12. Granek and Granek, Trans. Tran. Garments: current mammalian usage and future applications, *JONOMAS*, **6,** 271–278 (1985).
13. Y. Kim, J. B. Fahs and B. S. Tupper, Optimal electrode design for electrode surgery, defib., and extern. cardiac pacing, *IEEE Trans. Biomed. Eng.*, **33,** 845–853 (1986).

Automedica, 1989, Vol. 11, pp. 25–42
Reprints available directly from the publisher only
Photocopying permitted by license only

ELECTROCUTANEOUS SENSATIONS ELICITED USING SUBDERMALLY LOCATED ELECTRODES

R. R. RISO, A. R. IGNAGNI and M. W. KEITH

*Rehabilitation Engineering Center, Case Western Reserve University
Cleveland, Ohio*

(Received September 1988)

This paper describes techniques by which an electrocutaneous communication display can be made implantable so as to eliminate the need for mounting externally fastened electrodes on the skin surface. Subdermal electrical stimulation provides consistent and reliable evoked tactile sensations that may be used for sensory substitution systems.

INTRODUCTION

Electrocutaneous communication involves the application of electrical pulses to an individual's skin to elicit sensations that are typically described as having a tapping or vibratory quality which is localized to the skin surface in contact with the electrode. By appropriately coding the electrical stimulation, an individual's intact skin sense may be used as an input channel to receive augmentative or substitute sensory information to lessen some existing sensory deficit or disability condition. Applications for electrocutaneous systems have included sensory feedback for neuromotor prostheses [1]; asensory hands due to Hanson's disease [2]; artificial limbs [3–7]; vocoders to assist deaf persons in the perception of speech [8, 9]; and reading and mobility aides for the blind [10–14]. The applications of sensory feedback in limb prostheses has recently been reviewed in references [15] and [16].

An electrocutaneous display may be formed from either one or several separately located electrodes. With a single electrode display the most common feedback codes employ frequency and/or amplitude modulation of the activity of the cutaneous electrode. The provision of multiple electrodes in the form of an array or matrix allows the additional dimensions of the spatial location of the active elements of the display to be used in the sensory coding scheme.

Tactile substitution systems, whether based on direct electrical stimulation of the skin afferent nerve fibers as discussed above, electrical stimulation of peripheral trunk nerves [17, 18] or on mechanical stimulation of the skin mechanoreceptors [19–21], have the inherent advantages over auditory and visual displays that the signals are private to the user, to provide good cosmetic acceptability, and the signals are not difficult to attend to in noisy environments.

Despite these positive qualities, few tactile based sensory substitution systems have gained widespread acceptance by the intended users. For electrocutaneous systems this can be traced to the inconvenience of having to "don and doff" the display electrodes each day, to poor consistency in the qualities of the sensation provided and to occasionally uncomfortable sensations being evoked. These problems are mainly caused by difficulties in placing the electrodes in the exact same location each day; in maintaining the same level of hydration of the skin; and in securing the electrodes against the skin in a reliable manner.

Differences in the sensation thresholds and perceived intensity of electrical stimuli may occur even with very small (ca. 1–2 mm) repositioning errors in electrode mounting. To compensate for this, means for independent amplitude control must be available for the user to adjust the stimulus intensity for each electrode channel of the sensory display. This increases the complexity and bulk of the system hardware, and where large numbers of electrode channels are involved, the system may become too cumbersome for the user to operate.

These deficiencies can be overcome by designing sensory substitution systems which may be totally implanted within the body. The success of this approach has previously been demonstrated for sensory feedback systems in limb prostheses in which electrical stimulation was applied directly to an amputee's median or sciatic [17, 18] nerve by means of an implanted stimulator and epineural electrodes. As part of the effort to achieve this goal for electrocutaneous based sensory systems, techniques are being developed to elicit electrocutaneous sensations by means of implanted subdermal electrodes. This paper reviews the status of that development. Aspects of this work have been previously reported in abstract form [22, 23].

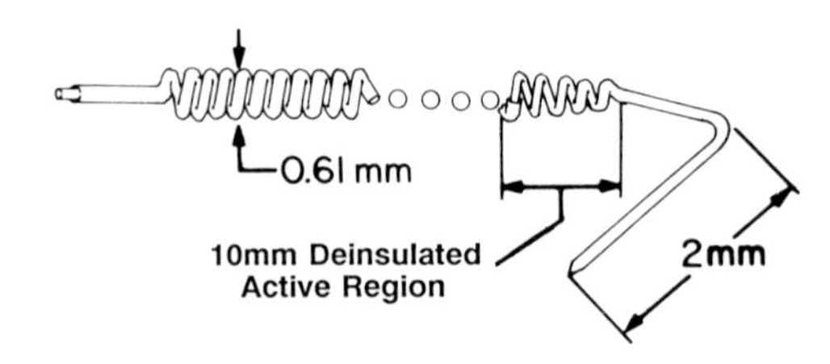

Figure 1 Schematic drawing showing structural details of the multistrand coiled wire percutaneous electrode used for subdermal electrocutaneous stimulation.

METHODS AND MATERIALS

Electrode Description

Two types of electrode have been used for subdermal stimulation:

Coiled wire electrode—The coiled wire type of electrode has been most extensively studied (Figure 1 and 3). It is similar to the type originally designed by Caldwell and Reswick [24] and is identical to the percutaneous coiled wire electrodes being used to stimulate muscle [25, 26]. The electrode is formed by winding 76 μm, Teflon insulated, 10 strand 316L stainless steel wire about a 0.15 mm mandrel. Before winding, the insulation is stripped off for a distance of 25 mm. After winding, this deinsulated region forms an active electrode area of about 10 mm^2 and overall length of 10 mm. Impedance of the electrode measured at 1 kHz is approximately 1.2 kohms. The completed electrode is loaded into a 19 gauge hypodermic needle and a hook or "barb" is formed at the electrode tip. The barb keeps the electrode within the needle during insertion into the skin and, afterwards, anchors the electrode within the skin. The complete electrode--needle assembly is sterilized using ethylene oxide.

To install the electrode, the needle is manually inserted into the skin a few centimeters from the intended location of the electrode active region, and then the tip of the needle is advanced in a parallel plane beneath the skin until the desired location is reached. The needle is then slowly withdrawn allowing the free end of the electrode to pass through the lumen of the needle and leaving the electrode behind in the subcutaneous tissue. The skin penetration site may be pretreated by infiltration of an anesthetic agent to minimize discomfort upon insertion of the needle.

The experience obtained from studying a cumulative total of 16 electrodes that have been installed in six normal volunteer subjects and one upper extremity amputee subject is reported in this paper. Once installed, the percutaneous electrodes may be left in place for protracted studies of several months to years. A molded rubber cap may be fastened using double faced tape over the site of entrance of the electrode lead wires to provide mechanical protection and a convenient means of attaching a lightweight connector and cable. Each electrode consists of a monopolar lead only, and several such electrodes may exit from the skin in an area of only a few square centimeters.

The percutaneous site should be monitored for signs of irritation and cleaned with alcohol daily. The skin forms a tight seal about the electrode lead at the entrance site and thus prevents the entrance of bacteria. In the experience of investigators who have implanted hundreds of this type of percutaneous electrode for activation of paralyzed muscle over the passed decade, infection of the skin site has been extremely rare and has not been encountered in any of the sensory stimulation applications. It is speculated that the exact dimensions of the helical coil after winding may play an important role in allowing the skin seal to be effective. The percutaneous electrodes may be easily removed intact at any time simply by applying traction to the exposed lead.

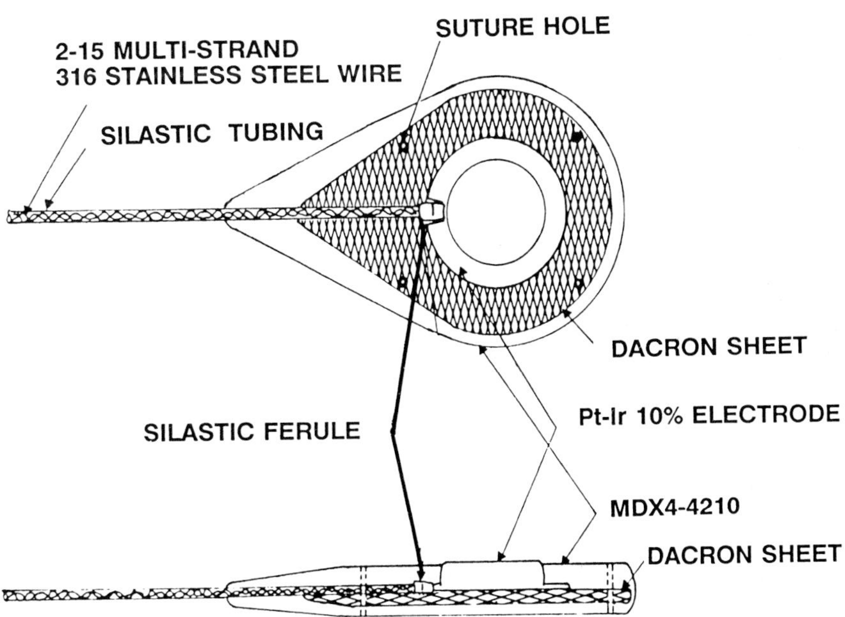

Figure 2 Schematic drawing depicting the structure of the disk type of electrode used for electrocutaneous sensory feedback in an implanted Functional Neuromuscular Stimulation Grasp Restoration Neuroprosthesis System. The electrode requires surgical installation and is driven by an implanted stimulator.

Disk electrode—Another implantable electrode that has been demonstrated to be effective for subdermal stimulation of the skin consists of a 3 mm diameter platinum-iridium disk attached to a molded backing of dacron reinforced silastic rubber (Figures 2 and 3). This monopolar electrode was first developed for epimysial stimulation of skeletal muscle for use with implanted functional neuromuscular stimulation systems and has been deployed for electrocutaneous sensory feedback purposes as well in one quadriplegic individual. The disk electrode is surgically attached to the underside of the skin using sutures so that the active surface of the electrode contacts the underside of the skin. The insulated back surface of the electrode faces the underlying tissue and helps to prevent current leakage which might otherwise excite nearby muscles or nerve branches [1, 27, 28].

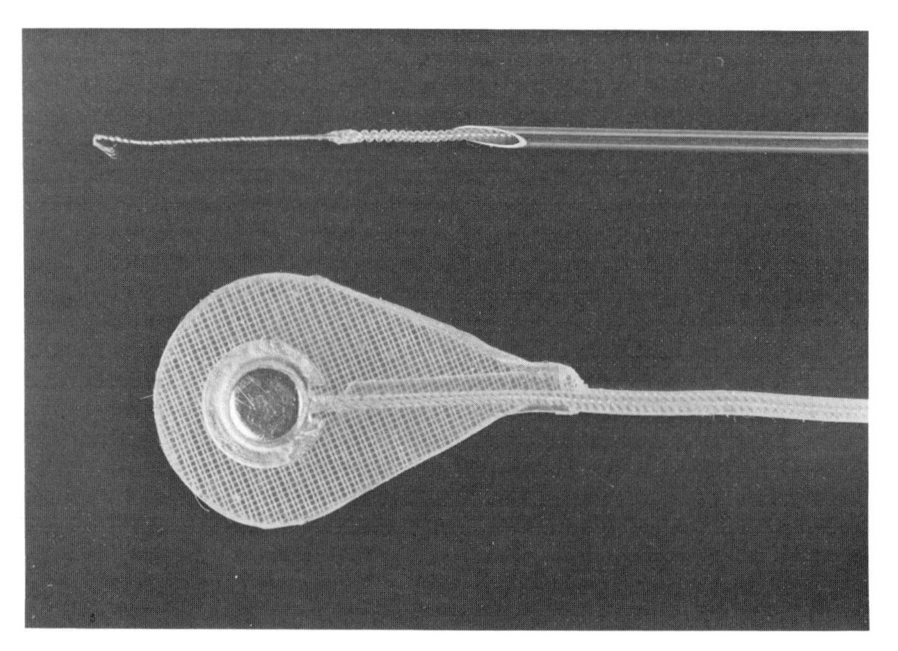

Figure 3 Photograph showing the two types of electrodes used for subdermal electrocutaneous stimulation. Upper—coiled wire electrode loaded into a 19 gauge hypodermic needle which is used during the percutaneous installation procedure. Lower—disk type of electrode used for permanent installation on the underside of the skin and connected to an implanted stimulator.

Location of Active Sensory and Indifferent Electrodes

For the studies involving the subdermal electrocutaneous stimulation, the skin in the region of the upper arm has been the primary site for location of the electrodes with two exceptions as follows: In the case of the disk electrode the location was above the chest near the clavicle bone, and in the case of one of the electrodes in an arm amputee subject, the location of the coiled wire electrodes was on the upper forearm skin. Where multiple subdermal electrodes were installed in the same subject, the active tips were separated by at least 35 mm to insure that the subject could easily discriminate the activation of each electrode site from the others.

The indifferent electrode used in conjunction with the coiled wire subdermal electrodes consisted of a 41 mm × 88 mm, self-adhesive disposable carbon impregnated rubber pad (3M Myocare) which was placed over the deltoid region of the shoulder. For the case of the disk subdermal electrode a portion of the metal case of the stimulator package (implanted into a pacemaker location in the chest) served as the indifferent electrode. In both cases the indifferent electrode provided a large diffuse current path to minimize the current density at the indifferent electrode site. This assured that the effective stimulation of the skin was confined to the immediate vicinity of the subdermal electrode active tip, thus affording good localizability of the evoked electrotactile sensations.

Stimulus Waveform

The choice of stimulus waveform is important when using implanted electrodes to maintain the integrity of the stimulated tissue and the electrodes. Biphasic capacitively coupled constant current pulses are used to prevent any net accumulation of charges within the tissue. An asymmetric waveform is employed which has an initial cathodic phase (50 μs duration) of sufficiently high current amplitude to excite the skin afferent nerve fibers. This primary stimulus phase is followed by a secondary anodal phase which is limited in amplitude to be much lower than the primary phase. This is done to maintain the voltage at the active electrode tip below the stainless steel corrosion potential during the time that it is driven anodically and is therefore vulnerable to corrosion. To maintain charge balance, the duration of the secondary phase is increased as necessary to compensate for the reduced amplitude [29].

An additional consideration for the stimulus waveform is the introduction of an interphase delay of 100 μs. This has the effect of allowing the initial (primary) cathodic stimulus phase to be maximally effective in exciting the afferent nerve fibers before the introduction of the secondary anodic stimulus phase. The action of the secondary stimulus phase is to repolarize the membrane. If the secondary phase were delivered without any delay, it would serve to counteract the action of the primary stimulus phase. Therefore, the net effect of purely biphasic stimulation without an interphase delay would be that higher stimulus intensities would be needed. This would increase the possibility of producing tissue damage from the stimulation [30, 31].

Psychophysics—Just Noticeable Differences of Frequency

The efficacy of subdermal electrocutaneous stimulation was evaluated by measuring the ability of subjects to discriminate changes in stimulus frequency and comparing those results to the case where the stimulation was applied using more conventional concentric surface mounted electrodes. Measurement of these performance parameters consisted of making psychophysical determinations of "just noticeable differences" (JNDs) using a computer adaptation of Cornsweet's "Dual Staircase Method" [32]. This technique is an abbreviation of the Method of Constant Stimuli in which a predetermined assortment of comparison stimuli are successively paired with the standard stimulus. The subject was required to indicate on each trial (stimulus pairing) if a change was perceived or not. For the case of frequency discriminations, for example, the staircase method began with pairings using comparison stimuli that were successively just slightly higher in frequency than the standard. An ascending series was continued until a level of comparison stimulus was reached which was sufficiently higher than the standard, so that the subject was able to detect the difference. After that event occurred, the direction of the staircase was reversed, and the next stimulus pairings that were given differed less and less from the standard until the subject could no longer discern any difference. At that point the descending series was terminated and another ascending series was begun. With this scheme repeated over and over the only stimulus pairings which were given (except at the very start of the procedure) were those in which the comparison stimulus was equal to or just bracketed the JND value. This resulted in a highly efficient experimental technique. To be certain that the subject did not learn the control strategy behind the staircase stimulus presentations, two independent staircase experiments were given simultaneously with their respective trials interspersed (thus the name "dual-staircase").

At least 55 trials were given per test to determine the JND at each test frequency. For these studies the size of the JND was defined as the minimum stimulus change that could be detected for 80% of the test trials. As a control for haphazard guessing a percentage of the trials was arranged so that no change in the stimulus occurred. Data were rejected and the respective test repeated if a subject had more than one false positive response for any control trial within the set of 55 trials. Subjects controlled the onset of each test trial by depressing a switch. "Self-pacing" helped to assure that the subjects were attentive for each trial. A time delay of 750 ms was introduced between the depressing of the button and the delivery of the trial to ensure that the motor act of depressing the button did not interfere with the subject's performance of the discrimination task. A PDP-1123 minicomputer generated the frequencies, and when a shift was made, it was executed instantaneously so that the very next interpulse interval was that which was appropriate for the comparison stimulus. During the frequency discrimination tests, the subjects were free to adjust the intensity of the stimulus at any time. Also, if they felt that any accommodation effects were becoming severe enough to compromise their judgments, they could interrupt the testing and resume after a brief rest period during which the stimulus was turned off. Subjects were well practiced at the JND task before the data were recorded.

RESULTS

Sensation Thresholds and Stability

Thresholds for electrocutaneous sensation using coiled-wire subdermal electrodes are typically found to require levels of current in the range of 0.3–6.0 ma when measured using stimulus pulses of 50 μs duration and 5 Hz repetition frequency. The current threshold associated with each different electrode is dependent on the particular location of the electrode active tip after installation. Figure 4 gives an example of the relative stability of the current at the sensation threshold as a function of time after implantation for three subdermal electrodes implanted in the same subject.

The stability of the threshold current is excellent for the two most sensitive electrodes. For the remaining electrode (labeled No. 3 in Figure 4), there was a change in behavior between the 3rd and 5th weeks after which this electrode also showed good stability.

The histogram of Figure 5 shows the distribution of the threshold currents for 16 electrodes at approximately 7 days following implantation. The current range is

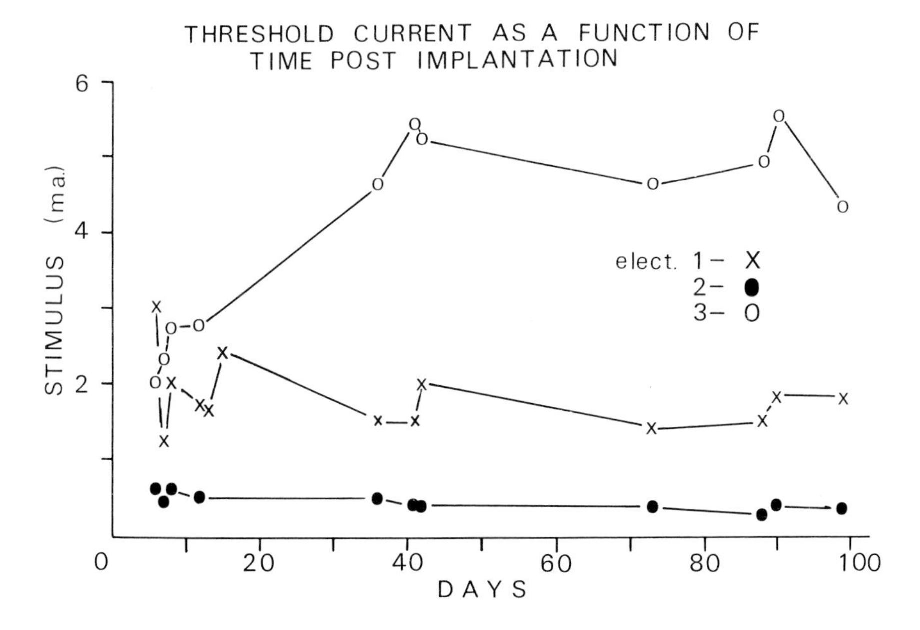

Figure 4 Showing stability of the current level at the sensation threshold for three different coiled wire type subdermal electrodes installed in the upper arm of the same subject. Thresholds were obtained using biphasic capacitively coupled pulses from an adjustable constant current stimulator. The pulse duration and pulse repetition rate were held constant at 50 μs and 5 Hz, respectively.

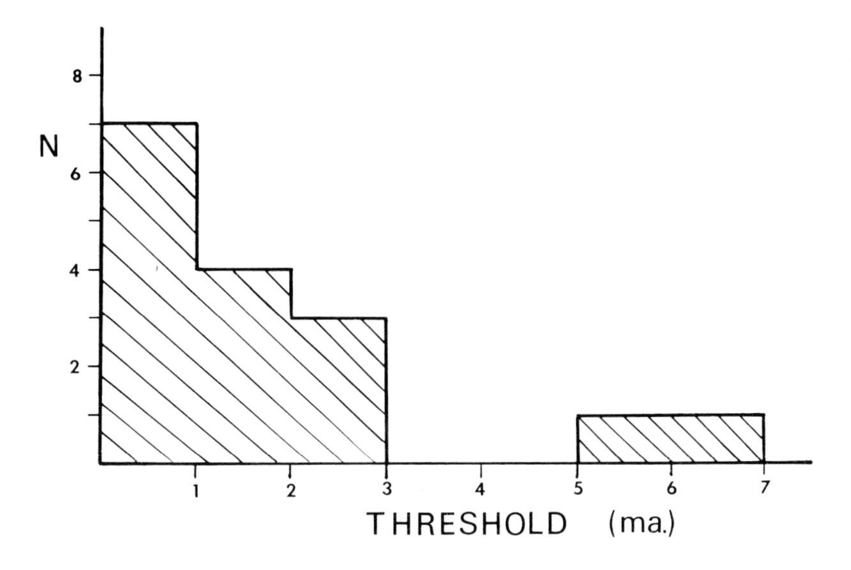

Figure 5 Histogram showing the sensation current thresholds for 16 different coiled wire subdermal electrodes implanted in 7 subjects. The threshold measurements shown were measured approximately seven days following implantation. Current range was 0.3–6.0 mA with mean = 1.4 mA.

0.3–6.0 mA and the mean value is 1.4 mA. For 8 of these electrodes that were studied for at least 30 days, the mean current was 1.2 mA at day 7 and this value was unchanged at day 30. These data indicate that the electrodes retain their ability to stimulate the skin within reasonable limits of current. Mechanical fixation of the electrodes within the tissues appears to be adequate to provide this performance stability.

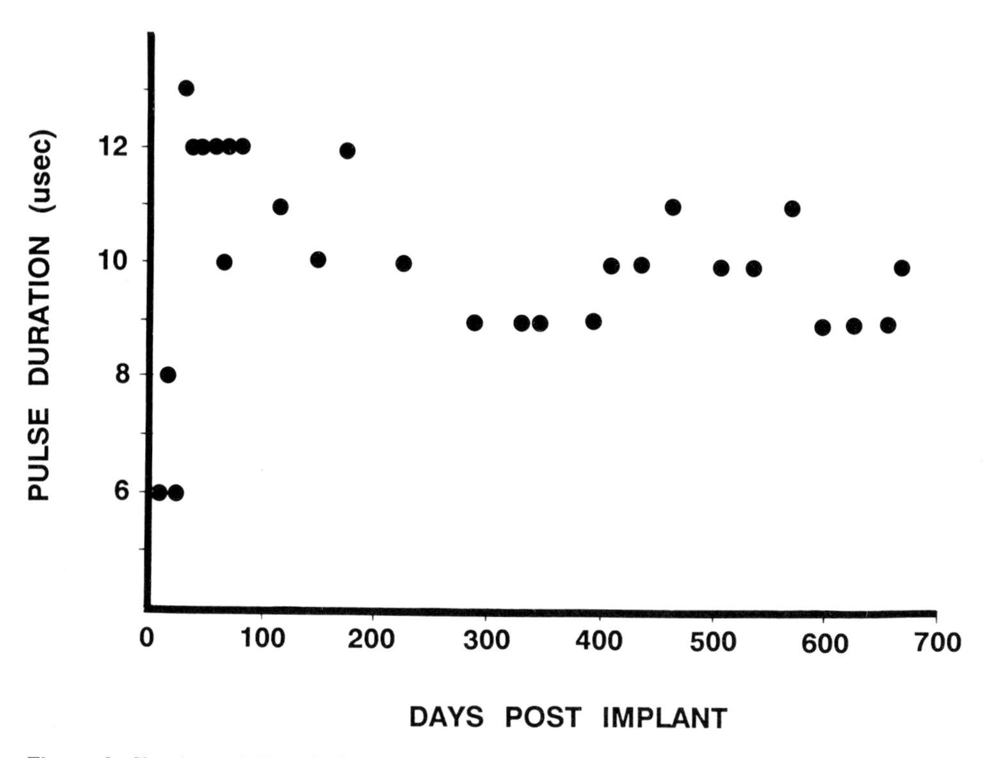

Figure 6 Showing stability of stimulus pulse duration at the sensation threshold for a disk type subdermal electrode. The electrode was implanted in the upper chest region of a quadriplegic individual to provide electrocutaneous sensory feedback as part of an implanted Functional Neuromuscular Stimulation Hand Neuroprosthesis system. Stimulus pulses were all constant current with a fixed amplitude of 20 mA.

Threshold Stability for Disk Type Electrode

The functional stability associated with using the disk type of electrode for subdermal electrocutaneous stimulation has been excellent for the single subject in whom this has been implemented to date. A history of threshold measurements collected over nearly two years are presented in Figure 6. The ordinate of the graph gives the duration of the stimulus pulses needed to attain the sensation threshold. The reason for this is that the implanted stimulator hardware operates with a fixed current intensity of 20 mA. Modulation of the stimulation delivered to the electrodes is either by changing the pulse duration or pulse repetition frequency, and depends on whether the electrode is intended to stimulate the skin or muscles. For the case of the muscle stimulation, the frequency is held constant at

12.5 Hz and the pulse duration is modulated. In the case of cutaneous stimulation, the pulse duration is selected to elicit a moderately strong, yet comfortable sensation, and the feedback information is frequency encoded. We have found that a pulse duration of 31 μs elicits a satisfactory level of sensation for this particular user over the frequency range of 4–55 Hz that is used, regardless of whether he has just turned on his neuroprosthesis system or has been using it protractedly.

It may be said, therefore, that for the one user of an implanted sensory system with whom we have experience, the sensory stimulation parameters were adjusted once only, with no further change required during the two years hence. Figure 6 shows some initial decrease in electrode sensitivity during the first few weeks after implantation when tissue encapsulation was taking place. Following this period, however, no systematic changes in the threshold pulse duration are evident. Most likely, the variations of approximately 2–3 μs shown for the remainder of the history, reflect variability inherent in the user's subjective responses in reporting when the threshold was reached. Rigorous psychophysical tests of sensation threshold were not attempted with this individual because of the tedium that this would engender. (For purposes of monitoring the sensory threshold, 1 Hz pulses were always utilized).

Dynamic Range

The dynamic range between the threshold for sensation and the level of stimulus current which causes discomfort was measured on several occasions for the various electrodes and was within 3 to 5 times the threshold current. This range is adequate for many applications and may be expected to increase as subjects receive stimulation day after day as has been reported by Saunders [8] and by Solomonow and Prados [33] for studies involving surface electrocutaneous stimulation. The increase in dynamic range is thought to derive from a learning effect.

Quality of Evoked Sensation

Subjects report that the sensations elicited by subdermal electrodes using pulse widths between 10 and 100 μs and pulse repetition rates between 2 and 100 Hz are comfortable, clear and consistent, and generally are preferable to those evoked using surface applied stimulation. In three cases, testing of a subdermal coiled wire electrode immediately after installation led to an unacceptable evoked sensation: In one of these cases a subject reported that the stimulation felt indistinct or "cloudy" in terms of the discriminability of the stimulus frequency. In the other two unacceptable cases the stimulation was not comfortable and seemed to produce only a dull "aching" feeling. In these instances the electrodes were later withdrawn and replaced by others. In our experience an electrode which elicits a comfortable sensation immediately after insertion continues to do so. Similarly, an electrode which initially evokes an uncomfortable sensation continues to do so.

Just Noticeable Differences of Frequency

Successive JNDs of increases in frequency were determined over a range of 2–100 Hz for several subjects to determine the useable bandwidth and the theoretical resolution that would be available for an electrocutaneous sensory feedback system that utilized a frequency modulation code. Figure 7 shows the results obtained for one subject. The tests were also performed using a surface mounted electrode which was positioned directly over the location of the subdermal electrode. The data were collected by starting at a base frequency of 2 Hz and determining the frequency change that corresponded to the first JND level. That value was then taken as the new standard frequency and the second level JND was sought. The process was repeated until the upper level of 100 Hz was reached. The two curves in the figure show that the JND versus frequency ratio is not constant but varies considerably with frequency.

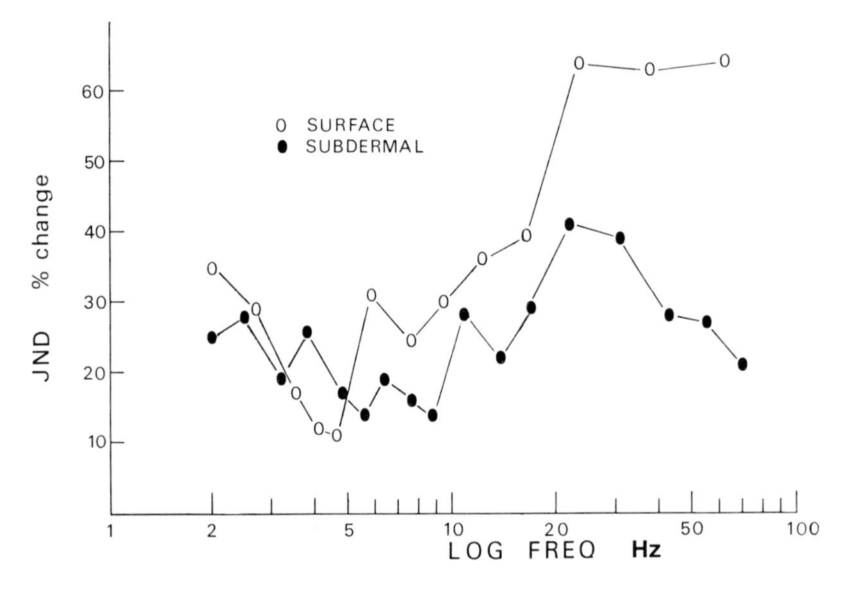

Figure 7 Showing the magnitude of successive JNDs of frequency for electrocutaneous stimulation over the range of 2–100 Hz obtained with one subject using a coiled wire subdermal electrode and a concentric surface mounted electrode (center electrode element was 3 mm diameter) as indicated. The surface stimulation was given in the location directly over the implanted electrode. Note that the subject was able to discriminate substantially smaller increases of frequency, particularly above 20 Hz, when the subdermal stimulation was used versus surface stimulation.

With the use of subdermal stimulation, the subject was able to discriminate at least 16 JNDs of frequency in the range 2–55 Hz, whereas only 11 JNDs could be discriminated when surface stimulation was employed. In terms of the average size of the JNDs (expressed as percent change), the subdermal data yielded 24.3%, whereas this was increased to 28.5% for the use of the surface stimulation.

To further quantify the distinctness of the sensation from the subdermal stimulation in comparison with conventional surface applied stimulation, the size of JNDs of frequency were measured in another series of tests involving three subjects. Specifically, JNDs were determined for increases of frequency starting from six arbitrary standard values of 2, 5, 10, 20, 30 and 50 Hz. Twelve experiments in all were conducted on each of the subjects. There were two electrode locations studied for each subject, and a subdermal and surface mounted electrode was tested at each of these two locations. Three independent evaluations of the JND tests for each electrode type-location pair were performed. A controlled order for the implementation of the tests was maintained as follows:

Tests 1–3 Position one—surface electrode
Tests 4–6 Position two—subdermal electrode
Tests 7–9 Position two—surface electrode
Tests 10–12 Position one—subdermal electrode

Approximately 1.5 to 2 h were required to obtain the JND values for the full set of 6 standard frequencies, and generally only a single such test was held on a single day.

Examples of the psychometric functions which were obtained for discriminations of the frequency for one of the subjects are shown in Figure 8. Each of the curves represents the mean value that was derived from combining the results of the three separate evaluations for the respective electrode type and electrode position indicated. As can be seen from the data for frequencies that are beyond 20 Hz, this subject was able to discriminate smaller changes in frequency when the stimuli were presented with the subdermal electrodes as compared to when surface electrodes were used. This finding was demonstrated with the data of the other two subjects as well. When the averages for the size of the JNDs at all of the tested frequencies were compiled separately for the subdermal electrodes and for the surface electrodes of each of the three subjects, the average JNDs were found to be lower for the subdermal electrodes than for the surface electrodes by 25%, 10% and 23%, respectively, for the three subjects.

Ability of Subjects to Discriminate Absolute Frequencies of Stimulation

Three subjects who were already well familiar with the task of discriminating frequency transitions using electrocutaneous codes participated in a study to determine the ease with which they could learn to discriminate six frequencies on an absolute basis. Each subject was assigned a list of six frequencies beginning with 2 Hz and spaced 2 JNDs apart according to that person's own personal JND data (obtained in the same manner as the data shown in Figure 7).

Each subject was presented with the various frequencies in a pseudorandom sequence that consisted of 50 trials. On each trial the subject was required to identify the frequency that was presented as being either 1, 2, 3, 4, 5 or 6. The experimenter stated the correct response after each trial to provide performance feedback that could facilitate learning. Three blocks of 50 trials were given each day to each subject, and the testing was continued for each subject until the subject achieved an average performance score of 90% correct responses for the 3 blocks of trials given that day. Results showed that the three subjects achieve this criterion of performance within 3, 6 and 9 days of training, respectively.

Energy Considerations

Computations of the electrical energy needed to elicit electrocutaneous sensations via the coiled wire subdermal electrodes were made with the aid of a LeCroy waveform analyzer (sampling frequency 32 MHz). Pulse repetition rates of 5 and 25 Hz were studied in combination with pulse durations of 10, 50, 100, 150 and 250 μs. The level of stimulus current was adjusted to be just at the threshold for sensation as determined using a psychophysical testing paradigm, and the sensation current and voltage characteristics were recorded. The stimulus current was then increased to an arbitrary multiple of 2.5 times the threshold level, and the current and voltage waveform were again recorded. The area under the initial "active"

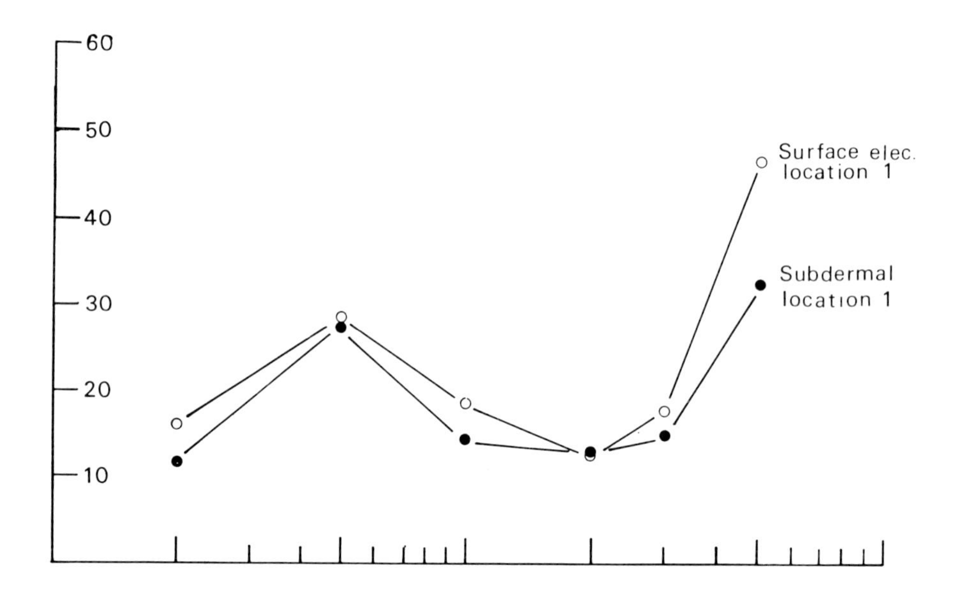

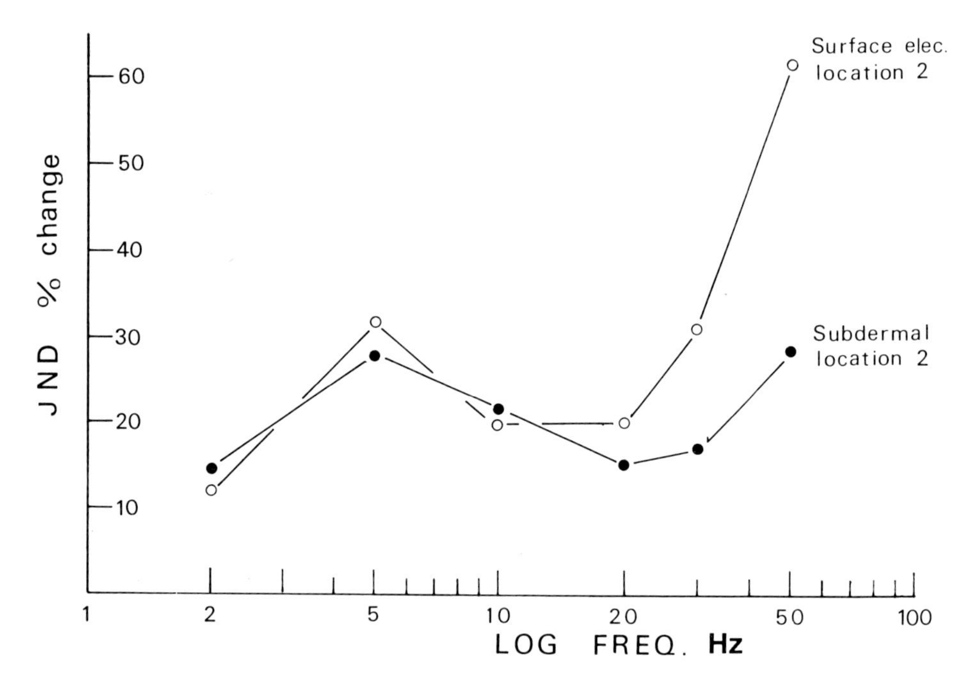

Figure 8 Showing size of JNDs for increases in stimulus frequency for two skin locations and for subdermal and surface stimulation as indicated. Each data point represents the mean value obtained from three separate evaluations of the JND size, and each such evaluation was performed on a different test day. Note the increased discriminability of the subdermal stimulation in comparison to the surface stimulation for the 30 and 50 Hz test frequencies.

phase of the biphasic stimulus waveform was computed by numerical integration and an average voltage obtained. These results were then applied to Equation 1 below:

$$\text{Power} = (I)(V)(\text{Duty Factor}). \tag{1}$$

In the equation, I represents the amplitude of the constant current pulse; V represents the average voltage during the active phase of the stimulus pulse; and the Duty Factor represents the fraction of time that the active phase of the stimulus current is flowing (i.e. for a 50 μs pulse repeated at 25 Hz the duty factor is equal to 50×10^{-6} s/pulse $\times$ 25 pulses/s or 1250×10^{-6}).

Three electrodes were studied (two were installed in one subject and the third was installed in a different subject). An example of the results obtained is given in Figure 9. The data show that the most efficient pulse duration to use is within the range of 50–100 μs. The power requirements increase substantially if pulse durations as short as 10 μs are employed, and to a lesser extent if pulse durations of 150 μs or greater are utilized. The power dissipated at the electrode–tissue interface appears to be of the order of 10 μW for 25 Hz stimulation and would thus be expected to be about 22 μW if the stimulation frequency is further increased to 55 Hz which, in our experience, is a practical upper bound for frequency modulation electrocutaneous codes.

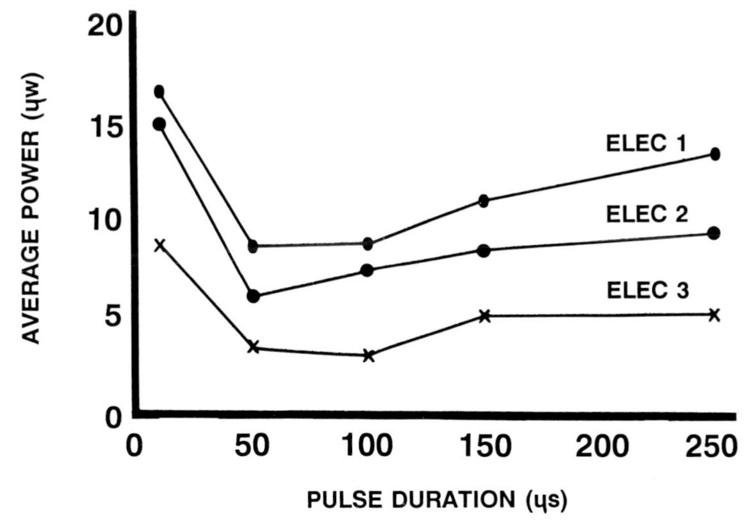

Figure 9 Average energy consumed for sensation threshold stimulation at the electrode–tissue interface for three different subdermal electrodes implanted in two subjects. Note the improved efficiency for the 50 and 100 μs pulse duration in comparison to the shorter (10 μs) and longer pulses. Stimulation frequency was held constant at 25 Hz throughout the investigation.

DISCUSSION

Location of Subdermal Electrodes

Electrodes for electrocutaneous displays should be placed in a region of skin that is free of mechanical disturbances from body movements, and where the sensory stimulation is unlikely to stimulate underlying trunk nerves or muscles. The most appropriate locations for the placement of the electrodes are the skin of the upper arms, trunk, abdomen or thighs.

Activation of Skin Afferent Nerve Fibers by Electrocutaneous Stimulation

Geldard [34] referred to electrical stimulation as the "great non-adequate stimulus" because it is able to excite skin afferent nerve fibers without regard for their specific modalities.

Nevertheless, some control of the evoked sensations is afforded by using pulses of short duration (e.g. 10–100 μs) so that large diameter (A-beta) sensory fibers that are associated with sensations of mechanical events on the skin are recruited preferentially to smaller diameter (A-delta and C) fibers that evoke sensations of pricking pain, burning pain and dull ache [31, 35–37].

The complement of large afferents in the hairy (i.e. non-glabrous) skin that are responsible for electrotactile sensations is unknown. A set of large afferents which innervate the hair follicles and are readily activated by movement of the hair shafts represent the best candidates, however. These afferents are thought to subserve the sense of flutter elicited by low frequency mechanical stimulation (10–40 Hz). It is interesting to speculate that the poor discriminability reported by subjects in our studies for stimulation above 55 Hz is related to the lower preferred stimulation frequency of these hair shaft afferents. Also innervated by large myelinated fibers and present in hairy skin are Pacinian Corpuscles and two groups of slowly adapting afferents. Of these slowly adapting afferents, the Pinkus corpuscles located at the skin upper surface are not very abundant in human skin. The second type of afferent, the Ruffini structure, is sensitive to stretching of the skin, but activity present in Ruffini afferents is thought not to reach consciousness [38].

The subdermal electrodes described in this paper are relatively large so that they stimulate many different cutaneous afferents simultaneously. The sensations evoked are similar to what is obtained using surface stimulation with the major advantage of affording greater consistency.

A study that initially suggested the high efficiency that is possible by using subdermal instead of surface stimulation is that of Saunders and Saunders [39]. They used a bipolar microelectrode implanted within the skin directly beneath a concentric surface stimulating electrode to monitor the electric events *in situ* during electrocutaneous stimulation. Their results indicated that "90% of the applied pulse was dissipated in the stratum corneum" and that "... the intradermal charge required to attain threshold was extremely small and highly stable".

Aside from the obvious savings from reduced power consumption, a further advantage of the use of subdermal electrocutaneous stimulation is that the smaller stimulus currents required help minimize stray electrical fields which could interfere with feeble EMG signals that are sometimes used for prosthesis control.

Discriminability of Electrocutaneous Frequency

Frequency modulation electrocutaneous codes are advantageous for sensory feedback applications where absolute levels must be communicated, because accommodation effects that can severely degrade intensity coding schemes have much less effect [40–42]. We have found that it is comparatively easy for subjects to learn to identify at least 6 absolute levels of frequency, and it is worthwhile to point out that about 7 items (code words in this case) is usually considered to be a practical limit for what the human mind can comfortably deal with regardless of the type of coding scheme used [43].

Our results indicate that the shape of the psychometric function obtained with any given subject is somewhat unique for each individual and that the ratio of the size of the JND and the base frequency is non-constant. These observations confirm previous reports: Our findings of 24.3% (subdermal stimulation) and 28.4% (surface stimulation) for the average frequency JND size agrees well with the 28% value reported by Szeto *et al.* [44] for surface stimulation.

Subjects stated subjectively that frequencies above about 55 Hz felt very much the same and caused a considerable amount of accommodation. Furthermore, these effects were worse with the surface stimulation as compared to the subdermal stimulation. These comments agree with the experimental results and confirm recommendations by others that the upper range for frequency stimulation should be limited to below 100 Hz [45].

Implanted Sensory Feedback System

The disk type of subdermal electrode is being used to give a user of a functional neuromotor hand prosthesis system feedback about the functional status of his stimulation system. A frequency modulation coding paradigm is used that consists of 5 frequencies (4, 7, 10, 20, 35 and 55 Hz). The five frequencies provide feedback of the output of the system proportional control signal which the user generates by voluntary movements of his shoulder, and which is monitored by means of a chest mounted transducer that extends to the user's shoulder. In addition, the feedback system assists the user in switching the system ON and OFF, and in selecting one of two grasp coordination patterns. The electrocutaneous sensory system replaces a previous system that was based on auditory signals.

Tissue Safety Considerations

Studies that directly involve the electrical safety of subdermal stimulation have not been reported. An estimate of the maximum current density that is likely to be encountered at the tissue interface when using the coiled wire electrodes can be made using 6 mA as an upper limit for the current obtained for the sensory threshold, a pulse duration of 50 μs, and a stimulus repetition frequency of 55 Hz. Additionally, studies have demonstrated that it is usually desirable to provide the stimuli in bursts of pulses rather than as single pulses and to modulate the burst repetition rate as the basis for the frequency coding [46].

If such bursts contain a maximum of 4 pulses, and if the current intensity during usage of the sensory system is typically adjusted to be about 2.5 times the threshold current level, then the charge density will be about 0.33 $\mu C/mm^2$. This level is below the value of 0.4 $\mu C/mm^2$ that has been reported by Mortimer *et al.* [29] to be safe for the stimulation of muscle tissue using the same coiled wire electrode design.

The development of implantable stimulators that provide multiple output channels is ongoing in several laboratories [28, 47–49] and some implantable stimulators are available from commercial suppliers. With the continued development of implantable electrocutaneous interfaces, many of the neuroprostheses applications for which these implantable stimulators were conceived may begin to incorporate components of sensory feedback which hopefully will enhance their functional utility in rehabilitating individuals with disabilities.

CONCLUSIONS

1) Subdermal stimulation of the skin to elicit electrotactile sensations has been demonstrated to be safe, reliable and comfortable, and provides a means whereby sensory communications displays may be made implantable.

2) Threshold currents for sensation using 50 μs pulses at 5 Hz are in the range of 0.3–6.0 mA for coiled wire subdermal electrodes.

3) The average size of JNDs for increases in stimulus frequency using subdermal coiled wire electrodes is found to be about 24% of the base frequency over the range of 2–50 Hz base frequency tested.

4) Using subdermal stimulation, subjects can learn to identify at least 6 levels of frequency (90% correct responding) in the range 2–55 Hz with modest amounts of training.

5) Maximal energy efficiency for subdermal stimulation using coiled-wire subdermal electrodes appears to be obtained using a stimulus pulse duration of about 50–100 μs.

6) A disk type of subdermal electrode that has been in use for nearly 2 years by a user of a functional neuromotor hand prosthesis system functions well and hasn't required any adjustment of its stimulation parameters.

ACKNOWLEDGEMENTS

The authors would like to express their gratitude to the engineers and other staff of the Rehabilitation Engineering Center of Case Western Reserve University for their contributions in designing the hardware and fabricating the electrodes used during the reported studies and to the subjects and patients who participated in the investigations. A note of particular thanks is extended to Mr. Scott Banks, Mr. Brian Smith and Mr. Ronald Hart for obtaining the sensation threshold data reported for the disk subdermal electrode. This work was supported by National Institute for Disability and Related Research grant No. G001005815 and by NIH Neural Prostheses Program contract NO1-NS-3-2345.

REFERENCES

1. R. R. Riso, A. R. Ignagni and M. W. Keith, Cognitive feedback for use with FNS upper extremity neuroprostheses, *Proc. 8th Meeting on External Control of Human Extremities*, ETAN, Dubrovnik, Yugoslavia, August (1987).
2. C. C. Collins and J. Madey, Tactile sensory replacement, *Proc. San Diego Biomed. Symp.*, **13,** 15–26 (1974).
3. I. Kato, S. Yamakawa, K. Ichikawa and M. Sano, Multifunctional myoelectric hand prosthesis with pressure sensory feedback system, *Proc. 3rd Int. Symp. on External Control*, 155–170, Dubrovnik, Yugoslavia (1970).
4. R. E. Prior, P. A. Case and J. Lyman, Supplemental sensory feedback for the VA/NU myoelectric hand: background and feasibility, *Bull. Prosth., Res.*, **BPR-10-26,** 170–190 (1976).
5. G. F. Shannon, A comparison of alternative means of providing sensory feedback on limb prostheses, *Med. Biol. Eng. Comput.*, **14,** 289–294 (1976).
6. R. N. Scott, R. H. Brittain, R. R. Calwell, A. B. Cameron and V. A. Dunfield, Sensory-feedback system compatible with myoelectric control, *Med. and Biol. Eng. Comp.* **18,** 65–69 (1980).
7. J. Kawamura, O. Sweda, H. Kazutaka, N. Kazuyoshi and S. Isobe, Sensory feedback systems for the lower-limb prosthesis, *J. Osaka, Rosai Hospital*, **5,** 104–112 (1981).
8. F. A. Saunders, W. A. Hill and C. A. Simpson, Speech perception via the tactile mode: progress report, *Sensory Aids for the Hearing Impaired* (Eds. H. Levitt, J. M. Pickett and R. A. Houde), IEEE Press, New York (1980), pp. 278–281.
9. D. W. Sparks, P. K. Kuhl, A. E. Edmonds and G. P. Gray, Investigating the MESA (multipoint electrotactile speech aid): the transmission of segmental features of speech, *J. Acoust. Soc. Am.*, **66,** 246–257 (1978).

10. R. D. Melen and J. D. Meindl, Electrocutaneous stimulation in a reading aid for the blind, *IEEE Trans. Biomed. Eng.*, **BME-18,** 1-3 (1971).
11. C. C. Collins, Tactile Television: Mechanical and electrical image projection, *IEEE Trans. Man-Machine Systems*, **MMS-11(2),** 85-91 (1970).
12. C. C. Collins and P. Bach-y-Rita, Transmission of pictorial information through the skin, *Advs. Biol. Med. Phys.*, **14,** 285-315 (1973).
13. C. C. Collins, Electrotactile visual prosthesis, in *Functional Electrical Stimulation* (Eds. F. T. Hambrecht and J. B. Reswick), Marcel Dekker, New York (1977), pp. 189-301.
14. S. Tachi, K. Tanie, K. Komoriya and M. Abe, Electrocutaneous communication in a guide dog robot (MELDOG), *IEEE Trans. Biomed. Eng.*, **BME-32,** 461-469 (1985).
15. C. A. Phillips, Sensory feedback control of upper and lower extremity motor prostheses, *CRC Critical Reviews*, Boca Raton, Fl. (in press).
16. A. Y. J. Szeto and R. R. Riso, Sensory feedback using electrical stimulation, in *Rehabilitation Engineering* (Eds. R. V. Smith and J. H. Leslie), CRC Press, Boca Raton, Fl. (in press).
17. F. A. Clippinger, A system to provide sensation from an upper extremity amputation prosthesis, in *Neural Organization and Its Relevance to Prosthetics* (Eds. W. Fields and L. Leavitt), Intercontinental, N.Y. (1973), pp. 165-176.
18. F. A. Clippinger, A. V. Seaber, J. H. McElhaney, J. M. Harrelson and G. M. Maxwell, Afferent sensory feedback for lower extremity prosthesis, *Clin. Ortho. Related Res.*, **169,** 202-206 (1982).
19. R. W. Mann and S. D. Reimers, Kinesthetic sensing for the EMG controlled "Boston Arm", *IEEE Trans. Man-Mach. Syst.*, **MMS-11 (1),** 110-115 (1970).
20. P. Clark and R. Savoie, Development of an advanced optical character recognition speech-output accessory for blind people, *Bull. Prosth. Res.*, **BPR 10-35,** 199-200 (1981).
21. C. A. Phillips and J. S. Petrofsky, Cognitive feedback as a sensory adjunct to functional electrical stimulation (FES) neural prosthesis, *J. Neurological and Orthopaedic Med. and Surg.*, **6(3),** 231-238 (1985).
22. R. R. Riso, A. Y. J. Szeto and M. W. Keith, Comparison of subdermal versus surface electrocutaneous stimulation, *IEEE Frontiers in Engineering in Health Care*, 343-347, Philadelphia, Pa. (1982).
23. R. R. Riso, M. W. Keith, K. R. Gates and A. R. Ignagni, Subdermal stimulation for electrocutaneous communication, *Proc. 6th Annual Conference on Rehabilitation Engineering*, 321-323, San Diego, Ca. (1983).
24. C. W. Caldwell and J. B. Reswick, A percutaneous wire electrode for chronic research use, *IEEE Trans. Biomed. Eng.*, **22,** 429-432 (1975).
25. P. H. Peckham, J. T. Mortimer and E. B. Marsolais, controlled prehension and release in the C5 quadriplegic elicited by functional electrical stimulation of the paralyzed forearm musculature, *Annals Biomed. Eng.*, **8,** 369-388 (1980).
26. E. B. Marsolais and R. Kobetic, Functional electrical stimulation for walking in paraplegia, *J. Bone Joint Surg.*, **69-A,** 728-733 (1987).
27. P. A. Grandjean and J. T. Mortimer, Recruitment properties of monopolar and bipolar electrodes, *Annals of Biomed. Eng.*, **14,** 53-66 (1986).
28. B. Smith, P. H. Peckham, M. W. Keith and D. D. Roscoe, An externally powered, multichannel, implantable stimulator for versatile control of paralyzed muscle, *IEEE Trans. Biomed. Eng.*, **BME-34,** 499-508 (1987).
29. J. T. Mortimer, Motor prostheses, in *Handbook of Physiology: The Nervous System*, Vol. II, Motor Control (Ed. V. B. Brooks), Am. Physiol. Soc., Bethesda, MD., (1981), pp. 155-187.
30. C. van den Honert and J. T. Mortimer, The response of the myelinated nerve fiber to short duration biphasic stimulating currents, *Ann. Biomed. Eng.*, **7,** 177-125 (1979).
31. P. H. Gorman and J. T. Mortimer, The effect of stimulus parameters on the recruitment characteristics of direct nerve stimulation, *IEEE Trans. Biomed. Eng.*, **BME-30,** 407-411 (1983).
32. T. N. Cornsweet, The Staircase Method in psychophysics, *Amer. J. Psychol.*, **75,** 485-491 (1962).
33. M. Solomonow and D. Prados, Further evidence of learning in the tactile sense, *Proc. 4th Ann. Conf IEEE EMBS*, 339-342 (1982).
34. F. A. Geldard, Adventures in tactile literacy, *Amer. Psychologist*, **12,** 115-124 (1957).
35. L. Vodovnik, C. Long, E. Regenos and A. Lippay, Pain response to different tetanizing currents, *Arch. Phys. Med. Rehab.*, **46,** 187-192 (1965).
36. R. G. Hallin and H. E. Torebjork, Electrically induced A and C fiber responses in intact human skin nerves, *Exp. Brain Res.*, **16,** 309-320 (1973).
37. J. C. Willer, F. Boureau and D. Albe-Fessard, Role of large diameter cutaneous afferents in transmission of nociceptive messages: electrophysiological study in man, *Brain Res.*, **152,** 358-364 (1978).
38. M. M. Merzenich and T. Harrinton, The sense of flutter-vibration evoked by stimulation of the hairy

skin of primates: comparison of human sensory capacity with the responses of mechanoreceptive afferents innervating the hairy skin of monkeys, *Exp. Brain Res.*, **9,** 236–260 (1969).
39. F. A. Saunders and V. F. Saunders, Intradermal recording during electrocutaneous stimulation, *Federation Proceedings*, **33,** No. 3, Part 1 (1974).
40. A. Y. J. Szeto and J. Lyman, Comparison of codes for sensory feedback using electrocutaneous tracking, *Ann Biomed Eng.*, **5(4),** 367–383 (1977).
41. A. Y. J. Szeto and F. A. Saunders, Electrocutaneous stimulation for sensory communication in rehabilitation engineering, *IEEE Trans on Biomed Eng.*, **BME-29(4),** 300–308 (1982).
42. A. Y. J. Szeto and L. Mao, Dermal effects of electrocutaneous stimulation, in *Biomedical Engineering I: Recent Developments* (Ed. S. Saha), (1982), pp. 121–124.
43. G. A. Miller, The magical number seven, plus or minus two: some limits on our capacity for processing information, *Psychological Review*, **63,** 81–97 (1956).
44. A. Y. J. Szeto, J. Lyman and R. E. Prior, Electrocutaneous pulse rate and pulse width psychometric functions for sensory communications, *Human Factors*, **21(2),** 241–249 (1979).
45. F. A. Saunders, Recommended procedures for electrocutaneous displays, in *Functional Electrical Stimulation* (Eds. F. T. Hambrecht and J. B. Reswick), Marcel Dekker, New York (1977), pp. 303–309.
46. R. R. Riso, A. R. Ignagni and M. W. Keith, Effects of stimulus parameters on the discriminability of frequency modulation codes for electrocutaneous sensory substitution systems, *Proc. 2nd Int. Conf. on Rehabilitation Engineering*, Ottawa, Canada (1984), pp. 435–436.
47. J. Holle, H. Thoma, M. Frey, H. Gruber, H. Kern and G. Schwanda, Functional electrostimulation of paraplegics: experimental investigations and first clinical experience with an implantable stimulation device, *Orthopedics*, **7(7),** 1145–1160 (1984).
48. N. Donaldson, A 24-Output Implantable stimulator for FES, in *Proc. of the 2nd Vienna International Workshop on FES*, Vienna, Austria (1986), pp. 197–200.
49. G. S. Brindley, C. E. Polkey, D. N. Ruston and L. Cardozo, Sacral anterior root stimulators for bladder control in paraplegia: the first 50 cases, *J. of Neurology, Neurosurg. Psychiatry*, **49,** 1104–1114 (1986).

Automedica, 1989, Vol. 11, pp. 43–52
Reprints available directly from the publisher only
Photocopying permitted by license only

CHARACTERIZATION OF THE LOWER LIMB OF QUADRIPLEGICS FOR THE DESIGN OF A CONTROLLER FOR FUNCTIONAL ELECTRICAL STIMULATION

BERTRAM N. EZENWA, DAVID B. REYNOLDS and BLAIR A. ROWLEY

Department of Biomedical Engineering, Wright State University, Dayton, Ohio

and

PAUL T. DANSET

National Center for Rehabilitation Engineering

The movement of the lower limbs of quadriplegics can be performed by surface functional electrical stimulation of the quadriceps muscles. The design of a controller for functional electrical stimulation requires the characterization of the plant, in this case, the quadriceps muscle system. This paper is concerned with the identification and modeling of the lower limb of quadriplegics during functional electrical stimulation assisted exercise. The approximated nonlinear gain characteristics and the dependence of the complex poles on the level of stimulation were examined. The results suggest internal nonlinear summation of individual twitches of the muscle fibers. A least squares identification technique was used to determine the model order that provides the best fit for the system dynamics. Variations in the plant's parameters between subjects and for a subject under different occasions were observed. These variations suggest that it is better to determine an exercise profile based on a global model of the system and then use some form of online adaptive technique to control the electrical stimulation of the muscles during functional electrical stimulation. The results are used to establish design requirements for an appropriate controller.

Keywords: Functional electrical stimulation, control voltage, progressive weakness characteristic, fading memory characteristic.

INTRODUCTION

Functional electrical stimulation (FES) orthosis for paralyzed muscles is the application of controlled electrical stimulation to the intact peripheral nerve in an attempt to restore functional muscle contractions in paralyzed muscles. Small electrical current pulses activate motor axons to replace control signals lost as a result of spinal cord injury, cerebral stroke or brain injury.

Several orthoses have been developed for the improvement of hand function [1–3]. These methods have concentrated on achieving prehension/release by stimulating the finger flexor and extensor muscles and using a splint to stabilize the wrist. In one of these studies [2], the patient supplies the control signal by voluntary movement which in turn is measured and used to control the level of electrical stimulation of the paralyzed muscles. Other orthoses have been developed to enable standing, walking and exercise [4–7]. In [4], the torque at the hip, knee and ankle joints of a healthy individual were used to determine the amount of FES needed to compensate for these values in a paraplegic during standing up. While these studies

Address for correspondence: Bertram Nworah Ezenwa, Laboratory of Applied Physiology, 3171 Research Blvd. Kettering, OH 45420.

have achieved good results in an attempt to use FES to aid the use of paralyzed muscles, however, it has been reported [8–12] that stimulated muscles possess nonlinear and possibly time varying properties, which makes the use of data from one individual less adequate as setup parameters for another. It also makes the use of classical control methods less ideal in general.

It has been reported [13] that variations in the characteristics of paralyzed muscles have resulted in the development of various models for the response of paralyzed muscles to FES, but that few of these models have been useful for efficient design of a digital control system. This observation raises the question about the choice of an appropriate controller for the paralyzed muscles during FES exercise.

The purpose of this study is to determine the general trend of the static and dynamic response characteristics of the paralyzed muscle due to functional electrical stimulation and to use the determined characteristics to investigate an appropriate controller for the paralyzed muscles during FES.

METHOD

Five quadriplegic subjects who had been screened for exercise by the use of FES were used. As Shown in Figure 1, electrodes are placed generally at the motor points of the quadriceps muscle, the motor point being the point at the quadriceps muscle where a given electrical stimulus produces the largest contraction. Two

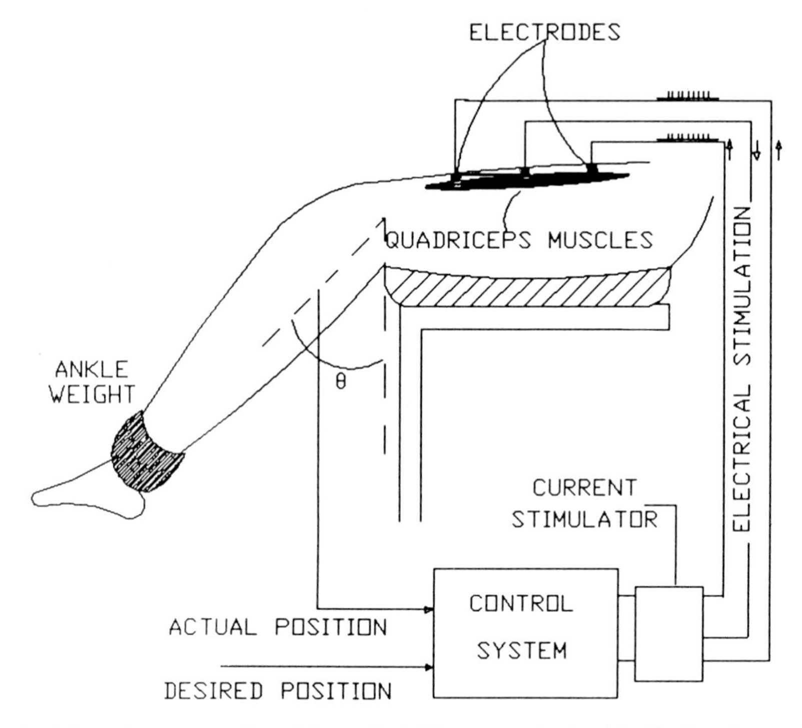

Figure 1 Schematic representation of the method. The operator's visual feedback and manual control substitute for the control system.

active electrodes referenced to an indifferent electrode in the middle are used to stimulate the muscles. An operator who controls the electrical pulses replaces the control systems. The stimulator used to supply electrical stimulation has the following characteristics: (a) is capable of generating biphasic constant current pulses, (b) has pulse repetition rate of 33.33 Hz, (c) has adjustable output current level to a maximum of 150 mA, (d) has adjustable pulse width to a maximum of 300 μs.

Experimental Setup

The experimental setup is as shown in Figure 2(a) which was later modified as in Figure 2(b). The modification was necessary for reasons such as cost and to enable direct calibration of effective workout of the paralyzed muscles. The operation of the setup is as follows: the quadriceps muscles are stimulated by electrical pulses from a constant current stimulator through two electrodes against an indifferent electrode. The amplitude of the current is controlled by the control voltage (CV) which is controlled by an operator's control of a potentiometer. A reference trajectory in the computer which determined the desired motion of the subject's leg consists of a raised sinusoid with a five seconds rise and five seconds fall. By following this reference trajectory, the quadriceps muscle is made to do work against a load in Figure 2(a) or ankle weight in Figure 2(b). The motion of the subject's leg is monitored by recording the knee joint angle. The knee joint angle is monitored indirectly by the rotation of a potentiometer, i.e. the knee sensor (KS) in Figure 2(a) or directly by the KS mounted on the knee brace in Figure 2(b).

Data for Dynamic Characteristics

The data for the determination of the dynamic characteristics of the system were collected by manually controlling the amount of electrical impulses to the quadriceps muscles. The quadriceps muscles are made to do work by having the subject's leg lift a load as shown in Figures 2(a) or 2(b). By visual feedback, the motion of the leg is made to track a reference trajectory by an operator who increases or reduces the electrical current input to the system in response to the error between the reference trajectory and the actual position of the leg. The position of the subject's leg is monitored by measuring the knee joint angle. The knee joint angles were sampled at 33.33 Hz just before the application of the next stimulus to the quadriceps muscles. This enabled the synchronization of the input/output to/from the muscle system. The input signals were also sampled at 33.33 Hz immediately following the sampling of the knee joint angle. The sampling occurred once again before the stimulus was sent to the muscles. Similar data were collected until the muscle was fatigued. The muscle was then allowed a rest period of five minutes before the process was repeated until the total number of lifts in an exercise session was more than 30. These input/output data were stored in the computer for off-line analysis.

Data for Static Characteristics

For the static characteristics, the amount of current stimulation (threshold currents) necessary to produce the onset contraction of the subject's muscle was recorded during the dynamic experiments. The aim of this measurement is to determine the

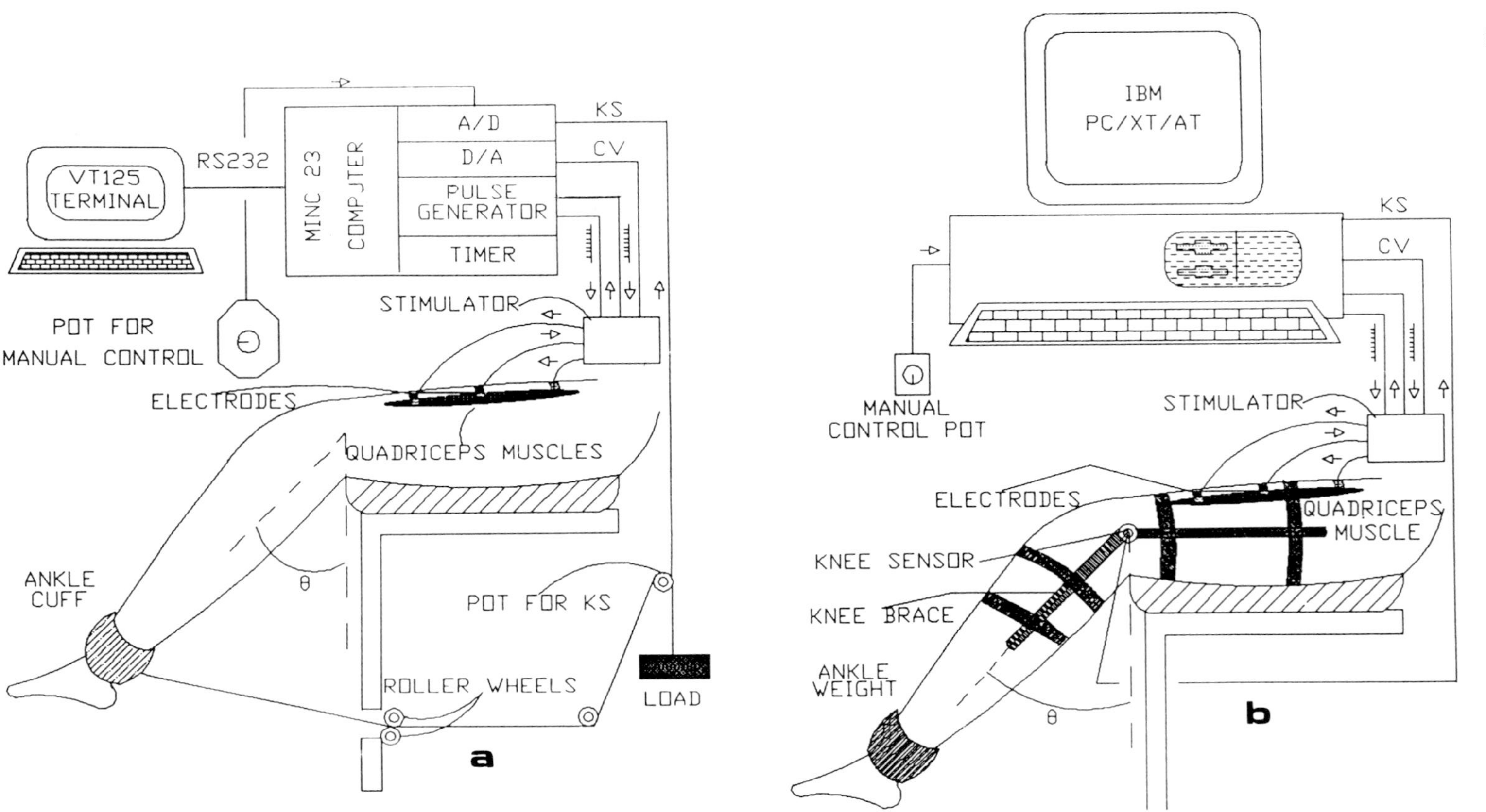

Figure 2 (a) Usual experimental setup. (b) Modification of (a). Note the advantages of (b) over (a), i.e., easier to use, direct calibration of effective workout of the muscles. KS is the knee sensor input signal to the computer, CV the control voltage, θ the knee joint angle, POT is the potentiometer. The indifferent electrode is at the middle of the two active electrodes.

changing impedance of the subject's muscles during exercise. This value was constantly determined at the beginning of each lift, until the muscle fatigued and for the entire exercise period.

ANALYSIS

Nonlinear Gain and Qualitative Studies

Curves of threshold currents versus lift number were plotted for each subject as shown by the square symbols in Figure 3. The curves for a subject were correlated qualitatively with the curves for that subject on different occasions and with those of other subjects. Two other parameters of the current demand were investigated. These include the mean stimulation current shown by cross symbols and the maximum current shown by triangles in Figure 3. A 3-D plot of the current demands for an exercise session of 35 lifts was made along with the error between the reference trajectory and the actual position for a subject. These plots are shown in Figure 4.

Dynamic Characteristics

The estimation of the parameters of a system demands that the input be rich and persistently exciting. From the response characteristics of the system, the input current provided such an input [13]. The system was characterized as a discrete time autoregressive moving average (ARMA) process. This process was written as a linear difference equation where process output is a linear combination of previous inputs and outputs. The general nth order difference equation relating the input $u(k)$ and the output $y(k)$ is given by

$$y(k)+a_1y(k+1)+\ \dots\ +a_ny(k-n)=b_0u(k)+b_1u(k-1)+\ \dots\ +b_nu(k-n) \quad (1)$$

where $a_1 \dots a_n$ and $b_0 \dots b_n$ are input and output coefficients respectively and k is the integer time index. The model order was sought by fitting the input–output data to different model orders and the goodness of fit compared. Two of the models considered will be discussed in this study. The first model (Model 1) has two input coefficients and two output coefficients and the second model (Model 2) has two input coefficients and three output coefficients. For each model, the parameters were estimated with a recursive least squares algorithm. The algorithm used is that of Hsia [14] where the parameter estimates are updated as each successive input–output pair is included. The parameter updating algorithm is given by

$$\phi(N+1)=\phi(N)+L(N+1)P(N)X(N+1)[y(N+1)-X^T(N+1)\phi(N)], \quad (2)$$

where

$$L(N+1)=1/[1+X^T(N+1)P(N)X(N+1)], \quad (3)$$

$$P(N+1)=(1/G)[P(N)-L(N+1)P(N)X(N+1)X^T(N+1)P(N)] \quad (4)$$

$$X(N+1)=[-y(N),\dots -y(N-k+1),\ u(N+1),\dots u(N-k+1)]^T \quad (5)$$

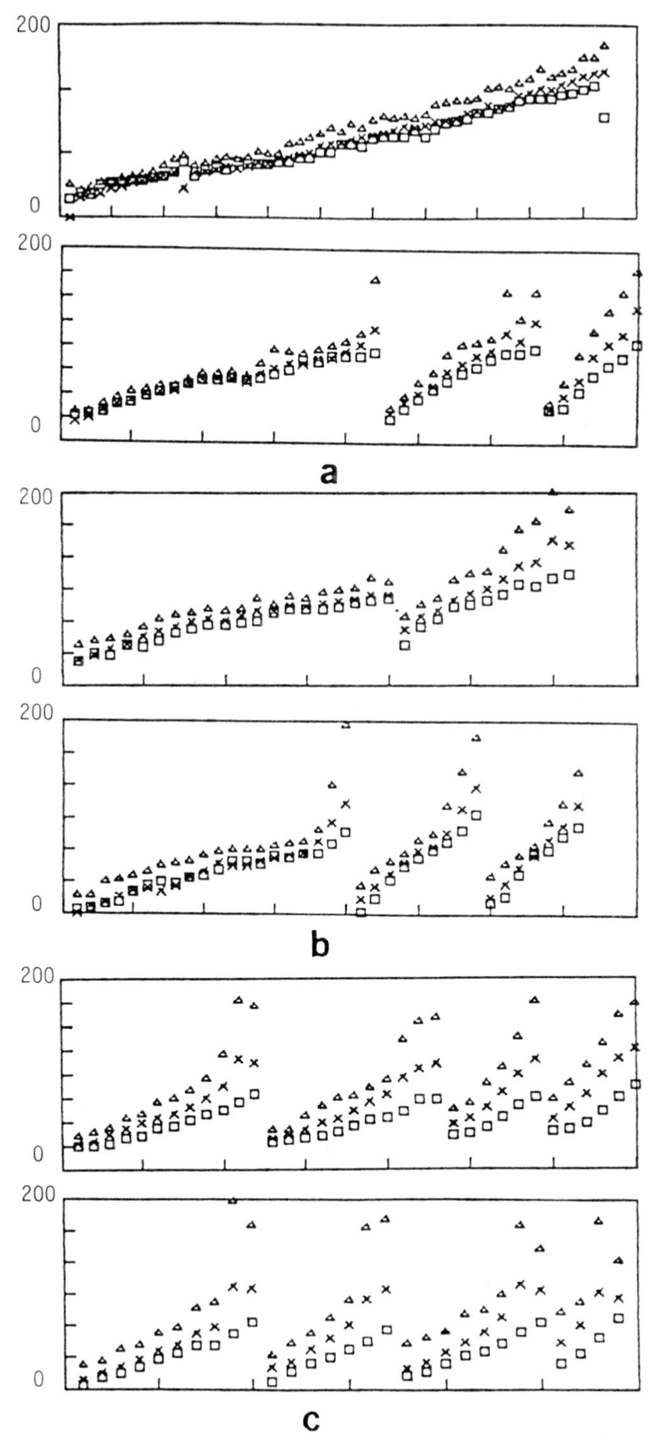

Figure 3 (a) Plot of threshold current in squares, mean stimulation currents in crosses and peak currents in triangles for one of the subjects. (b) and (c) similar results for two other subjects. Each tick mark in the horizontal axis represents five lifts. Note the variation in the number of lifts before fatigue in each subject and between subjects.

The algorithm could be started up with arbitrary values for $\phi(0)$, $P(0)$ as a large diagonal matrix, while $0 < G < 1$, specifically $G = 0.993$ to track slowly varying parameters.

RESULTS AND DISCUSSION

Nonlinear Gain and Qualitative Studies

As is shown in Figure 3, the threshold current increased with each increase in the lift number after a rest period. This increase is suggestive of increasing impedance of the muscle with subsequent lifts. If there is an increase in the impedance of the electrode/electrolyte/surface interface, the increase is negligible since after each rest period, the initial threshold current dropped back to the value at the beginning of the exercise. It is also worth noting that the increase in impedance is nonlinear in nature, and that the number of lifts before fatigue dropped off sharply after each rest period. The characteristics of the mean current and maximum current are very similar to the nonlinear gain characteristic, reinforcing the nonlinear characteristics of the system. Figure 4(a) is a pictorial view of the current demands of the system during lifts.

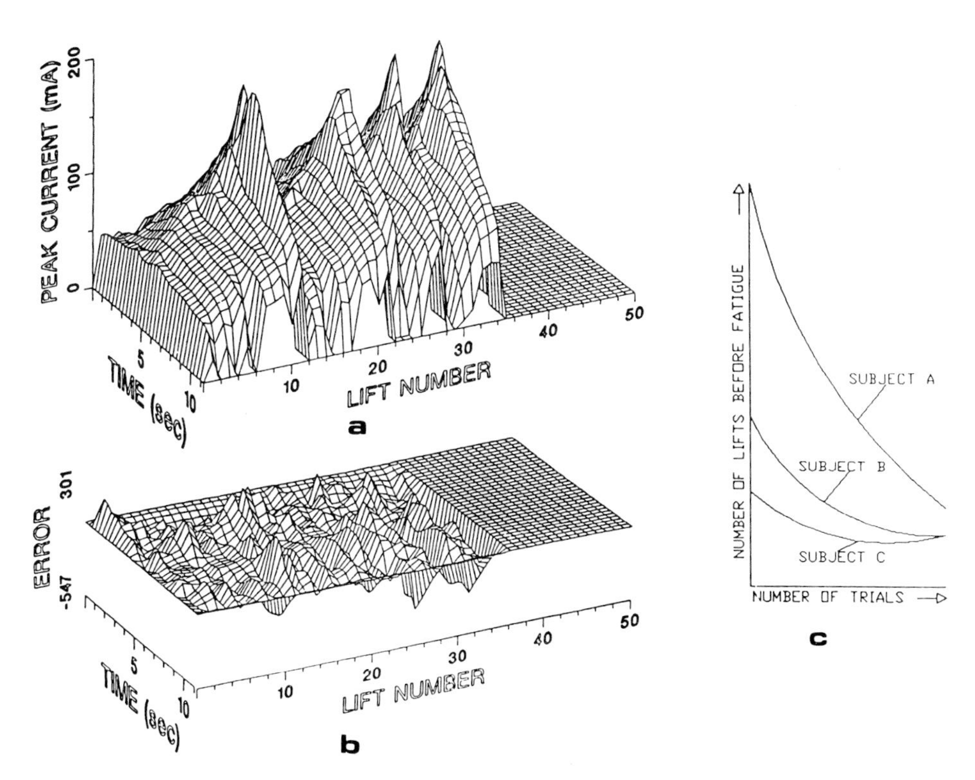

Figure 4 (a) Pictorial view of the current demands of the paralyzed muscles due to changing impedance of the system during functional electrical stimulation. (b) Error between the desired trajectory and the actual. (c) Variations in the progressive weakness characteristic between subjects.

From Figures 3 and 4(a), it can be seen that the muscles have a common property of progressive weakness or fading memory characteristics as the exercise session progressed. However, it is very evident that there are differences in the muscle characteristics between subjects and for the same subject on a different occasion. These differences include the number of lifts before fatigue and the progressive weakness characteristic (PWC), Figure 4(c). The PWC, which is inversely proportional to the number of times the muscle has fatigued, varied from subject to subject. In Figure 3(a), the number of lifts in between fatigues varied from 12 to 50; in Figure 3(b) it varied from 7 to 21, while in Figure 3(c) it varied from 6 to 12. It is equally noteworthy that manual control of the system to follow a reference trajectory is not easy as shown in Figure 4(b). This difficulty may be due in part to the time varying characteristics of the system.

Dynamic Characteristics

System identification by the method of least squares is only an approximation of the system dynamics. In general, as the number of the input-output parameters increase (increase in the model complexity), the approximation of the system dynamics approaches the explicit system dynamics, which will be very close to the dynamics of the system when the precise transfer function of the system is known. The input-output data collected during the experiment was fitted to two models with a recursive least squares algorithm [14]. Data from one of the subjects is shown in Table 1 and the variations in the parameters of the subject is shown in Figure 5. It can be seen in Table 1 that, as expected, the error of estimation (EE) was less for Model 2 than for Model 1. However, there are parameter variations for the subject between lifts in both models. The same variations were observed for the same subject under different occasions and between subjects.

Table 1 Table of statistics of parameter estimates

	Model 1	STDEV	Model 2	STDEV
a_1	1.8473	0.0438	1.3750	0.0061
a_2	−0.8491	0.4352	0.1619	0.8143
a_3			−0.5375	0.4800
b_1	0.132	0.0534	0.1088	0.5755
b_2	−0.1132	0.6177	−0.1011	0.5390
EE	1.9279	0.2542	1.3083	0.1873

Data obtained from one of the subjects for a total of 35 lifts. The values for a_1 to a_3, and b_1 and b_2 are the mean values. STDEV is the standard deviation, whereas EE is the mean of the mean squared estimation error less the common exponential factor.

Model 2 provided a better estimate for the system dynamics for the system; as a result, the dependence of the complex poles on the system gain was studied with Model 2. Figure 6 shows the movement of the complex poles in the z-plane. The poles of system became complex with a small increase in the gain of the system. When the gain was more than 5 the system became unstable. The location of the poles of the system in the complex plane depended on the gain of the system.

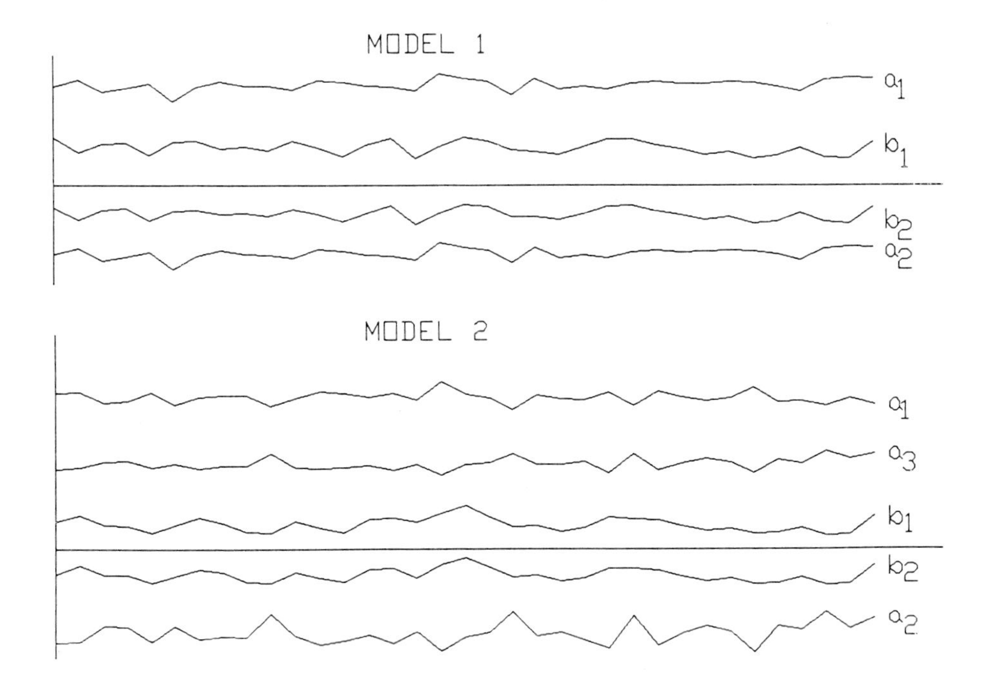

Figure 5 Choice of system characteristics. Data was obtained from the subject whose data is given in Table 1. Model 2 was chosen considering the dynamic characteristics.

It is known in control systems that a controller designed to operate with a set operating point fails when the operating point is changed. This brings up the question of the choice of the controller since from the static characteristics the threshold value changed with each lift thus changing the current demands to achieve the threshold. From the dynamic characteristics, the system parameters varied with each lift and had the potential of instability at high gains. These observations suggest that an online real-time adaptive control system will be an adequate controller for the paralyzed muscles during FES exercise.

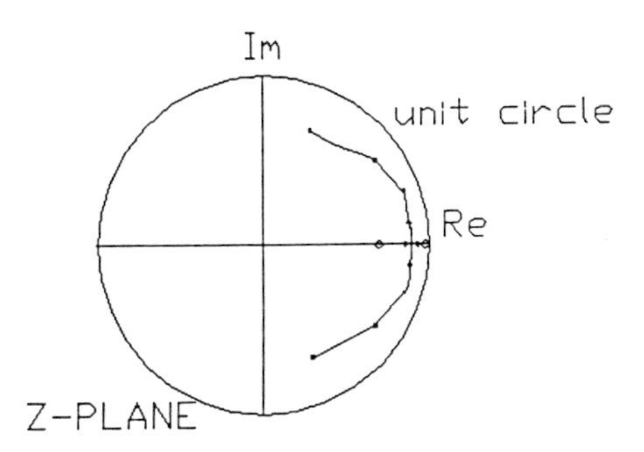

Figure 6 Dependence of nonlinear gain on complex poles for the data shown in Table 1. The gain was varied from 1 to 5, with an assumed natural frequency of 1 cycle per second. The system became unstable with the gain of more than 5.

CONCLUSION

The nonlinear gain characteristic of the system suggests internal nonlinear summation of individual twitches of the muscle fibers. This has been reported [9, 10]. The variation in the estimated parameters of a subject and between subjects together with the nonlinear summation of individual twitches makes the derivation of explicit models for the system unnecessary. Instead, it may be better to determine an exercise profile based on a global model of the system and then use some form of online adaptive technique to control the electrical stimulation of the muscles. This approach has been tried for the upper limb [15]. The results of this present study indicate that a linear difference equation with two input and three output coefficients provide a good representation of the response of the lower limb of quadriplegics to electrical stimulation. The use of a recursive least squares algorithm allows for real-time parameter estimation, making the use of adaptive techniques (model reference adaptive control), a better control strategy for the system.

REFERENCES

1. P. H. Peckham, E. B. Morsolais and J. T. Mortimer, Restoration of key grip and release in the C-6 tetraplegic through functional electrical stimulation, *J. Hand Surg.*, **5,** 462–469 (1980).
2. P. H. Peckham, J. T. Mortimer and E. B. Marsolais, Controlled prehension and release in the C-5 quadraplegic elicited by FES of paralyzed forearm musculature, *Ann. Biomed. Eng.*, **8,** 369–388 (1981).
3. A. Kraij and Vodovink, FES of extremities, *J. Med. Eng. Technol.*, **1, 2,** 12–15 and 75–80 (1970).
4. T. Bajd, A. Kralj and R. Turk, Standing up of a healthy subject and a paraplegic patient, *J. Biomech.*, **15,** 1–10 (1983).
5. E. B. Marsolais and R. Kobetic, Functional walking in paralyzed patients by means of electrical stimulation, *Clin. Orthop*, **175,** 30–36 (1983).
6. R. M. Glaser, J. R. Strayer and K. P. May, Combined FES leg and voluntary arm exercise of SCI patients, *PROC. Seventh Annu. Conf. IEEE Eng. in Med. and Biol. Soc.*, Chicago (1985), pp. 308-313.
7. J. S. Petrofsky, C. A. Phillips, R. Douglas and P. Larson, A computer-controlled walking system: the combination of an orthosis with functional electrical stimulation, *J. Clin. Eng.*, **11,** 121–133 (1986).
8. G. C. Joice, P. M. H. Rack, Isotonic lengthening and shortening movements of cat soleus muscle, *J. Physiol.*, **204,** 475–491 (1969).
9. G. C. Joice, P. M. H. Rack and D. R. Westbury, The mechanical properties of cat soleus muscle, *J. Physiol.*, **204,** 461–474 (1969).
10. R. B. Stein and F. Parmiggiani, Nonlinear summation of contraction in cat muscles: I early depressions, *J. Gen. Physiol.*, **78,** 277–293 (1981).
11. F. Parmiggiani and R. B. Stein, Nonlinear summation of contraction in cat muscles: II later facilitation and stiffness changes, *J. Gen. Physiol.*, **78,** 295–311 (1981).
12. P. M. H. Rack and D. R. Westbury, The effects of length and stimulus rate on tension in isometric cat muscle, *J. Physiol.*, **204,** 443–460 (1969).
13. L. A. Bernotas, P. E. Crago and H. J. Chizneck, A discrete-model of electrically stimulated muscle, *IEEE Trans. Biomed. Eng.*, **BME-33,** 829–838 (1986).
14. T. Hsia, *System Identification*, DC Heath and Company, Lexington, Toronto, Ont., Canada (1977).
15. J. Allin and G. Inbar, FNS parameter selection and upper limb characterization, *IEEE Trans. Biomed. Eng.*, **BME-33,** 809–817 (1986).

Automedica, 1989, Vol. 11, pp. 53–70
Reprints available directly from the publisher only
Photocopying permitted by license only

A FLEXOR TENDON MODEL OF THE HAND

DAVID J. GIURINTANO

Paul W. Brand Research Laboratory, Gillis W. Long Hansen's Disease Center, Carville, LA 70721

and

DAVID E. THOMPSON

Department of Mechanical Engineering, Louisiana State University, Baton Rouge, LA 70803

The hand has three major types of long flexor tendons: the Flexor Pollicis Longus of the thumb, the Flexor Digitorum Profundus and the Flexor Digitorum Superficialis of the fingers. These tendons cross multiple joints and traverse the length of the fingers. In order to mathematically model the kinematics of the long flexors, one's model must include the ability to allow for articulations to occur in any part of the chain. The flexor tendon model proposed allows for varying degrees of freedom at each of the three joints that the long flexors cross. The tendons were constructed from four primitive structures. The control points of these primitives are transformed with respect to the position of the bones of the hand. This interactive simulation is displayed on an Evans and Sutherland PS330 color graphics terminal and is one component of a comprehensive model for use in a hand biomechanics computer workstation.

INTRODUCTION

The construction of a physical analogue to the human hand and its muscle–tendon control system would be a valuable and difficult task. However, to understand the effects of variations in the lengths of the bones and in the placement of tendon paths, one would be required to construct a new analogue for each variation. An alternative method is to model the hand mathematically. In such a model, lengths of bones and placements of tendon paths are described symbolically. Modeling can be performed by simply assigning new numerical values to the symbols. Through the use of the rules of mathematics, additional information can be extracted from the model without the difficulties associated with the construction of a new physical analogue.

The model presented here is a mathematical representation of the kinematics of the long flexor tendons of the thumb and the fingers and is one component of a hand biomechanics workstation. The objective of the workstation is to allow a hand surgeon to interactively optimize a tendon transfer operation by modifying a simulation of the operation performed on the workstation. The surgeon will pick the path of the proposed donor tendon and its insertion point. The mechanical effectiveness of the transfer will be calculated. The surgeon will then evaluate the procedure and modify it if the results of the proposed surgical simulation are not satisfactory.

The flexor tendon model simulates the Flexor Pollicis Longus (FPL), the Flexor Digitorum Profundus (FDP), and the Flexor Digitorum Superficialis (FDS). The

Please direct all correspondence to David J. Giurintano.

tendon paths were assumed to be built from several primitives. Using these elementary building blocks, any flexor tendon path can be modeled. The paths of the FPL, FDP, and FDS were defined, constructed and integrated into an existing simulation by Buford and Thompson [1] depicting the joints of the thumb.

The model is incorporated into a computer graphics simulation so that the results can be displayed in a manner that is conducive to its understanding and correct interpretation. Otherwise, the only output of the model would be endless tables of tendon excursion versus joint angle (for five degrees of freedom of movement). The graphical output of the tendon's path with respect to the skeletal structure of the hand allows the viewer a sense of tendon kinematics. For this simulation to aid surgeons in performing improved tendon transfers, it must be very realistic.

BACKGROUND

In order to model the paths of the long flexors of the hand, a look at previous models is required. In 1961, Landsmeer [2] proposed his three models of a tendon crossing a joint. Model I was a tendon traveling over a trochlea. Model II was of a tendon running through a loop. Model III was of a tendon traveling through a tendon sheath. The very existence of three models is indicative of an incomplete knowledge of the geometric path of tendons crossing joints. It is probable that various models must be used to account for anatomical variations between joints.

In a workstation for hand biomechanics, one must consider the mechanical effects of proposed transfers. Brand [3] explained the effects of tendon transfers in the biomechanical realm. He was a proponent of the concept that the torques imposed on a joint by the tendons crossing that joint must be balanced. Brand discussed joint motion with respect to its axis of rotation and stressed the importance of observing the excursion required to restore function to the hand. If a transfer operation is performed using a muscle which does not meet the excursion requirement, then the range of motion of the joint will be reduced, not just at the joint to which the transfer was made, but for all of the joints the transfer crosses.

In Brand's exhaustive work on the mechanics of the hand [4], the muscles were described in terms of their fiber length and volume. A simple model of the muscle permitted the conversion of this data into relative tension capacity and excursions. Tendon excursion and its relationship to the range of motion was defined. The effect of external loads on the joints was described. Other factors, such as edema and friction, were discussed along with their effects on the joints. Relative to this study, Brand also presented data for moment arms versus joint angles, range of motion and required excursions for the FDP, FDS, and FPL tendons.

One must also use experimental data to verify the models proposed. An *et al.* [5] measured the tendon excursion versus joint angle for the tendons of the fingers. Landsmeer's models were discussed, and joint thickness versus excursion was correlated.

One of the first models of a tendon traversing an entire ray was proposed by Ou [6]. He developed a single joint model to predict the excursions of the tendons of the thumb. Ou's model allowed the carpometacarpal joint three degrees of freedom: rotation about the long axis of the metacarpal, flexion-extension, and abduction-adduction. The metacarpalphalangeal joint and the interphalangeal joint were assumed to be fixed, thereby eliminating any other degrees of freedom. For the model, the tendon's path was assumed to be a straight line from the pulley point to the attachment point. Although quite simple, Ou's model showed a good agreement with the experimental data for all eight muscles of the thumb.

Buford [7] extended Ou's work by developing a detailed interactive computer graphics simulation of the kinematics of the thumb. He also modeled the tendon paths as straight lines from pulley to insertion. The database for the bones of the thumb was obtained by digitizing X-rays of a cadaver specimen used by Ou. The simulation was displayed on an Evans and Sutherland PS330 color graphics system connected to a VAX host computer. A special graphics control program allows the user to interactively rotate, flex-extend, or abduct-adduct the thumb at the carpometacarpal joint. The metacarpalphalangeal and interphalangeal joints were similarly user-controlled and were modeled as simple flex-extend joints. Although only a database for the thumb exists, data for the bones of the hand and fingers could be readily added into the simulation. With Buford's simulation, one can observe the tendon paths, the moment arms at the carpometacarpal joint for the tendons, perform proposed tendon transfers, display the results, and observe the resultant excursions of the proposed transfer.

MODEL

The model of the finger flexor tendons is a kinematic simulation of the joints and the tendons of the hand. The model can describe the paths of the FPL, the FDP, and the FDS. In order to add realism to the simulation, the calculated path of the tendon is translated into four parallel paths thereby transforming the tendon from a line into a three-dimensional structure. Data for the model was obtained from X-rays and dissection of cadavers. Presently, data from scans of a cadaver's hand using CT images have been incorporated into the model.

An assumption of the model was that the axes of rotations of the fingers' joints were fixed with respect to the proximal bone segments over the full range of motion. The positions of the axes were determined through cadaver dissection and measurement. The inter-joint distances, the displacements of the rotation axes and the distances between the bones and tendon sheaths were also assumed to be constant for this simulation. The Carpometacarpal (CMC) joint of the thumb and the Metacarpophalangeal (MCP) joints of the finger were assumed to be three

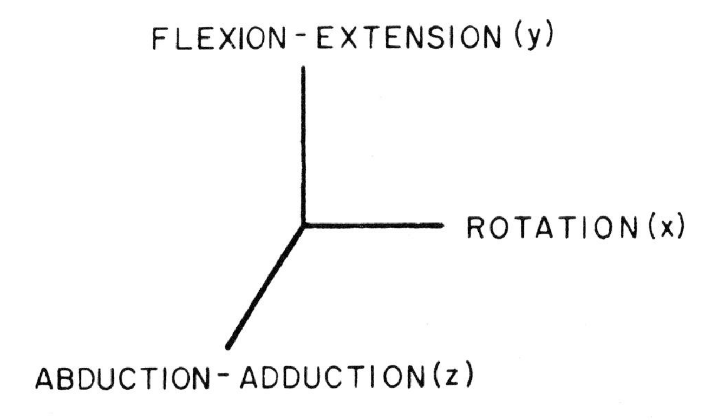

Figure 1 Orientation of the joint axes.

degree of freedom joints [rotation about the Rotation (x), Flexion-Extension (y), and Abduction-Adduction (z) axes]. The Metacarpophalangeal and Interphalangeal (IP) joints of the thumb and the Proximal Interphalangeal (PIP) and Distal Interphalangeal (DIP) joints of the fingers were modeled as one degree of freedom joints [rotation about the Flexion-Extension (x) axis] (Figure 1).

This model assumed that the flexor tendons of the hand may be derived from several primitives. These primitives are concatenated together to form the tendon's path. The model also allows for the ability to vary the moment arm of the tendon as it crosses the joint. Finally, the ability to incorporate soft tissue effects and tension-dependent models of the tendons exists in this preliminary model. After expressing the mathematical behavior of these primitives, they were combined to create the models of the tendons. Figures 3, 5, 6, and 7 depict the various primitives necessary to represent the flexor tendons of the hand. The reference position of a finger was defined as zero degrees of motion (neutral position) for all of the degrees of freedom of a joint of a bone with respect to the more proximal bone. The reference position of the thumb was described similar to the fingers except that the orientation of the CMC joint was not defined as zero degrees of motion. The orientation allowed the thumb to lie in the plane of the hand at rest. The angles of orientation that defined the position of the CMC joint were defined based on the data supplied by Cooney *et al.* [8].

The position of the control points of the primitives are mathematically described using the inter-joint distances and the degree of articulation of the joints of the kinematic chain. As depicted in Figure 2, mathematical matrix operations are used to move the control points from a reference position, P, to a new position, P', of the finger in space. This is described mathematically as

$$[P'] = [T_i] * [P].$$

Primitive A simulates a tendon traveling from control point 1 to control point 2 in a straight line. There are three cases of a tendon's path traversing a straight line. The first case is a tendon traveling through a tendon sheath (Figure 3A); the second case is a tendon crossing a joint and then entering a tendon sheath (Figure 3B); or the third case is a tendon exiting a sheath and inserting into a phalanx (Figure 3C).

Primitive B (Figure 5) simulates a tendon traversing a joint by following a circular path. Landsmeer's [2] III model is the basis of this simulation. This circular path is defined by three points—the distal point of the proximal sheath, B, the proximal point of the distal sheath, C, and an unknown point of closest approach, I, from the instant center of the joint, JC, to the tendon the arc crosses. The distance from JC to I is defined as the moment arm of the tendon, MA, and is obtained from experimentation. In order to determine point I, the problem is first transformed into a two dimensional system to simplify the solution (Figure 4).

The origin of the system is defined as the distal point of the proximal sheath, B. The x axis runs through the proximal point of the distal sheath, C. The joint center, JC, is then transformed to the local x–y plane. The tendon's radius of curvature, R, is calculated. Once the center of the radius of curvature, RC, is determined, the point of closest approach, I, is calculated. The point in the 2-D solution plane is then transformed back into the global system and the point of closest approach, I, is defined.

Primitive C (Figure 6) simulates a tendon bifurcating and then merging back into a single slip. The basis for this primitive is the bifurcation of the FDS to create a slip for the FDP to pass through.

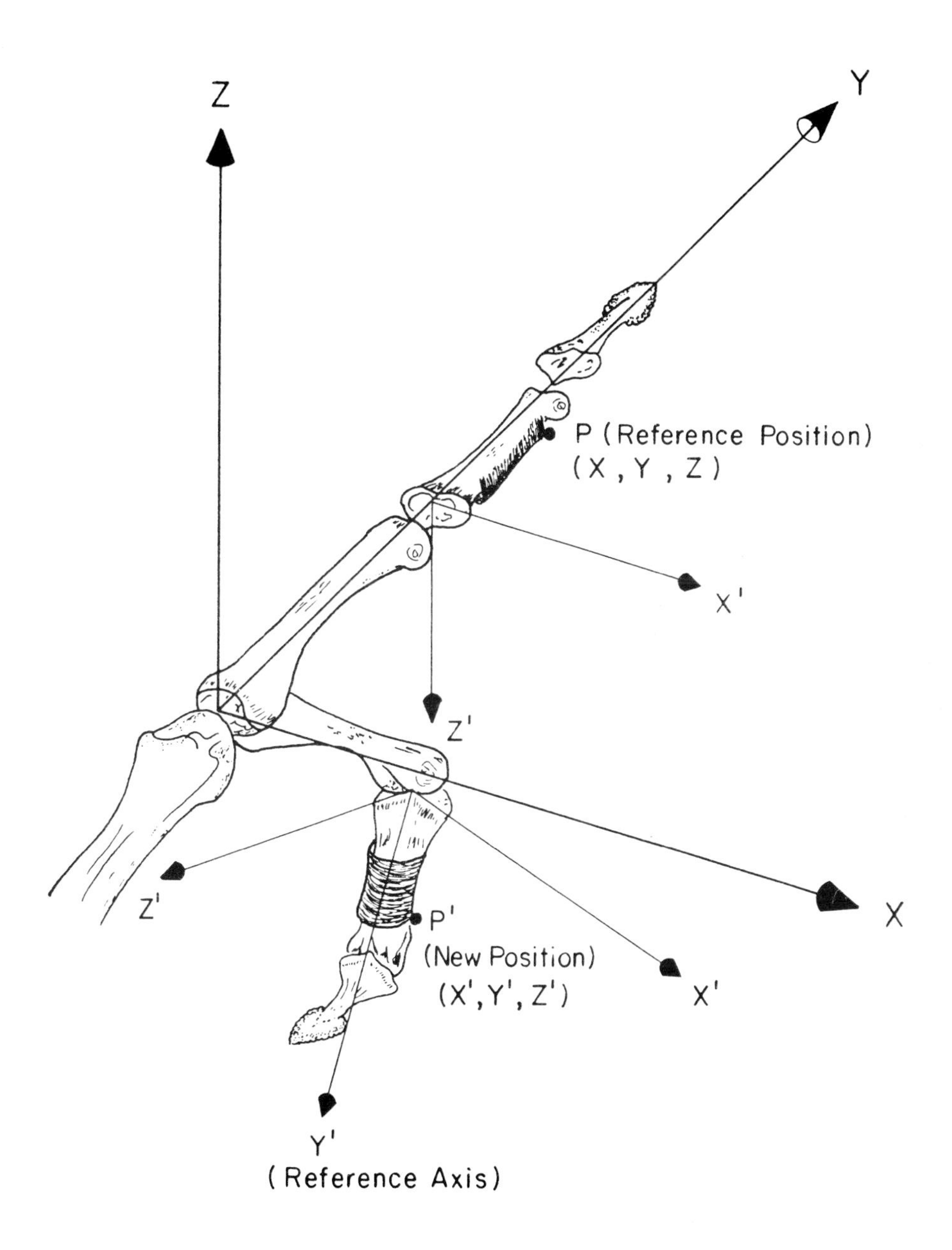

Figure 2 An example of the orientation of the axes of the transformed phalangeal segment.

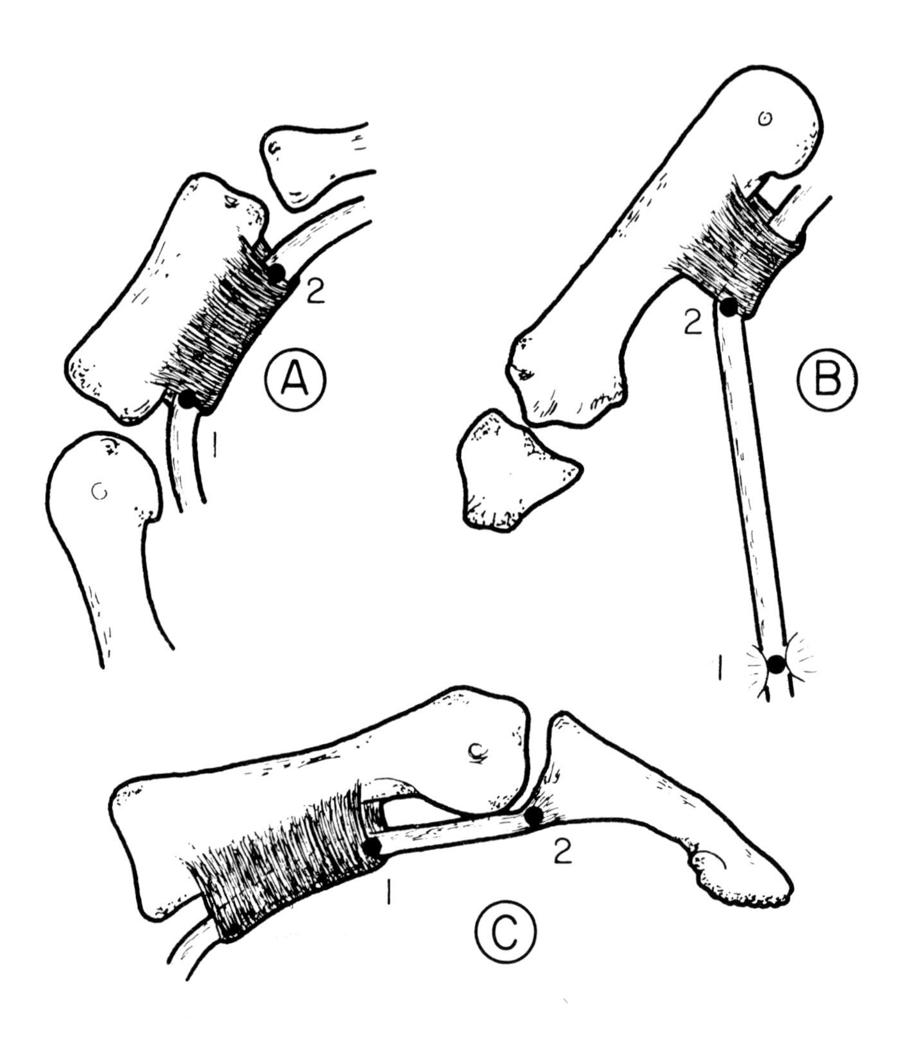

Figure 3 Tendon primitive *A* is a simulation of a straight line tendon path. *A* depicts a tendon traveling through a tendon sheath. *B* is a tendon bowstringing across a joint. *C* depicts the tendon exiting its sheath and attaching to the bone.

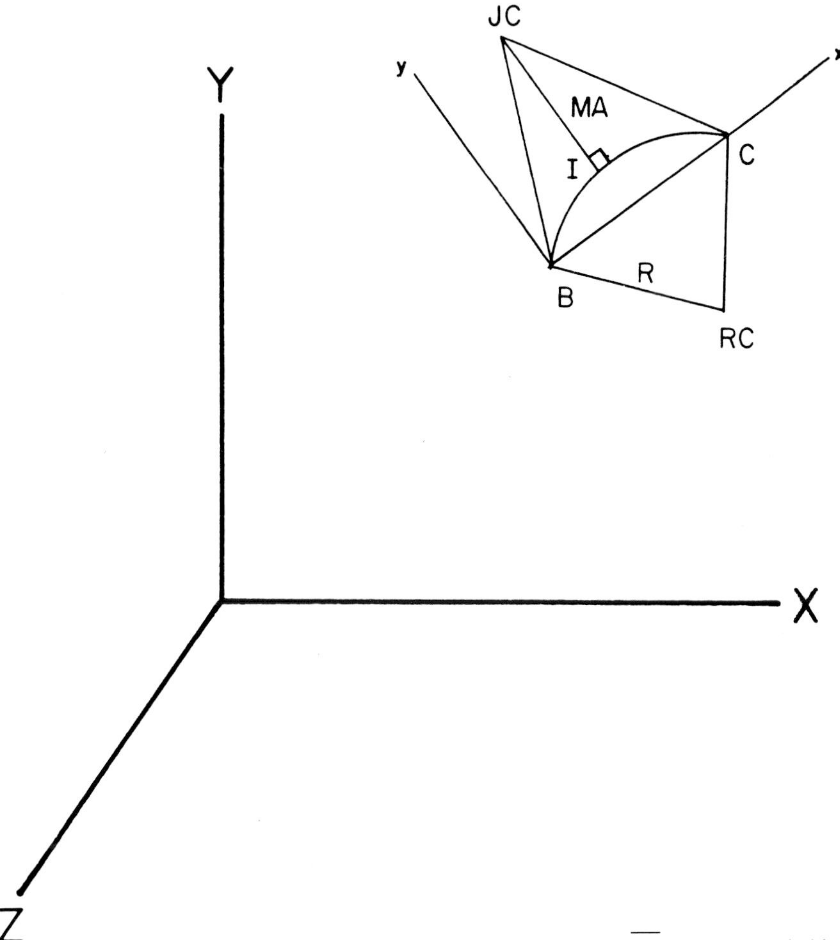

Figure 4 The 2-*D* description of the definition of the circular arc, $\overline{BC}$, for tendon primitive *B*.

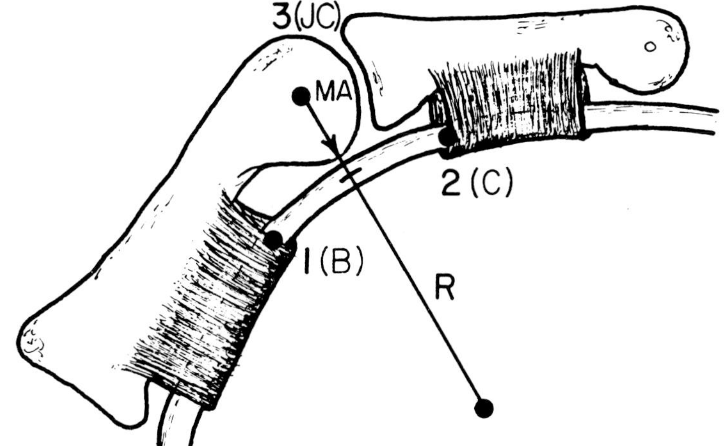

Figure 5 Tendon primitive *B* is a simulation of Landsmeer's third model for a tendon path, $\overline{BC}$, crossing a joint by traversing a circular path.

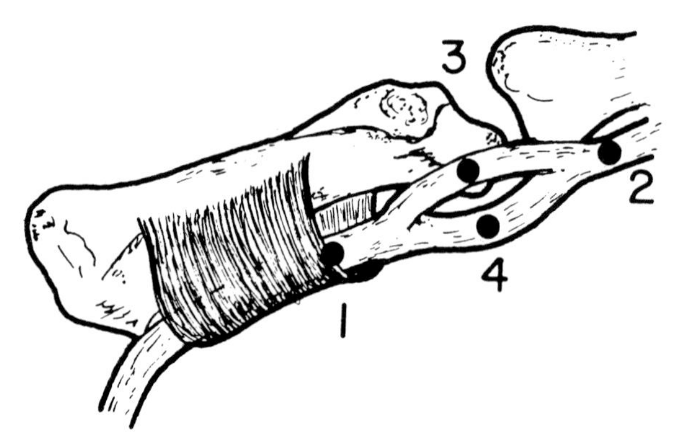

Figure 6 Tendon primitive *C* is a simulation of a tendon dividing and reconnecting to form a loop.

Primitive *D* (Figure 7) simulates a partial tendon path bifurcating and then inserting into a bone. The most distal path of the FDS is the basis for this primitive.

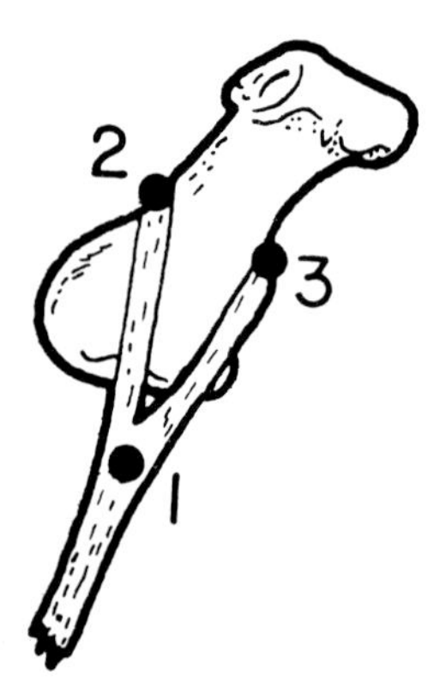

Figure 7 Tendon primitive *D* is the simulation of a tendon dividing and attaching to a bone with dual attachment points.

In order to model the FPL one must simulate its path from its exit of the carpal tunnel by bowstringing across the CMC joint, traveling through a tendon sheath and then traversing a circular path across the MCP joint, traveling through the sheath, and inserting into the distal phalanx. The FPL's path is defined by combing primitives *A* and *B* in the sequence A–B–A–A from the pulley point to the insertion point (see Figure 8).

The FDP's path is similarly defined by combining primitives *A* and *B*. However, the FDP traverses a circular path across the MCP joint instead of bowstringing across the joint. The FDP's building blocks sequence is B–A–B–A–A (see Figure 9). The path of the tendon proximal to the MCP joint of the finger has not been modeled in this work.

The FDS's path is simulated by the building block sequence B–A–C–A–D. The FDS traverses the MCP joint by a circular path similar to the FDP, then enters a

corresponding joint angle. The control points are calculated and the tendon's new path is displayed on the PS330's screen. The control points are also used to compute the excursions of the tendons relative to their lengths at the reference position. The total excursion, E_{total}, of the path is calculated by

$$E_{total} = d(P, A) + d(B, C) + d(E, AT),$$

where $d(P, A)$ is defined as the distance from point P to Point A.

The total excursion when the tendon is in its reference position is defined as its resting length, R. Therefore, the tendon excursion, T, is

$$T = E_{total} - R.$$

This information is then returned to the PS330 for immediate display. The model calculations are executed interactively on the host computer.

RESULTS

The resulting display of the simulation for the nine long flexors of the phalanges are depicted in Figures 11A to 11D. These displays simulate the hand performing specific motions, the hand at rest (Figure 11A), the hand grasping a cylinder (Figure 11B), tip pinch of the thumb and index finger (Figure 11C), and the hand in extension (Figure 11D). The tendons track a path through the range of motion near the bones to which the tendons' sheaths are attached. The positions were obtained by rotating the control dials of the PS330 (Figure 12), thereby simulating the position of the hand for the tasks described above.

The model not only displays the paths of the tendons but also calculates the tendons' excursions through the range of motion. Logically, the tendon's excursion decreased as the flexion angle increased. The model predicted a maximum FDP excursion of 2.3 cm for PIP joint flexion. This prediction compared favorably to An *et al.* [5] whose maximum excursion at the PIP joint was 2.1 cm. The maximum excursions of the model also approximated the magnitude of the profundus tendon excursions as specified in Brand [4].

Although the flexor tendon model was not rigorously verified with experimental data, the experimental data presented above does agree with the predictions of the model. The results of the simulation also met the goal of producing accurate and realistic interactive renderings of the flexor tendons of the hand.

CONCLUSION

The model of the long flexors of the hand was constructed to serve two purposes: to produce accurate realistic renderings of the flexor tendons and to predict accurate tendon excursions. The graphics realism of the flexor tendons portrayed here is immensely more satisfying than the previous two-point tendon model, and the model also improved the accuracy of the prediction of the tendon excursion. The two-point model would overestimate tendon excursion by 50% for maximum flexion of the joints. The real accuracy of this model still needs to be verified by comparison of the model's predicted tendon excursions to experimental data.

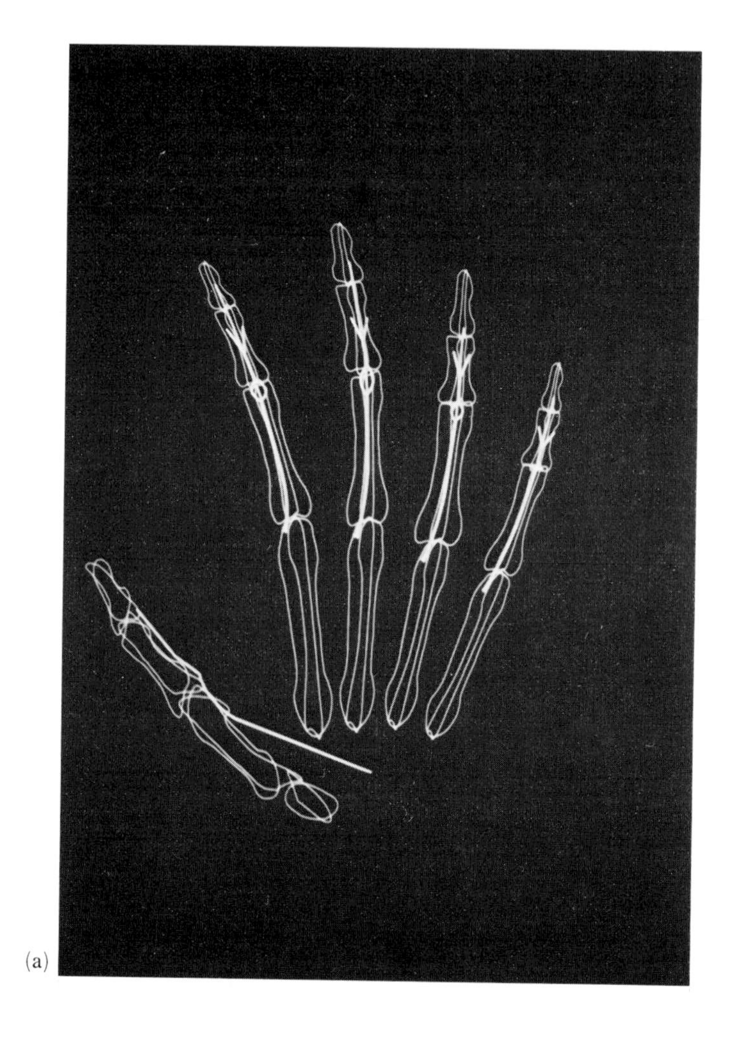

(a)

Figure 11 Photographs of the flexor tendon model simulation from the Evans and Sutherland display. (a) Hand in a *resting* position.

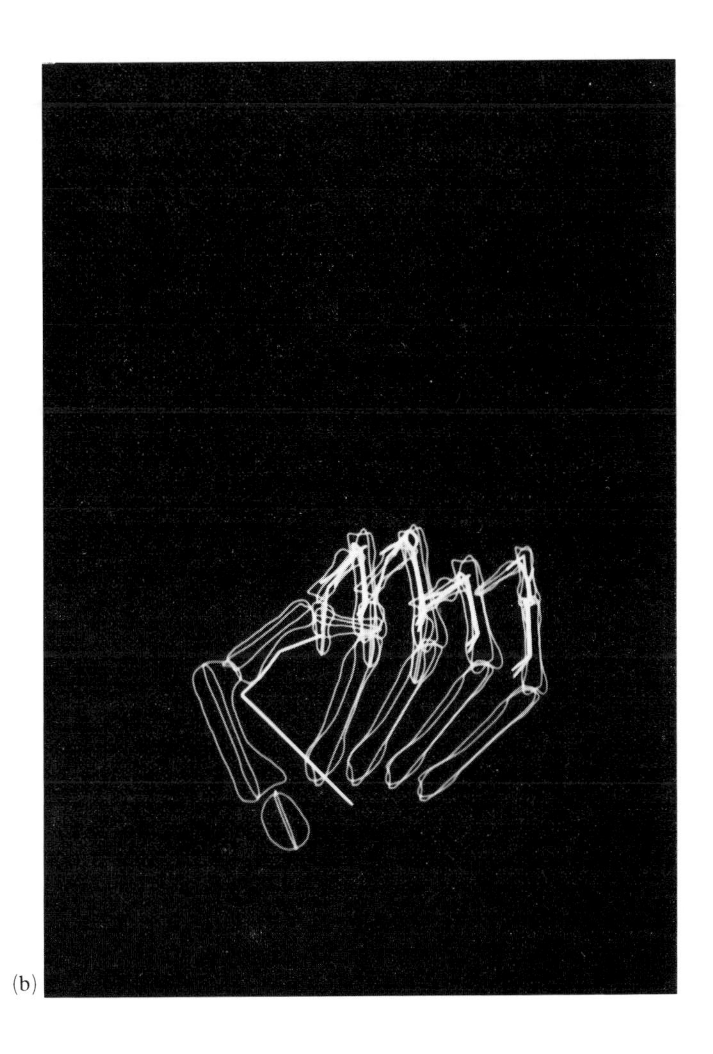

(b)

Figure 11 (continued) (b) *Cylinder grip.*

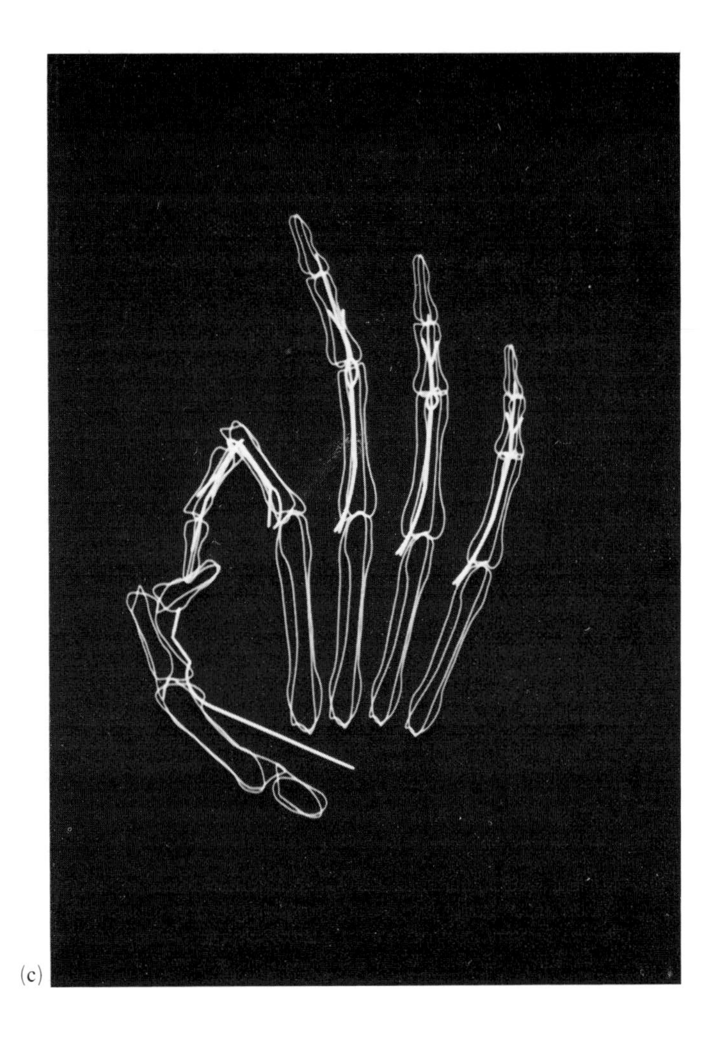

Figure 11 (continued) (c) Hand in the *tip-pinch* position.

(d)

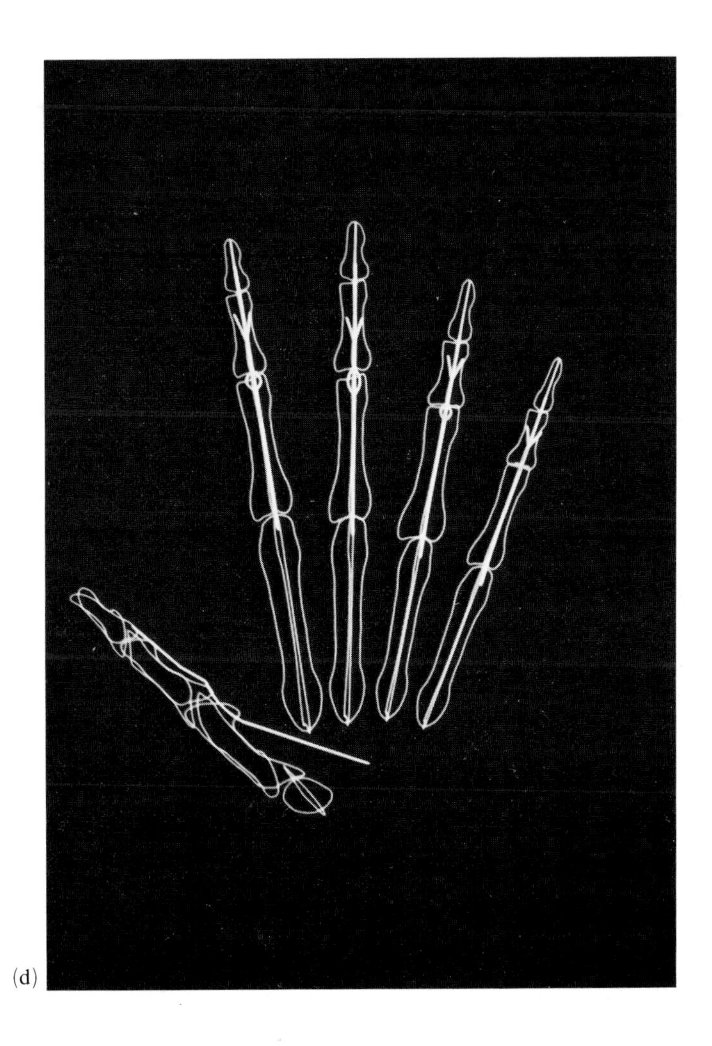

Figure 11 (continued) (d) Hand in a *fully extended* position.

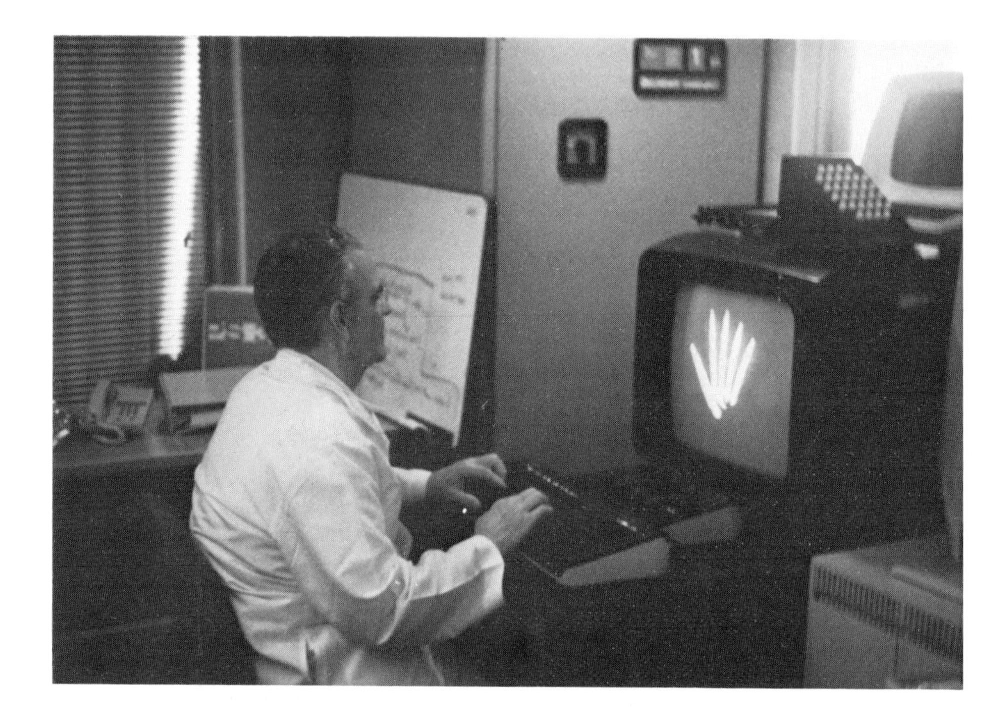

Figure 12 A user of the simulation at the PS330.

Presently, the model is being incorporated into a kinematic skeletal structure composed of bones created from CT images. Myers *et al.* [9] have developed a method to convert these images in segmented bone files. These images are displayed on an Evans and Sutherland PS390 color graphics terminal. The model software is also being converted so that its calculations will be performed by the PS390's processor. Figure 13 is a display of the new simulation.

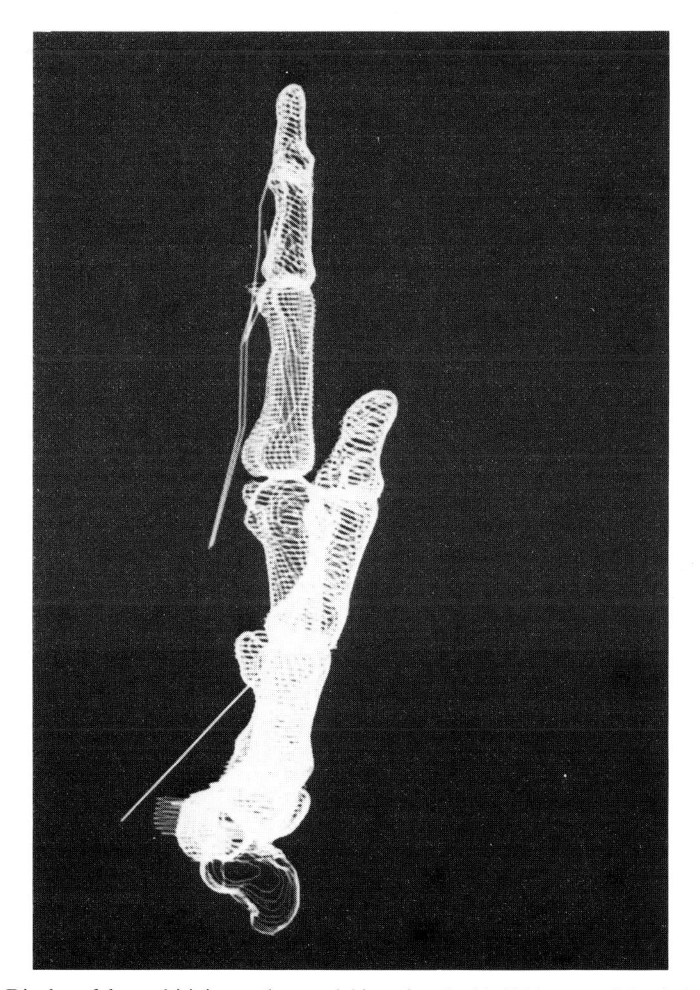

Figure 13 Display of the multi-joint tendon model interfaced with CT images of the skeletal structure.

ACKNOWLEDGEMENTS

The funding for this research was provided by the U.S. Public Health Service, Department of Health and Human Services under research contract 240–83–0060. Additional computing support was provided by the Computer Graphics Research and Applications Laboratory, Department of Mechanical Engineering, Louisiana State University through a joint research program with Digital Equipment Corporation. Evans and Sutherland has provided a PS 390 display system for use in this project. The CT images for this research were provided by Digital Diagnostics, Inc. of Baton Rouge, LA.

REFERENCES

1. W. L. Buford and D. E. Thompson, A system for 3D interactive simulation of hand biomechanics, *IEEE Trans. Biomedical Engr.*, **34** (6), 434–453 (1987).
2. J. M. F. Landsmeer, Studies in the anatomy of articulation, *Acta. Morphol. Neerlando-Scandinavia*, **3**, 287–321 (1961).
3. P. W. Brand, Biomechanics of tendon transfer, *Orthopedic Clinics of North America*, **5** (2), 205–230 (1974).
4. P. W. Brand, *Clinical Mechanics of the Hand*, C. V. Mosby, St. Louis (1985).
5. K. N. An, Y. Ueba, E. Y. Chao, W. P. Cooney and R. L. Linscheid, Tendon excursion and moment arm of index finger muscles, *J. Biomechanics*, **16** (6), 419–425 (1983).
6. C. J. Ou, The biomechanics of the carpometacarpal joint of the thumb, Doctoral Dissertation, Department of Mechanical Engineering, Louisiana State University (1979).
7. W. L. Buford, Interactive three dimensional computer graphics simulation of the kinematics of the human thumb, Doctoral Dissertation, Department of Engineering Science, Louisiana State University (1984).
8. W. P. Cooney, M. J. Lucca, E. Y. Chao and R. L. Linscheid, The kinesiology of the thumb trapeziometacarpal joint, *J. Bone and Joint Surgery*, **63-A** (9), 1371–1381 (1981).
9. L. M. Myers, W. L. Buford and D. E. Thompson, A graphics editor for 3-D CT-scan data for musculo-skeletal modeling, *Proceedings of the International Symposium CAR'*87 *Computer Assisted Radiology*, Springer-Verlag, Berlin (1987) pp. 476–483.

Automedica, 1989, Vol. 11, pp. 71-89
Reprints available directly from the publisher only
Photocopying permitted by license only

A PRELIMINARY STUDY OF RHYTHMIC MOVEMENT

H. HEMAMI, J. S. BAY and J. B. EVANS

Department of Electrical Engineering, The Ohio State University, Columbus, Ohio 43210

Two attributes of human movement that have common characteristics with physical man-made systems are stability and rhythmic movement. In this paper very simple rhythmic movements are studied within the confines of available basic engineering knowledge about physical systems. These simple movements merit study for two basic reasons:

1) They can shed light on how certain circuits of the central nervous system of living systems may be modeled.

2) They may eventually find application in functional neural stimulation for generation of movement patterns for rhythmic and nonrhythmic movement. The latter application would presume a second artificial control system to take care of postural stability and postural adjustments. However, this second system is not considered here.

The paper is devoted to the modeling and computer simulation of three systems that produce rhythmic movement:

1) coupled neural-type oscillators that may correspond to central pattern generators in living systems and how they may be used in synthesis of coordinated movement other than rhythmic movement,

2) limbs undergoing rhythmic movement brought about by reciprocal excitation and other forms of feedback, and

3) a combination of a limb and a neural circuit.

Previous research is studied for a general characterization of the behavior of central pattern generators in living systems. These generators (CPGs) are found to produce muscle stimulation signals with attributes amenable to reproduction by an artifical network of coupled nonlinear oscillators. It is shown that free parameters and coupling coefficients can be chosen to configure and tune the oscillators to result in outputs comparable to a wide range of observed CPG behavior. Furthermore, the artificial CPG model shows adequate structural consistency with neural networks to suggest its application in artificial neural stimulation.

A single limb planar arm with a pair of muscle models embodying the nonlinear spindle behavior is investigated by digital computer simulations. The effect of reciprocal excitation, leading to tremor-like oscillations, is demonstrated qualitatively.

Finally, linear models of a second order neural oscillator and the planar limb are coupled. The loci of the roots of the characteristic equation of the resulting fourth-order system are plotted to demonstrate that periodic movements can result from the couplings of both the musculo-skeletal and neural circuits.

Keywords: Artificial pattern generation, musculo-skeletal dynamics, neural prosthesis, neuromuscular stability, oscillations, reciprocal excitation, rhythmic motion, spindle feedback.

1 INTRODUCTION

Postural stability and adjustments and coordinated movements are two important and challenging areas of study for unravelling some of the vast control machinery of the central nervous system and for possibly helping the handicapped by artificial control systems and functional neuro-muscular stimulation. For ultimate success in helping the handicapped regain normal locomotion and manipulative function both areas are crucial and must be studied side by side. Postural adjustments have received a certain amount of scientific attention lately [1–6] and will not be further pursued here.

Rhythmic movements are a simpler class of coordinated movements and appear to provide efficient locomotion [7–9]. The origins of rhythmic movement have not been unequivocally established. Physiological evidence strongly indicates the presence of central pattern generators (CPGs) in the high level nervous systems of higher and lower animals. Grillner [10] has reported experiments with decorticate cats which exhibit normal locomotory function and normal gait transitions with increasing treadmill speed. Studies with other deafferented animals have suggested a source of locomotory pattern signals somewhere in the spinal column. A widely accepted conclusion is that groups of neurons act as coupled oscillating units which together constitute the CPG. Others believe that in the absence of cerebellar inhibitory signals, high gain devices in the spinal cord allow the musculoskeletal system to oscillate without resort to specific neural pattern generators. In generic control theoretical sense [11], an increase in the feedback loop gain of a linear system—either in the feedforward or feedback element of the loop—or the introduction of sufficient delay in the feedback loop can cause a system of third order or more to develop spontaneous oscillations or become unstable [11, 12]. In most man-made systems the same feedback machinery that brings about stability can also be used to cause periodic movements. It is possible therefore to combine the control circuitries needed for stability and feedback and by adjusting (or modulating) the gain of these circuits to maintain stability and produce periodic and rhythmic movement. Some generalizations and discussion of the additionally needed start-stop rule [13] are possible. In natural systems, however, it is not known whether or not the same circuits are used for coordination, rhythmic movement and stability, tracking and other skilled movements. Apart from the above considerations, stability plays a crucial role in constructing comprehensive models of the human or animal neuro-musculo-skeletal system that yield to functional and qualitative studies of movement. A generic approach to the stability problem, not strictly based on anatomical and physiological attributes of the neuromuscular system, has been to collectively model the visual, proprioceptive, and vestibular systems as a feedback system of angular positions and angular velocities of the skeletal segments. Digital computer simulation of such skeletal models with the above abstract angular position and angular velocity feedback have shown that indeed postural stability is guaranteed.

A further improvement in the models is to make them more physiologically accurate. Two factors of this improvement are replacement of moment of force (torque) actuators by systems of pairs of muscles, and use of more realistic models of the proprioceptive system, i.e., spindles, golgi, and joint sensors [4–16]. The present paper deals with the stability of a simple nonlinear reflex loop consisting of a planar arm (or hand), a pair of identical actuator muscles and a nonlinear spindle model. Certain simplifications and assumptions are made in order to keep the dimension of the overall system low and to render the programming and computations manageable. The major but modest goal is to be able to include large numbers of such actuators and sensors in computer models of the musculoskeletal system. As a result, the overall system becomes somewhat more physiologically relevant even though the individual muscle models may yet need improvements. The focus in this paper is on a nonlinear spindle model. The issues of delays, predictive mechanisms, golgi feedback, stiffness control and regulation [17] are not considered here. Stiffness regulation would be necessary only when the system is under additional persistent force disturbance. It is assumed the above system is not under the influence of such disturbances. Further background for this work may be found in the survey papers of Agarwal and Gottlieb [18] and Hulliger [19].

The problem of coupled oscillators as central pattern generators is studied first. The planar one-link torque muscle system is considered next. Finally, a coupling of a linear planar limb with a second order linear neural circuit is considered.

2 GENERATION OF ARTIFICIAL PATTERNS

Recently a mathematical model has been constructed from nonlinear oscillators which successfully simulates the CPG signals present during rhythmic locomotion. Coupled van der Pol equations were equipped with tunable parameters and coupling gains which enable the network to artificially generate leg trajectories for biped walking and hopping gaits [20–22].

In this paper we generalize the attributes of neural CPGs and demonstrate that the same network of oscillators reasonably models the structure and function of neural networks.

Many authors have shown that the most basic requirement of motor stimulating signals is the phasic relationship of bursts in each channel. Examples are the alternation of bursts in agonist-antagonist pairs, burst synchronization in interlimb coordination, and intersegmental lags in swimming fish. Digital computer simulations in [21] have verified the effectiveness of coupling strength tuning in control of the coupled oscillator's phasic relationship. A limited range of amplitude and frequency was also achieved through parameter variations within each oscillating unit.

2.1 *Formulation of Two Coupled Oscillators*

Consider the simple CPG represented by the two oscillator network of Figure 1. This system realizes the equations

$$\ddot{x}_1 - \mu_1\left(p_1^2 - \frac{k_1}{\mu_1}x_a^2\right)\dot{x}_1 + g_1^2 x_a = q_1,$$

$$\ddot{x}_2 - \mu_2\left(p_2^2 - \frac{k_2}{\mu_2}x_b^2\right)\dot{x}_2 + g_2^2 x_b = q_2, \tag{1}$$

where

$$x_a = x_1 - \lambda_{12}x_2,$$

$$x_b = x_2 - \lambda_{21}x_1,$$

and

$$\dot{x}_i = \frac{dx_i}{dt}.$$

Roughly speaking, μ_i is used to affect waveshape; wave amplitudes are approximately proportional to p_i; g_i affects frequency; and q_i represents static and

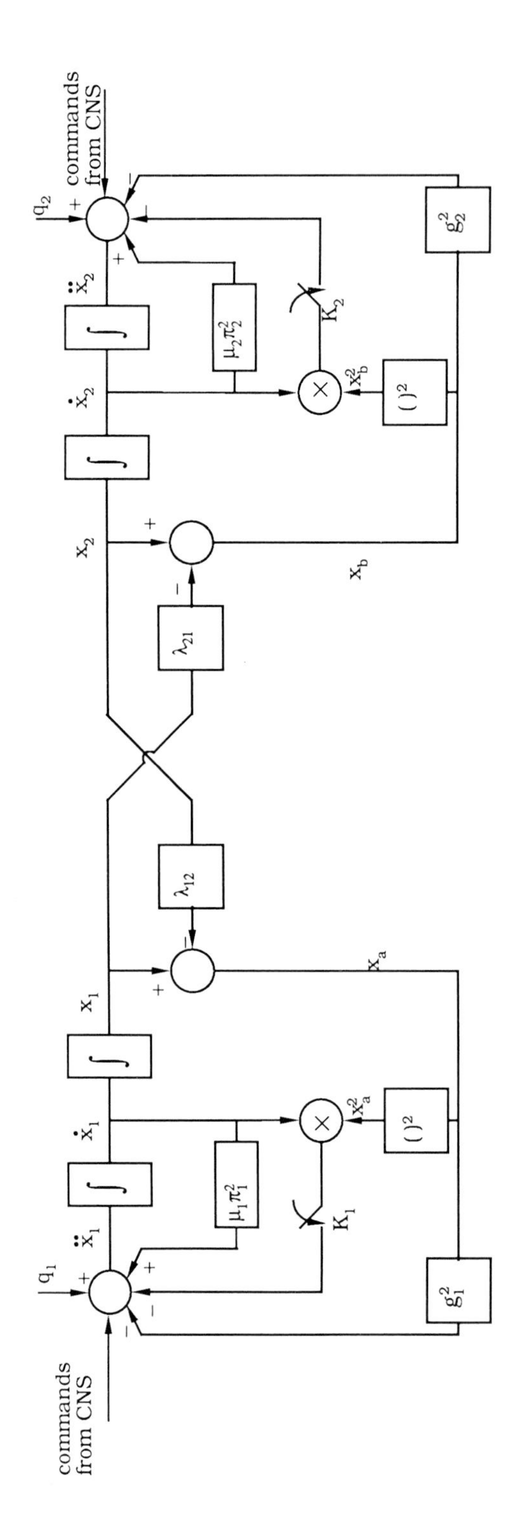

Figure 1 Schematics of a pattern generator with its feedforward and feedback gains.

dynamic offset, all for $i=1$, 2. These relationships are complicated by the interconnection coefficients λ_{ij}, which provide a path for each oscillator to perturb the others' characteristics. The algebraic sign of the λ_{ij}'s determines the excitatory or inhibitory nature of the coupling and thus controls primarily phase relationships. If the newly introduced gains k_i are constrained to have binary values 0 and 1, they can easily be seen to determine the nonlinearity of the oscillators. The variability of these and all the other parameters in Eq. (1) enables the system to exhibit the structure of a *hard-wired* neural network without the rigidity of fixed roles [23]. If it is assumed that a different network is dedicated to each synergist then reconfiguration of parameter values will accomplish gross changes in motor behavior. For example, if one wishes to wiggle his fingers while keeping his outstretched arm stationary, the normally periodic signals to shoulder and arm muscles can be tuned to drive isometric tension. The constant signals provided when p_i^2 is turned off (set to zero) circumvent any switching action in oscillator unit interconnections and provide complete functional disconnection.

2.2 *Pattern Generation*

All of the pattern behavior attributed to CPGs in [10] can be mimicked with the model (1). *Push Button* controls on coupling coefficients have been shown [22] to effect automatic gain transitions in bipeds and quadrupeds. Undulatory spinal motions can be elicited by coupled rings of oscillators which exhibit 360°/n phase differences in adjacent oscillators of an n oscillator ring. Intermediate phase differences which correspond neither to alternating nor synchronous coordination can be generated for use in, e.g., dancing, athletics, or pathological gaits.

Further versatility of the model is demonstrated by the extension of its application to nonperiodic point-to-point motion. Since the gains k_i can be set to zero, the oscillators, coupled or decoupled, can behave as linear fourth- or second-order systems. Thus the push-button command signals [10] can be processed into a wide variety of impulse or step responses. Figure 2 shows a typical schedule of parameters and its resulting waveforms.

The above coupled van der Pol model of a CPG is entirely functionally consistent with more popular abstract models of neural oscillators [10, 23]. Structurally, construction of the network was not intentionally guided by any knowledge of actual neural interconnections. However, if there exists a practical interest in the use of the artificial CPG in the laboratory, the consistency with physiological and anatomical data should be discussed.

In natural CPGs the mechanism for control of pattern characteristics and coupling strength is not fully understood. Studies indicate that the CPGs are activated by descending pathways, but that parameter control depends heavily on reflex effects. In cats, for example, afferent information about hip position, movement direction, and extensor load [1] may be processed to one or more channels of CPG input. Currently, there is not a preferred model for the neural connection of CPGs. Several authors have developed neural models of interconnected oscillating units with some regard to individual unit construction [24, 25]. In all such models mutual inhibitory and excitatory coupling is the primary mechanism for phasic control, as represented in the mathematical model (1). Although we cannot claim an exact correspondence between oscillator network gains and any particular neural signals, we can show consistency with popular theory of CPG structure.

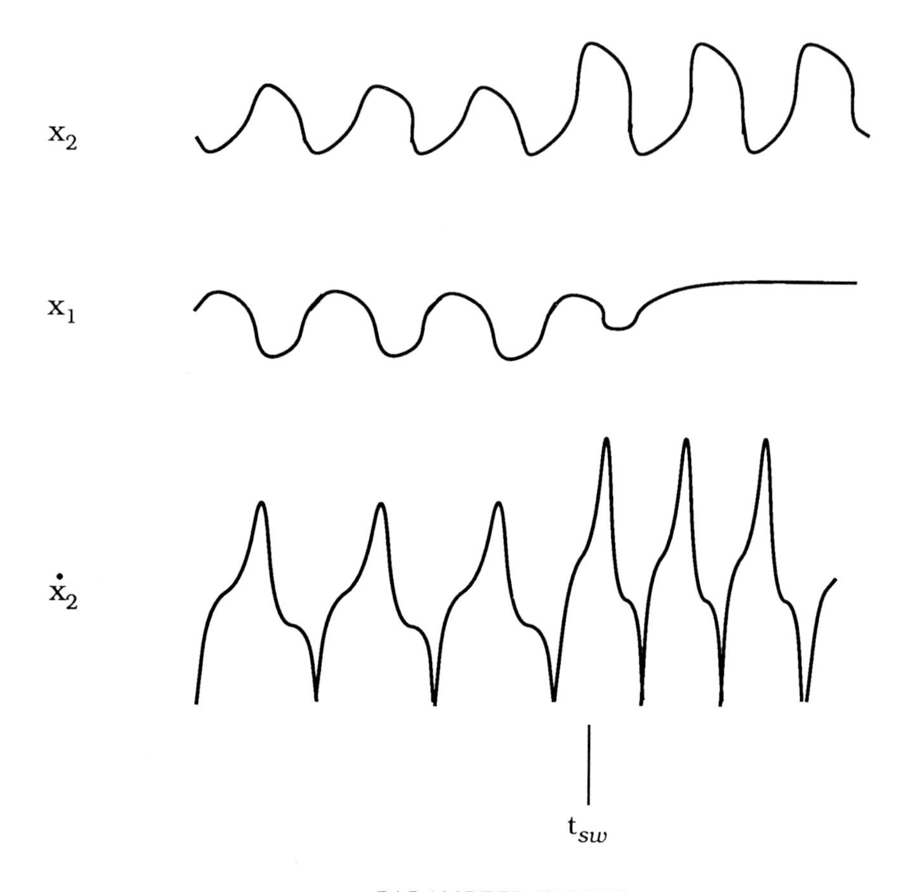

PARAMETER VALUES

$t < t_{SW}$	$t > t_{SW}$
$\mu_1 = \mu_2 = 2.5$	$\mu_1 = \mu_2 = 2.5$
$p_1^2 = p_2^2 = 2.0$	$p_1^2 = -4.0 \quad p_2^2 = 2.0$
$g_1^2 = g_2^2 = 4.0$	$g_1^2 = g_2^2 = 4.0$
$q_1 = q_2 = 0.0$	$q_1 = 8 \quad q_2 = 0.0$
$\lambda_{12} = \lambda_{21} = 0.5$	$\lambda_{12} = \lambda_{21} = 0.0$
$k_1 = k_2 = 1.0$	$k_1 = 0.00 \quad k_2 = 1.0$

Figure 2 The two sets of parameters need to gradually silence x after t_{sw} and to maintain rhythmic movement in x_2.

For laboratory implementation of artificial patterns, we admit the necessity of output signal processing. For example, if frequency modulated signals are required for muscle stimulation, extra hardware such as voltage controlled oscillators can be added to the circuit of Figure 1. Although two oscillators are shown here, the coupling scheme easily generalizes to arbitrarily large applications.

2.3 *Physiological Evidence*

As far as existence of natural oscillators in the CNS is concerned, it appears that integration (in time) or temporal summations as they are referred to in some literature exist in the auditory [26] and oculomotor systems [27, 28]. They are also referred to as neural integrators. Further, how such integrating circuits may naturally be designed by a neural network is discussed in [28]. It is conceivable that similar circuits may be involved in the spinal cord or higher control centers of the CNS for integration purposes.

3 SINGLE LIMB MOTION

The planar one-link system to be considered in this section is shown in Figure 3. Angle θ and $d/dt(\theta) = \dot{\theta}$ are the states of this system. Let J be the moment of inertia of the arm, m the mass of the arm, g the gravity constant and a moment arm of the identical flexor and extensor muscles, and r the distance of the center of gravity from the base.

The differential equations of the system are

$$J\ddot{\theta} - \mathrm{mgr} \sin \theta = +a(F_e - F_f), \tag{2}$$

where F_e and F_f are the extensor and flexor forces respectively.

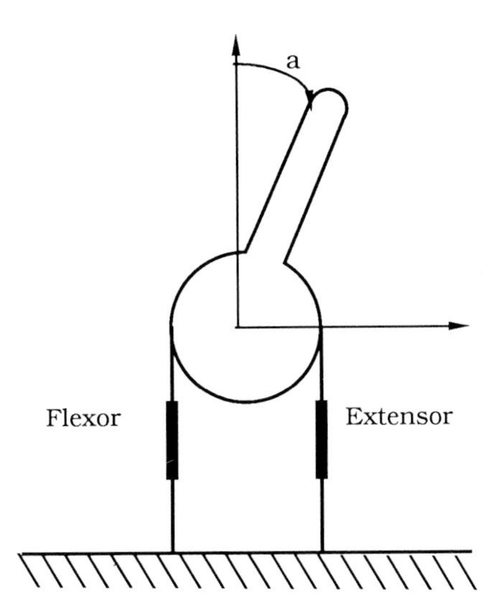

Figure 3 Arm with flexor and extensor actuators.

3.1 *The Muscle Model* [29]

The muscle model to be employed here does not have internal dynamics or states. This means the time scale of the tasks required from this muscle is relatively long compared to the time constants of the internal states so that for all practical purposes the muscle model has no internal memory.

It is further assumed that the length of the muscle does not change appreciably and consequently the forces generated both in contraction and in stretch depend on the rate of the change of the muscle length alone and not on the length itself.

The dependence of the force on the rate of the change of the muscle length is assumed to be a linearized version of Hill's equation [30, 31, 14 (Eq. 5)].

$$F = G_1 N - k_3 \dot{l}, \tag{3}$$

where F is the force produced by the muscle; N is the total neural input to the muscle; G_1 is a simple constant (adjustable gain); k_3 is a constant; and $\dot{l}$ is the contraction velocity (Figure 4).

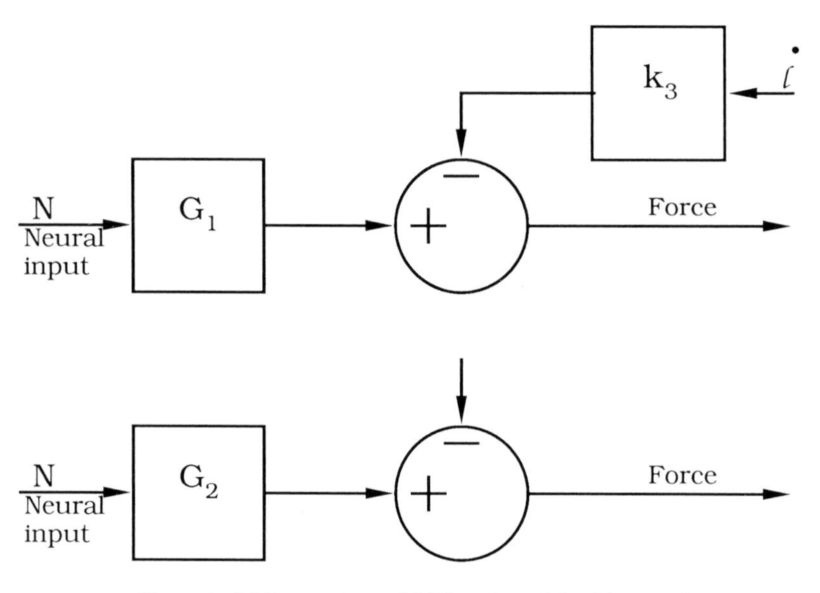

Figure 4 (a) Contraction and (b) Stretch models of the muscle.

Gielen and Houk [15] have proposed a force response when the muscle is stretched. If Eqs. 2 and 3 of Gielen and Houk [15] that respectively represent the output of the spindle and the force output in response to constant velocity stretch are divided, one arrives at a formula for the dynamics of the muscle in stretching,

$$F/N \approx G_2 \dot{l}^{(n_1 - n_2)}, \tag{4}$$

where G_2 is a constant, $\dot{l}$ is the velocity of stretch (negative contraction velocity) and n_1 and n_2 are exponents with estimated values of 0.17 and 0.30, respectively [15]. Therefore, in stretching, the muscle can be represented by a nonlinear amplifier whose gain is velocity dependent and is given by the right side of Eq. (4).

3.2 *The Spindle Model* [32–34]

According to Loeb and Hoffer [32] and Burke [33], spindles are sophisticated sensory organs that may perform a variety of functions depending on the task at hand. Specifically, it is proposed here to consider the spindle as a primary source of afferent feedback that performs servo-like action. Specifically, it is made of a gravity sensor, a position feedback sensor, and a velocity sensitive sensor. The latter will have a form proposed by Gielen and Houk in stretching and no activity in shortening [33]. Burke [33] has discussed the plausibility of no fusimotor action in contracting muscles when the external loads are small or the movements are fast. The block diagram of Figure 5 summarizes this model of the spindle.

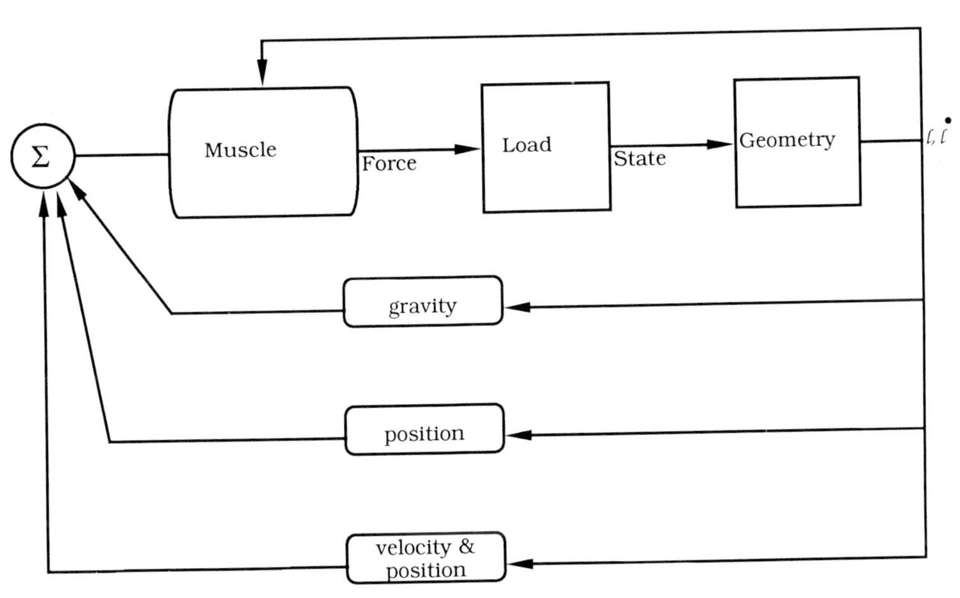

Figure 5 Three functional components of the spindle.

With the above models of the limb, the two muscles and the spindle, one can study the behavior of the one limb system in the phase plane of θ and $\dot{\theta}$. The phase plane can be divided into four quadrants.

In quadrant 1, $\theta > 0$ and $\dot{\theta} > 0$. In this region the flexor's spindles are active and the extensor spindles are passive. In the second quadrant in the phase plane $\theta < 0$, $\dot{\theta} > 0$. The extensor is contracting to bring the arm to its equilibrium and the flexor is passive. In the third quadrant $\theta < 0$ and $\dot{\theta} < 0$. This quadrant is analogous to region (1) except that the role of the flexor and extensor are exchanged. In quadrant four, $\theta > 0$, and $\dot{\theta} < 0$. Here the flexor is shortening to bring the arm to its vertical equilibrium and spindles for gravity and position feedback are active while, because of its shortening, the nonlinear flexor spindle is off. The extensor spindles are all considered to be off.

3.3 *Reciprocal Excitation*

Studies of reciprocal excitation as a differentiating feature in spasticity show that simultaneous activity exists in the shortening muscle during a stretch reflex experiment. The size of the EMG is directly proportional to the rate of the stretch. This means in the present model the following modifications must be made:

$$\begin{aligned} &\text{quadrant 1} \quad F_e = k_4\dot{\theta}, \\ &\text{quadrant 2} \quad F_f = 0 \\ &\text{quadrant 3} \quad F_f = -k_4\dot{\theta}, \\ &\text{quadrant 4} \quad F_e = 0. \end{aligned} \tag{5}$$

3. *Numerical Parameters*

We have arbitrarily selected the parameters of the system below. There are more systematic ways of selecting these parameters [36] and using similar models for functional neuromuscular control [3] and for analysis [38].

A mechanical linkage with the approximate parameters of a hand are considered:

$$\begin{aligned} m &= 0.5 \text{ kg} \\ g &= 10 \text{ m/s}^2 \\ \text{length of hand} &= 0.20 \text{ m} \\ \text{distance of c. of g. from hinge } r &= 0.10 \text{ m} \\ \text{moment of inertia about base} &= 0.01 \text{ kgm}^2. \end{aligned}$$

The ranges of the state variables θ and $\dot{\theta}$ are respectively

$$\begin{aligned} -\pi/2 \leqslant \theta \leqslant +\pi/2 &\text{ radians}, \\ -3 \leqslant \dot{\theta} \leqslant +3 &\text{ radians/s}. \end{aligned}$$

The system parameters of the muscle G_1, k_3, *and* G_2 and the gain parameters k_5 and k_6 of the spindles are all positive numbers correspond to position and velocity feedback—Eq. 7 in [29].

These parameters are selected here very grossly to correspond to transient responses of several seconds duration.

$$\begin{aligned} G_1 = G_2 &= 10, \\ k_3 &= 0.1, \\ k_5 &= 0.025, \\ k_6 &= 0.024. \end{aligned}$$

The amount of reciprocal excitation is varied over a range from $0 \leqslant k_4 \leqslant 1$.

4 DIGITAL COMPUTER SIMULATIONS

The arm is given an initial position of one radian and the angular velocity of one radian per second. The amount of reciprocal excitation is zero. The phase plane response of the system (θ along the horizontal axis, and $\dot{\theta}$ along the vertical axis) is plotted in Figure 6. The response tends to the origin in about 10 s, and proves that

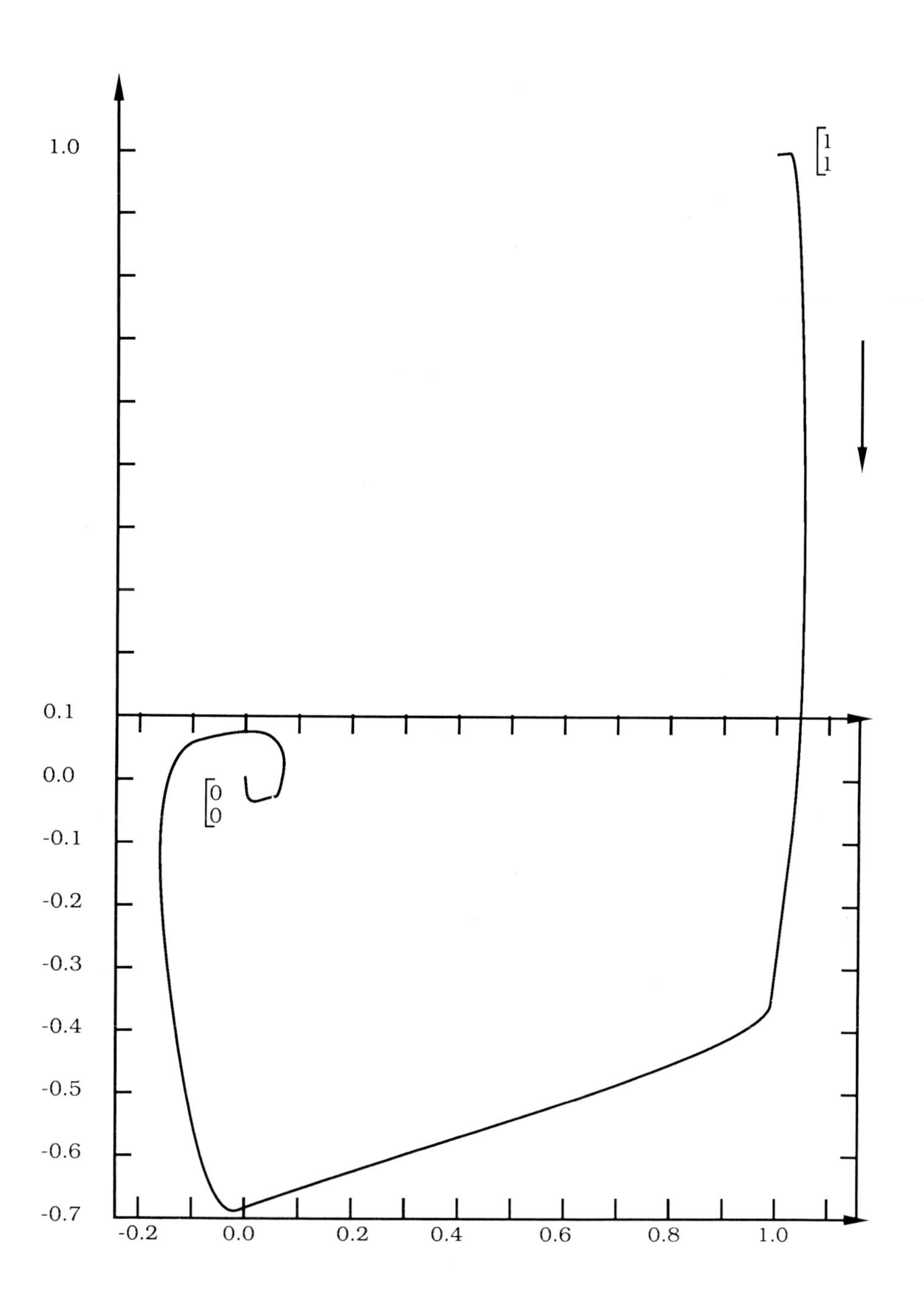

Figure 6 Phase plane response of the system.

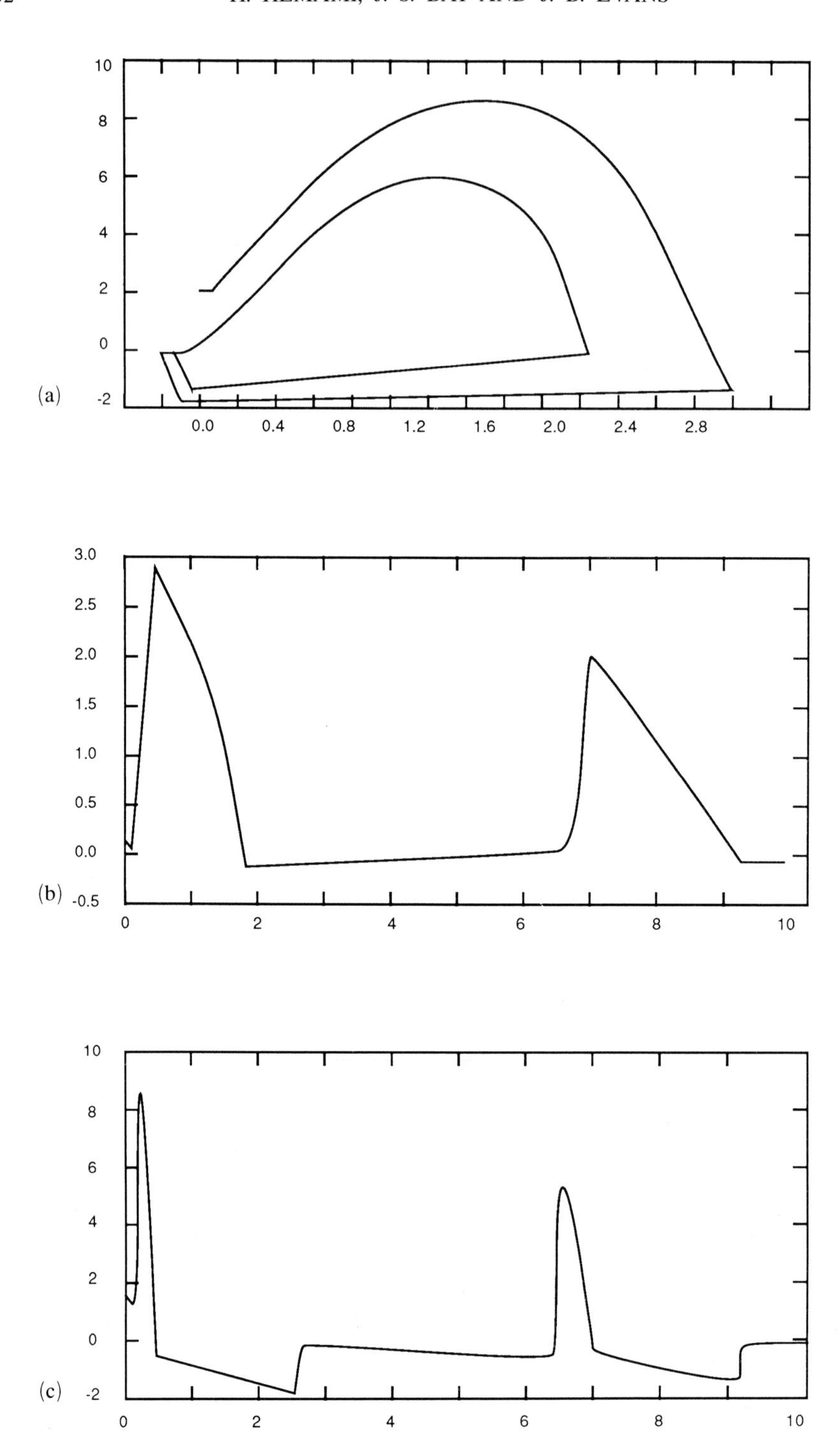

Figure 7 (a) Phase plane, (b) angle and (c) angular velocity for reciprocal excitation.

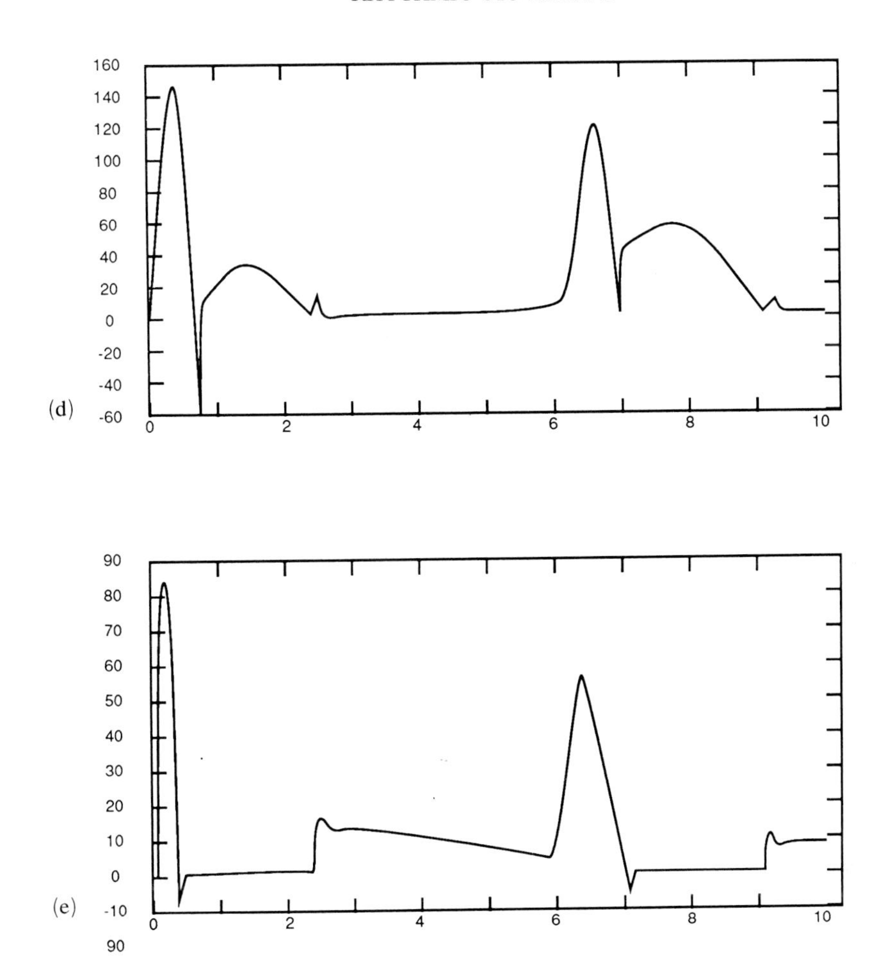

Figure 7 (continued) (d) Flexor and (e) extensor forces for reciprocal excitation.

the nonlinear spindle model in the stretch part plus the natural viscosity assumed for the muscle in contraction together stabilize the postural loop. Figure 7 shows the beginnings of oscillatory behavior when reciprocal excitation exists ($k_4 = 1$). It appears from these simulations that only when the gain of reciprocal excitation (k_4) is about 4 to 5 times of the spindle gain ($k_2 = 0.24$) these rhythmic oscillations begin to occur.

An alternative circumstance [10, p. 1202] that can lead to oscillations is discussed by Grillner. It is termed "position dependent negative feedback" by Grillner and it becomes active in the extreme positions of flexion and extension to ensure switching from flexor to extensor and vice versa. A qualitative phase plane trajectory of the oscillation, also called limit cycles in engineering literature, can be constructed as given in Figure 8. The trajectory is marked by ABCDEFA where points A and D correspond to extreme positions of extension and flexion, respectively. Around point A the flexor becomes active and the flexion force pulls

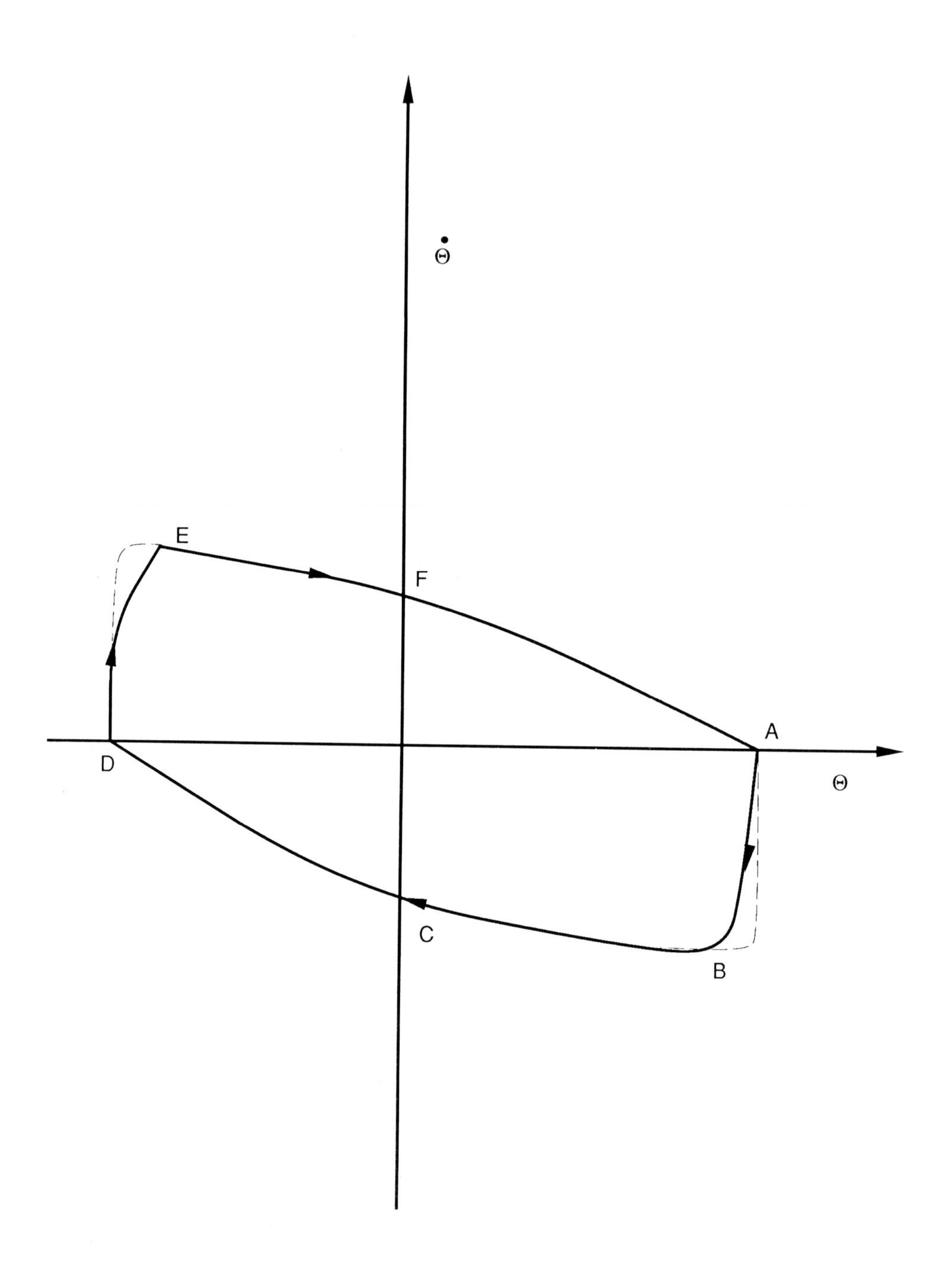

Figure 8 A simple oscillation trajectory in the phase plane.

the limb back, i.e., $\theta<0$. The activity of the flexor stops at B. The energy imparted to the limb during AB swings the limb beyond the vertical ($\theta=0$, point C) and brings it to the extreme flexion positions (D). At this point the extensor becomes active and imparts a velocity ($\dot{\theta}>0$) over the segment DE. This energy swings the limb and gets dissipated until the limb comes back to zero velocity at A where the flexor becomes active again and the cycle repeats.

Based on these results, it appears that the mechanical limb with reciprocal excitation that is not controlled from higher centers can fall into oscillations. Nonlinear spindle feedback in stretch coupled with the natural viscosity of the muscle in contraction, is sufficient to stabilize the one-link system. It is further shown that the system is immune to moderate amounts of reciprocal excitation.

4 COUPLED NEURO-MUSCULO-MECHANICAL SYSTEM

In this section we consider a coupled, linearized, fourth-order system consisting of a single system from Eq. (1) and a one limb mechanical system.

At the onset it must be assumed that the model here is physiologically very tenuous on two grounds:

1) There is no support phase here and consequently any feedback of support forces, motion of the skin, or surface [10] is ignored.

2) The feedback paths and parameters are not justified for either intact living systems or living systems with induced deficits. Grillner [10] has discussed in some detail the large number of feedback signals that can be active even in a simple system as we have considered here.

As one example of the complexity of the feedback paths Grillner [10, p. 1200] discusses continuous extensor activity in one limb during fictitious locomotion of a quadruped if the limb is stopped during the support phase. The prolonged extensor activity continues even when the limb is unloaded.

With these limitations of our present model, the linear neural circuit is assumed to be

$$\ddot{x}_3-\mu\dot{x}_3+g^2(x_3-x_4)=q_1. \tag{6}$$

Let the linearized musculo-mechanical system be

$$J\ddot{x}_4+b\dot{x}_4+h(x_4-x_3)=0. \tag{7}$$

Equation (7) is a simpler form of the single link (2) with gravity ignored and the effect of natural viscosity of the system [38] included as a linear term in velocity, i.e., $b\dot{x}_4$. Parameter h is analogous to the gain of the reflex loop, x_4 is analogous to the γ signal, and (x_4-x_3) is analogous to the position (length) dependent spindle signal. For simplicity we have combined the effects of the extensor and flexor in one equation. The coupling between the neural and mechanical system is by signal q_1:

$$q_1=-1.8125k\,\dot{x}_4, \tag{8}$$

i.e., the velocity of the limb is fed back to the input of the neural circuit. The gain k is controlled by higher centers.

Equations (6), (7), and (8) describe the behavior of this fourth-order system. The physical parameters of this system and the behavior of the loci of the roots of its characteristic equation [12, 37] are given in Figure 9. For this figure, we have used as parameters $\mu=0.25$, $g^2=4$, $J=0.1$, $b=0.2$, and $h=1$. One root of the characteristic equation always remains at the origin, i.e., the system has one integrator mode. The other three roots are in the left half of the complex plane at

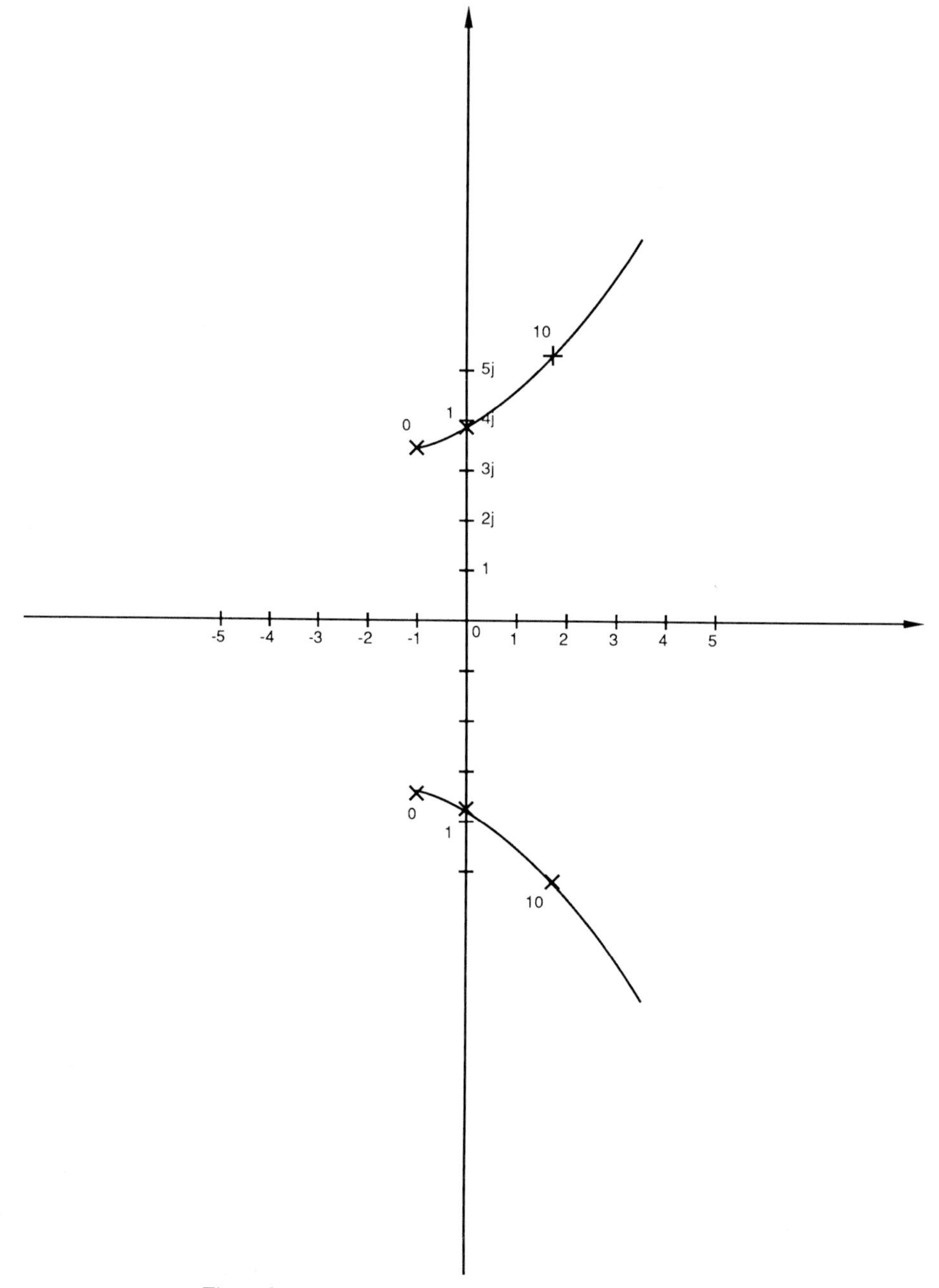

Figure 9 Loci of the three nonzero roots of the spinal cord.

-0.425 and $-0.66 \pm j\ 3.5$. As k increases, the real pole migrates further left on the real axis and the two complex poles migrate to the right.

For $k=1$ the four poles are 0.0, -1.75, and $\mp 3.67\ j$. For $k=10$ the four poles are 0.0, -5.5, and $1.87 \pm 5.5\ j$. Therefore, for $k=1$ the system oscillates with a period of about 2s.

5 DISCUSSION AND CONCLUSIONS

We have demonstrated here that coupled simple neural circuits autonomously oscillate as central pattern generations. We have also shown that the one-limb musculoskeletal system, under certain reciprocal excitation or extreme range feedback, can oscillate without any periodic input. The last section also shows that under favorable conditions combinations of a neural circuit and one-limb musculoskeletal system can oscillate and produce rhythmic movement. These results can be extended to more elaborate neural circuits and to multi-linkage skeletal systems with many muscles. One major difficulty is making such mathematical models physiologically more accurate. Because of the richness of the sensory and control machinery of the central nervous system impinging on such neuro-musculoskeletal models, and because of the elaborate feedback and feedforward paths, it should be very difficult to identify quantitatively all the gates and their functions. Nevertheless, this analysis will be valuable for designing artificial gait and artificial patterns that may eventually have to be interfaced with natural systems in functional neuromuscular stimulation [39].

A major modification required is that all the designs have to be repeated with frequency modulation or pulse code modulation so that these artifical circuits may be coupled more easily to natural nerves and muscles. Alternatively, one may introduce such transformations from amplitude to frequency only at the interfaces to and from the natural system. The question of stability and postural adjustment still remains and must be separately addressed.

ACKNOWLEDGEMENTS

This work was supported in part by the National Science Foundation under Grant ECS-820-1240, and in part by the Department of Electrical Engineering of The Ohio State University. The authors would like to thank Professor H. C. Ko, Chairman of Electrical Engineering, for his continuous support and sustained encouragement of this work.

REFERENCES

1. J. Massion, *Role of Motor Cortex in Postural Adjustments Associated with Movements in Integration in the Nervous System*, Igaku Shoin, Tokyo, Japan (1979), pp. 239.
2. L. Nashner and G. McCollum, The organization of human postural movements: a formal basis and experimental synthesis, *Behavior and Brain Science*, **8,** 135 (1985).
3. K. Mauritz, J. Dichgans and A. Hufschmidt, Quantitave analysis of stance in late cortical cerebellar atrophy of the anterior lobe and other forms of cerebellar ataxia. *Brain*, **102,** 461–482 (1979).
4. J. Yang, Biomechanical strategies of postural control in standing walking, Ph.D. Dissertation, University of Waterloo (1987).
5. H. Hemami, Some physical attributes of postural adjustments, report BC-87-HH-161, The Ohio State Univ. Dept. of Elec. Eng.
6. B. Maki, P. Holliday and G. Fernie, A postural control model and balance test for the prediction of relative postural stability, *IEEE Trans. Biomed. Eng.*, **BME-34,** 797–810 (1987).

7. D. Winter, Kinematic and kinetic patterns in human gait: variability and compensating effects, *Human Movement Science*, **3,** 51 (1984).
8. D. Winter, Energy generation and absorption at the ankle and knee during fase, natural and slow cadences, *Clinical Orthopaedics and Related Research*, **175,** 147 (1983).
9. D. Gordon, E. Robertson and D. Winter, Mechanical energy generation, absorption and transfer amongst segments during walking, *Journal of Biomechanics*, **13,** 845 (1980).
10. S. Grillner, *Control of Locomotion in Bipeds, Tetrapods, and Fish in Handbook of Physiology, Section, The Nervous System, part* 1, Volume II, Bethesda, Maryland: American Physiological Society (1981), pp. 1179, Chapt. 26.
11. H. Hemami and C. L. Golliday Jr., The inverted pendulum and piped stability, *Mathematical Biosciences*, **34,** 95–110 (1977).
12. S. Yurkovich, S. Hoffman and H. Hemami, Stability and Parameter Studies of a Stretch Reflex Loop Model, *IEEE Trans. Biomed. Eng.*, **34,** 547–553 (1987).
13. R. Jagacinski, B. Plamondon and R. Miller, Describing movement control at two levels of abstraction, in *Human Factors Psychology* (Ed. P. A. Hancock), Elsevier Science Publishers B.V. (North Holland) (1987), pp. 199–247.
14. J. Houk, P. Crago and W. Rymer, Function of the spindle dynamic response in stiffness regulation—a predictive mechanism provided by nonlinear feedback, in *Muscle Receptors and Movement* (Eds. H. Taylor and A. Prochazka), McMillan (London) (1981), pp. 199–309.
15. C. Gielen and J. Houk, Nonlinear viscous properties of the wrist motor servo in man, *J. Neurophysiol.*, **52,** 533–569 (1984).
16. R. Stein and M. Oguztoreli, Tremor and other oscillations in neuromuscular systems, *Biological Cybernetics*, **22,** 147–157 (1976).
17. N. Hogan, Adaptive control of mechanical impedance by coactivation of antagonistic muscles, *IEEE Trans. Automatic Control*, **AC-29,** 681–690 (1984).
18. G. Agarwal, C. Gyan and G. Gottlieb, Mathematical ·modeling and simulations of the postural control loop: parts i—iii, *CRC Critical Reviews in Biomedical Engineering* Vol 8, issue 2, 1982, pp 93–134, Vol 11, issue 2, 1984, pp 113–154, and Vol 12, issue 1, 1985, pp 49–93.
19. M. Hulliger, *Recent Advances in Muscle Spindle Physiology*, (*Habilitationschrift*), Zurich: Institute fur Hirn Forschung der Universitat (1982).
20. J. S. Bay, Coupled nonlinear oscillators as central pattern generators for rhythmic locomotion, M. S. Thesis, The Ohio State University (1985).
21. J. S. Bay and H. Hemami, Modeling of a neural pattern generator with coupled nonlinear oscillators, *Biomed. Eng.* **34,** 297–306 (1987).
22. J. S. Bay, J. B. Evans and H. Hemami, Generation of Artificial Patterns for Autonomous Motor Stimulation, *Proc. Ninth IEEE Conference of the Engineering in Medicine and Biology Society*, Vol. 3, November (1987), pp. 1581–1582.
23. A. Selverston, J. P. Miller and M. Waldepuhl, Neural mechanisms for the production of cyclic motor patterns, *IEEE Trans. Systems, Man, and Cybernetics*, **SMC-13,** 749–757 (1983).
24. W. Friesen and G. Stent, Generation of a locomotory rhythm by a neural network with recurrent cyclic inhibition, *Biological Cybernetics*, **18,** 27–40 (1977).
25. M. Egelhaaf and P. Benjamin, Coupled neuronal oscillators in the snail lymnaea-stagnalis-endogenous cellular properties and network interactions, *J. Exp. Biol.*, **102,** 93–113 (1983).
26. J. Zwislocki, Theory of temporal auditory summation, *Accoustical Society of America*, **32,** 1046–1060 (1960).
27. S. Cannon and D. Robinson, Loss of neural integration of the oculomotor system from brain stem lesions in monkey, *J. of Neurophysiology*, **57,** 1383–1409 (1987).
28. S. Cannon, D. Robinson and S. Shamma, A proposed neural network for the integrator of the oculomotor system, *Biol. Cybernetics*, **49,** 127–136 (1983).
29. H. Hemami, Stability of a Simple Nonlinear Reflex Loop, in *Proc. of the* 8*th Annual Conference of the IEEE Engineering in Medicine and Biology*, Fort Worth, TX (Ed. V. Kondraske and C. J. Robinson), vol. 3 (1986), pp. 1586–1590.
30. A. Hof and J. Van den Berg, EMG to force processing I: An electrical analogue of the Hill muscle model, *J. Biomechanics*, **14,** 747–758 (1981).
31. A. Hof and J. Van den Berg, EMG to force processing II: Estimation of parameter of the Hill muscle model for the human triceps surae by means of a calfergometer, *J. Biomechanics*, **14,** 759–770 (1981).
32. G. Loeb and J. Hoffer, Muscle spindle function during normal and perturbed locomotion in cats, in *Muscle Receptors and Movement* (Eds. A. Taylor and A. Prochazka), MacMillan (London) (1981), pp. 219–228.
33. D. Burke, Muscle spindle function during movement, *Trends in Neuroscience*, **3,** 251–253 (1980).
34. I. Boyd, The isolated mammalian muscle spindle, *Trends in Neuroscience*, **3,** 158–265 (1980).

35. B. Myklebust, G. Gottlieb, R. Penn and G. Agarwal, Reciprocal excitation of antagonistic muscles as a differentiating feature in spasticity, *Ann. Neurology*, **12,** 367–374 (1982).
36. J. Allin and G. Inbar, FNS—parameter selection and upper limb characterization, *Technion I.I.T.*, no. EE Pub. 477, January (1984).
37. R. Jaeger, Design and simulation of closed-loop electrical stimulation orthoses for restoration of quiet standing in paraplegia. *J. Biomechanics*, **19,** 825–835 (1986).
38. S. Lehman and L. Stark, Perturbation analysis applied to eye, head and arm movement models, *IEEE Trans. Systems, Man, and Cybernetics*, **SMC-13,** 972–979 (1983).
39. D. Peterson and H. Chizeck, Linear quadratic control of a loaded agonist-antagonist muscle pair, *IEEE Trans. Biomed. Eng.*, **BME-34,** 790–796 (1987).

Automedica, 1989, Vol. 11, pp. 91–98
Reprints available directly from the publisher only
Photocopying permitted by license only

DYNAMICS AND CONTROL OF A MUSCLE-LOAD SYSTEM DURING ELECTRICAL NERVE STIMULATION

P. H. VELTINK

Biomedical Engineering Division, Faculty of Electrical Engineering

and

J. A. VAN ALSTÉ

Coordination Centre for Biomedical Engineering

(Received September 1988)

The dynamics of a muscle connected to a second order linear load were investigated during artificial nerve stimulation. Results of simulations with a third-order non-linear model describing the combined muscle-load system were compared with recordings of animal experiments on loaded muscles which were artificially stimulated.

Muscle length control by electrical nerve stimulation was investigated. Compensation of only non-linear recruitment, low-frequency load characteristics and mean maximal force appeared to be insufficient for proper length control.

Keywords: Muscle length control, nerve stimulation, muscle dynamics.

1 INTRODUCTION

Contraction of skeletal muscles can be controlled by artificial nerve stimulation. The objective of our research in this field is to obtain functional movements in paralyzed limbs of stroke or spinal-cord-injured patients.

A joint angle in a paralyzed limb can be controlled by stimulating one or several muscles influencing the joint torque. The mechanical properties of the limb and human body act as a mechanical load to the active muscles.

Most research in the control of skeletal muscle contraction using artificial stimulation concern isometric contractions [1–3] controlling muscle force. However, when controlling a joint angle in a dynamic skelton-muscle-load system, the muscle contraction is not isometric and the load characteristics must be accounted for. Therefore, we investigated control of muscle length, where the muscle was connected to a linear second order mechanical load [4].

2 METHODS

2.1 *Experimental Methods*

Experiments were performed on rats (Wistar, male, 3–4 months old, 0.30–0.35 kg), anaesthetized intraperitoneally with pentobarbital sodium. In the right hind limb the peroneous communis nerve was stimulated.

Correspondence to: Dr. J. A. van Alsté, Coordination Centre for Biomedical Engineering, University of Twente, P.O. Box 217, 7500 AE Enschede, The Netherlands, Telefax: 31–53–356490.

The tibialis anterior muscle was connected to a programmable dynamic load [5], consisting of a strain gauge based force transducer, a real-time load processor and a servo-controlled linear motor system. A second order load was used in the experiments:

$$F = -M \cdot \ddot{x} - D \cdot \dot{x} - C^{-1} \cdot x, \tag{1}$$

where the load parameters are M (mass), D (damping) and C (compliance). F represents the muscle force, and x is the position of the muscle tendon connected to the load (x increases for lengthening muscle).

Monopolar cathodic stimulation was performed, using monophasic constant current pulses. The tip of a 200 μm diameter stainless steel wire electrode was placed directly at the nerve surface for stimulation. An indifferent electrode was placed in between muscles outside the stimulation region. Muscle contraction was modulated by motor neuron recruitment through variation of the stimulation pulse amplitude. Stimulation frequency and pulse width remained constant (30 Hz and 60 μs respectively).

2.2 *Model of the Muscle-Load System*

The muscle was modeled using the classical model of muscle dynamics (Figure 1) [6, 7], consisting of a contractile component CC, a series elastical component SEC and a parallel elastical component PEC.

The contractile component is velocity dependant [6, 7]. During shortening the contractile component is described by the hyperbolic Hill-relation [6, 7]:

$$v_{cc} = -b \cdot \frac{\dfrac{-F_{cc}}{F_0} + 1}{\dfrac{F_{cc}}{F_0} + n}, \quad \frac{F_{cc}}{F_0} < 1 \tag{2}$$

where b and n are constants. Parameter b can vary considerably among muscles and determines the maximal shortening speed of the CC. Parameter n varies much less between muscles: Hof [8] determined $n = 0.12$ in the human calf muscle. F_{cc} is the force the CC exerts, v_{cc} is the velocity of the CC. F_0, called active state, is a function of time which gives the force of the CC when the CC would contract isometrically. The CC shortens ($v_{cc} < 0$) when $F_{cc}/F_0 < 1$.

During lengthening the muscle force quickly follows the active state F_0, with little dependancy on velocity v_{cc} [6, 7]. We used a parabolic relation between F_{cc} and v_{cc} during lengthening ($F_{cc}/F_0 > 1$):

$$v_{cc} = \left(\frac{F_{cc}}{F_0} - 1\right) \cdot \left[s \cdot \left(\frac{F_{cc}}{F_0} - 1\right) + \frac{b}{n+1}\right], \quad \frac{F_{cc}}{F_0} > 1, \tag{3}$$

where s is a constant. The relations between F_{cc} and v_{cc} (2), (3) and their derivatives connect continuously for $v_{cc} = 0$.

The time function of the active state F_0 was based on Hof [6]. The active state response to a submaximal stimulus at $t = 0$ was modeled as:

$$F_0(u;t) = u \cdot F_m \quad 0 > t > t_1$$
$$= u \cdot F_m \cdot \{(1-\beta)e^{-(t-t_1)/t_2} + \beta e^{-(t-t_1)/t_3}\} \quad t > t_1. \tag{4}$$

F_m is the maximum amplitude of the active state. The fraction of the maximal active state F_m, which is obtained when stimulating submaximally is represented by u and depends on the stimulus amplitude. We did not explicitly incorporate the length dependance of the CC in our model. This length dependance was accounted for by considering F_m at the resting length used (when the muscle is not stimulated). Parameters β, t_1, t_2 and t_3 were taken such that the simulated isometric muscle twitch resembled the measured isometric muscle twitch close.

The series and parallel elastical components were given an exponential length–force relationship [6]:

$$F_{\text{sec}} = F_{s0} \cdot (e^{\alpha_s \cdot (x - x_{cc})} - 1),$$
$$F_{\text{pec}} = F_{p0} \cdot (e^{\alpha_p \cdot x} - 1), \tag{5}$$

where F_{s0}, F_{p0}, α_s and α_p are constants. The position associated with the CC (Figure 1) is x_{cc}.

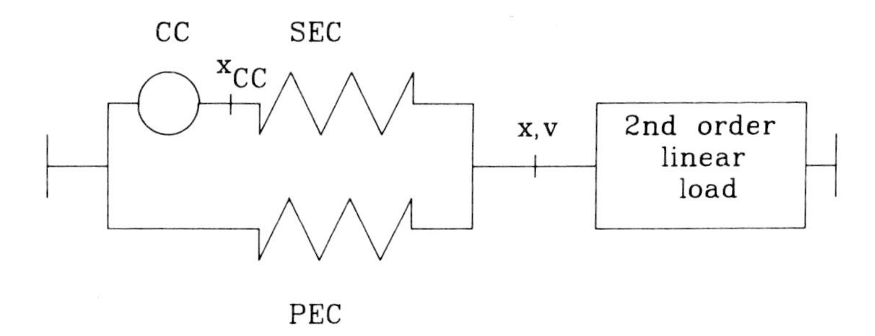

Figure 1 Classical model of muscle dynamics, incorporating a contractile component CC, and series and parallel elastic components (SEC and PEC). The muscle is connected to a second-order mechanical load.

The muscle-load system can be described in a state space description of the following form:

$$\dot{\mathbf{x}} = \mathbf{f}(\mathbf{x}, u). \tag{6}$$

The time derivative of the state vector $\mathbf{x}$ is a non-linear function $\mathbf{f}$ of the state vector and of an input vector u. This model is of the third order, with state variables muscle length x and velocity $v = \dot{x}$, and length of the contractile component x_{cc} (Figure 1). Input to the system is the recruitment fraction u. Two of the three state variables correspond to the second-order linear load. The length of the contractile component x_{cc} corresponds to the first-order non-linear muscle model. Using (2)–(5) and $F_{cc} = F_{sec}$, the state space description (6) can be written as:

$$\dot{x} = v \tag{7a}$$

$$\dot{v} = -\frac{1}{M}\{C^{-1}x + Dv + F_{s0}(e^{a_s(x-x_{cc})} - 1) + F_{p0}\cdot(e^{a_p \cdot x} - 1)\} \tag{7b}$$

$$\begin{aligned}\dot{x}_{cc} &= -b\frac{-F_r+1}{F_r+n}, \quad F_r < 1 \\ &= (F_r - 1)\left[s(F_r-1) + \frac{b}{n+1}\right], \quad F_r > 1\end{aligned} \tag{7c}$$

where

$$F_r = \frac{F_{cc}}{F_0} = \frac{F_{s0}(e^{a_s(x-x_{cc})} - 1)}{F_0(u;\, t)}. \tag{8}$$

2.3 *Open-loop Control System with Compensation Block*

As a first approach to control of muscle length control we used a simple compensation block to compensate for some of the non-linear characteristics of stimulation and muscle dynamics (Figure 2). In the compensation block the non-linear recruitment properties of nerve stimulation, the low-frequency load characteristics, and the mean muscle force were taken into account. The algorithm of the compensation block determines the actual stimulation current pulse amplitude A from the desired muscle length x_d as:

$$A(t) = R\left[\frac{x_d(t)}{C \cdot F_{am}}\right]. \tag{9}$$

Function R is the inverse of the recruitment curve, which we defined as the relation between stimulus amplitude and force amplitudes of single isometric muscle twitches. The force amplitudes of the twitches were normalized on the twitch amplitude during supramaximal stimulation. The recruitment curve must be measured before using the compensation block for length control. C is the load compliance, which characterizes the low-frequency load characteristics. F_{am} is the mean muscle force during maximal stimulation at 30 Hz.

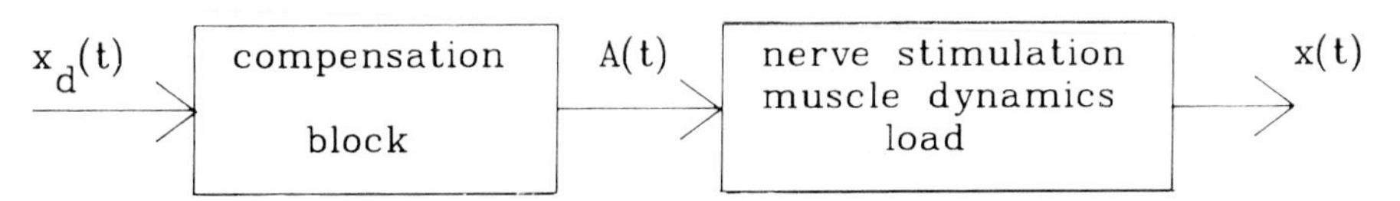

Figure 2 Concept of length control using a compensation block. $x_d(t)$, desired muscle tendon position; $x(t)$; actual muscle tendon position; $A(t)$, stimulation pulse amplitude.

3 RESULTS

3.1 *Preliminary Model Verification and Parameter Identification*

The model (7) was implemented in a simulation program and simulation results were compared with experimental results. In this study only muscle lengths were considered for which the passive muscle force was low. Therefore, the force contribution of the PEC could be neglected.

Preliminary parameter identification and model verification was done by matching simulation results with experimental results in the case of isometric twitches, and in the case of supramaximal stimulation of the muscle-load system for several load characteristics.

Figure 3 gives an example of a measured isometric twitch and a simulated twitch. The isometric twitch was simulated with only the first order non-linear muscle model, which is represented by Eqs. (7c) and (8), with $\dot{x}=0$. The parameters of the active state (t_1, t_2, t_3, β, F_m), the SEC (F_{s0} and α_s) and the CC (b, n, s) were given a value such that the form of the measured and simulated twitches resembled well, and a realistic twitch–tetanus ratio was obtained.

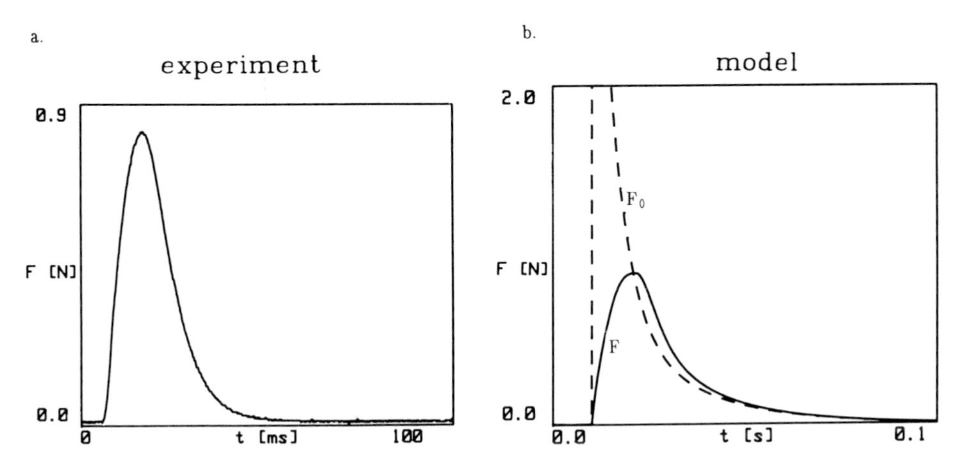

Figure 3 Measured (a) and simulated (b) isometric muscle twitch during submaximal stimulation. F: muscle force, F_0: active state. *Model parameters*: $b=10$ mm/s, $n=0.15$, $s=50$ mm/s, $F_{s0}=1.0$ N, $\alpha_s=4$ mm^{-1}, $t_1=5$ ms, $t_2=5$ ms, $t_3=20$ ms, $\beta=0.3$, $F_m=4$ N, $u=0.5$.

Two seconds of maximal stimulation of the loaded muscle at 30 Hz was simulated and compared with experimental results for several loads. Figure 4 gives an example. The parameters of the muscle model (active state F_0, CC and SEC) were further identified by matching of the simulated and measured force and length registrations.

3.2 *Performance of the Open-loop Compensation Block Controller*

First, the parameters of the compensation block (9) were identified: The mean muscle force during 2 s supramaximal stimulation F_{am} was measured and the recruitment curve was determined by measuring isometric twitch amplitudes at several stimulation pulse amplitudes. The obtained information was used for identifying the parameters of the compensation block. Then, length control experiments were performed. This procedure was repeated several times for several rats. Figure 5 gives a typical result. During shortening the error in the muscle length appeared to be rather high, although the shortening velocity was only 1 mm/s. During lengthening this error was much smaller. The velocity dependance of the muscle force may explain the error during shortening.

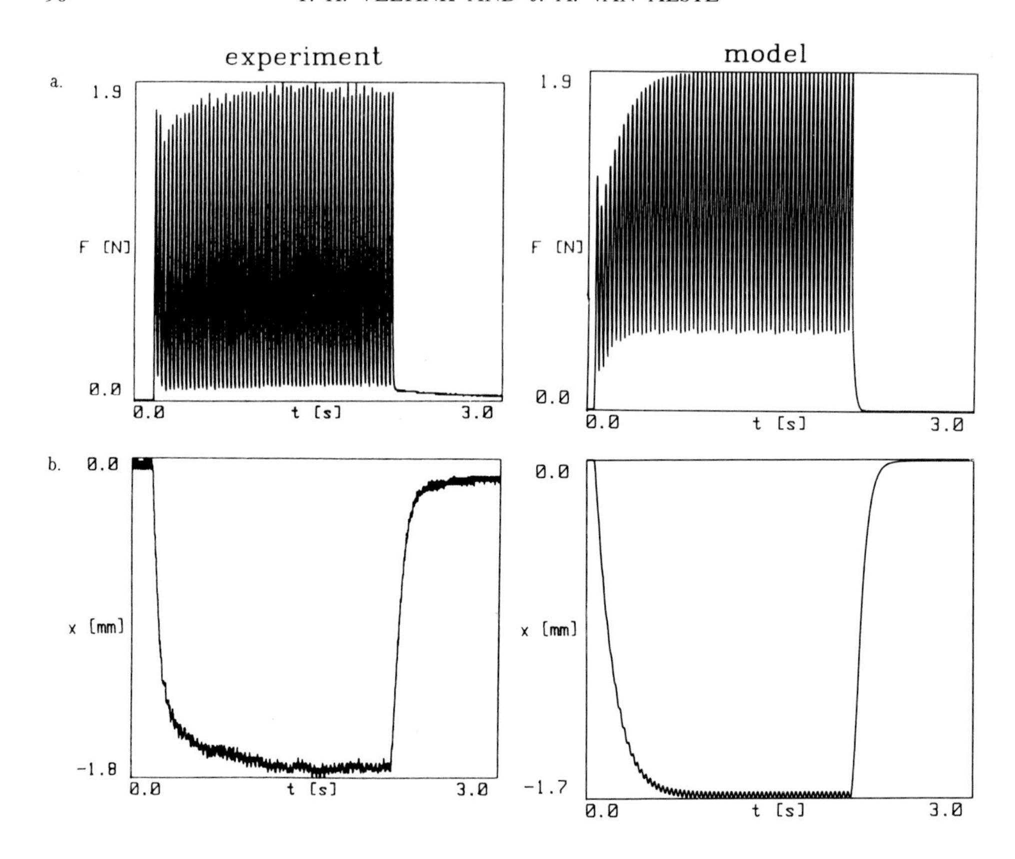

Figure 4 Typical experimental and simulation results of 2 s supramaximal stimulation. A critically damped load was used with a resonance frequency of 4 Hz: $m = 1.05$ kg, $C^{-1} = 663$ N/m, $D = 53$ Ns/m. The parameters of the muscle model were the same as in Figure 3, except $u = 1.0$. a, force signal $F(t)$; b, length signal $x(t)$.

4 DISCUSSION

Preliminary model verification was satisfying. However, further research is needed for a final model verification and for independent identification of the parameters.

In order to improve the simple open-loop compensation control block [(9) and Figure 2] account must be taken of the velocity dependent characteristics of muscle force. Therefore, we are now developing non-linear control strategies based on the model of the muscle-load system described in this paper (7).

ACKNOWLEDGEMENTS

We would like to thank J. E. van Dijk and J. Put for assisting in the preparations and performance of the animal experiments, and H. Nijmeijer for helpful discussions concerning the state-space model of the muscle-load system and non-linear control strategies.

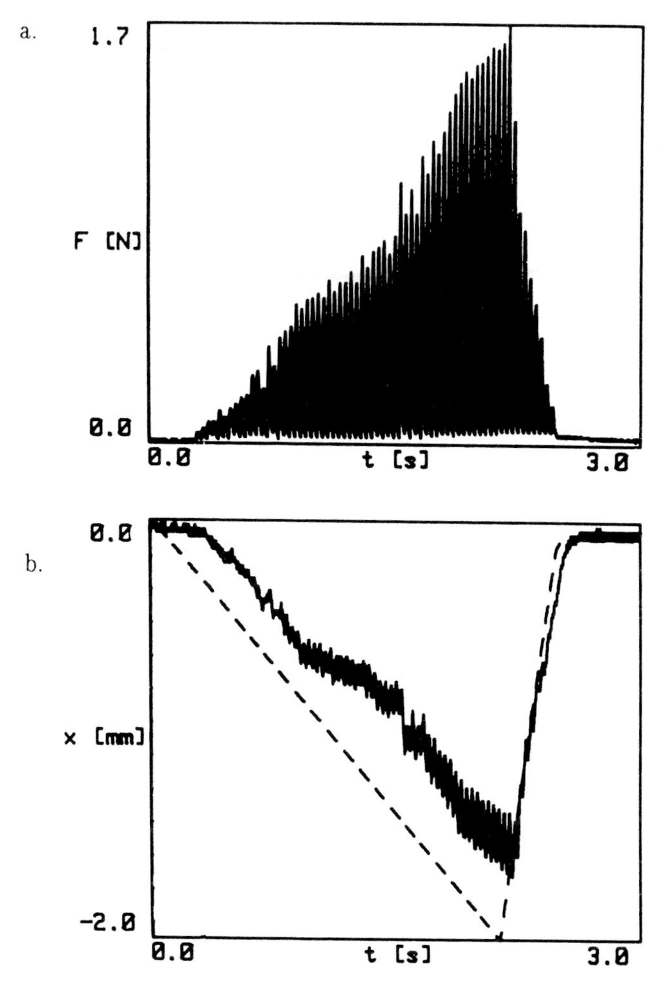

Figure 5 A typical registration of an open-loop control experiment using the compensation block of Figure 2. A critically damped load was used with a resonance frequency of 7 Hz: $m = 0.2$ kg, $C^{-1} = 391$ N/m, $D = 17.8$ Ns/m. a, force signal $F(t)$; b, length signal $x(t)$.

REFERENCES

1. L. A. Bernotas, P. E. Crago and H. J. Chizeck, A discrete-time model of electrically stimulated muscle, *IEEE Trans. Biomed. Eng.*, **33,** 829–838 (1986).
2. L. A. Bernotas, P. E. Crago and H. J. Chizeck, Adaptive control of electrically stimulated muscle, *IEEE Trans. Biomed. Eng.*, **34,** 140–147 (1987).
3. B. H. Zhou, R. Baratta and M. Solomonow, Manipulation of muscle force with various firing rate and recruitment control strategies, *IEEE Trans. Biomed. Eng.*, **34,** 128–139 (1987).
4. J. A. van Alsté, P. H. Veltink and H. Nijmeijer, Muscle length control by electrical nerve stimulation, *Proc. of the IEEE 9th Ann. Conf. of the Engineering in Medicine and Biology Society*, **3,** 1372–1373 (1987).
5. P. H. Veltink, J. E. van Dijk and J. A. van Alsté, Programmable dynamic muscle load for animal experiments, *Med. Biol. Eng. Comput.*, **26,** 234–236 (1988).

6. A. L. Hof and J. W. van den Berg, EMG to force processing I: an electrical analogue of the Hill muscle model, *J. Biomechanics*, **14,** 747–758 (1981).
7. H. Hatze, A. myocybernetic control model of skeletal muscle, *Biological Cybernetics*, **25,** 103–119 (1977).
8. A. L. Hof, Jw. van den Berg, EMG to force processing II: estimation of parameters of the Hill muscle model for the human triceps surae by means of a calfergometer, *J. Biomechanics*, **14,** 759–770 (1981).

Automedica, 1989, Vol. 11, pp. 99–109
Reprints available directly from the publisher only
Photocopying permitted by license only

SIMULATION AND MODELLING OF A MICROCOMPUTER CONTROLLED ABOVE-ELBOW PROSTHESIS

J. S. PHILIPPE-AUGUSTE and D. T. GIBBONS

Department of Electrical Engineering, University of Ottawa, 770 King Edward Avenue, Ottawa, Ontario, Canada K1N 6N5

and

M. D. O'RIAIN

Department of Rehabilitation Engineering, The Rehabilitation Centre, 505 Smyth Road, Ottawa, Ontario, Canada K1H 8M2

(Received September 1988)

This paper describes an externally powered above-elbow prosthesis which uses the technique known as *extended physiological proprioception* and which has a microcomputer control system allowing for selectable input-output relationships. A bench model of this prosthesis has been constructed and basic tests have been done using normal subjects. Based on the results of these tests, a computer simulation has been developed together with a complete data gathering and processing system. The simulation permits precise investigations of prosthesis performance.

Keywords: Prosthesis, extended physiological proprioception, myoelectric control, computer simulation.

1 INTRODUCTION

The recent advances of technology in the fields of electronics, computers, and mechanical engineering, have raised hopes for improving the performance of externally powered upper extremity prostheses. The availability of powerful microprocessors and digital signal processors (DSPs) allows the use of pseudo-intelligent controlling units that can easily be incorporated into a prosthesis. With this amount of computing power available at low cost, investigators are no longer limited in the complexity of the controlling algorithm they wish to incorporate into their prostheses. Hence many research groups have come up with a variety of control strategies for their externally powered prostheses. The majority of these systems use myoelectric (EMG) control [1–4]. The popularity of this control strategy comes from the fact that it is a very "elegant" one, since the control signals going to the prosthesis may come from the myoelectric activity of the muscle which formerly controlled the lost limb. Therefore, in some sense, the brain's own signals are used to control the motion of the prosthesis [5]. Although the validity of myoelectric control based techniques seems reasonable, in practice the performances are very disappointing. There are a number of reasons for this. In the cases of high levels of amputation, the electrical activity of the remaining muscles have only minor correlations with the normal arm movements and this makes coordinated control of several functions very difficult. A more important problem is that any EMG controlled prosthesis is inherently open-loop [6]. The absence of position proprioception and other related problems of reliability in executing movements [7] has contributed to the poor success rate of myoelectrically controlled prostheses. These facts have encouraged us to investigate other available control strategies for externally powered multifunctional prostheses. Our research was aimed at finding a viable control strategy, implementing it and performing suitable evaluations. This was done and an initial evaluation was performed. The results established the usefulness and viability of the new prosthesis design [7]. The next major step, which we are presenting in this paper, involved developing a computer simulation incorporating all the features of the new prosthesis. This allows rapid and accurate investigations of different aspects of the prototype without having to change components each time.

2 THE CONTROL STRATEGY

An interesting approach to overcoming the problems encountered with EMG control has been proposed by Simpson [6]. The control strategy he proposed, and which he termed: "extended physiological proprioception" (e.p.p.), uses the positions of intact joints as the controlling input signals to a prosthesis. This strategy is based on a recognition of the fact that intact joints possess inherent position feedback. By establishing a one-to-one relationship between the position of the intact joints and the position in space of the terminal device of the prosthesis, the natural feedback of the intact joints can be "extended" to the prosthesis and thereby provide it with proprioceptive (position) information. Based on the ideas by Simpson, we have designed an above-elbow prosthesis and implemented a bench model for testing (see Figure 1). The positioning input of the system is the intact shoulder and the outputs are the elbow and wrist joints. A one-to-one relationship is always maintained between inputs and outputs. At any instant in time, by knowing the position of the shoulder, the amputee knows exactly where the prosthetic arm lies in space.

In Simpson's designs, the mechanical construction of the prosthesis guaranteed the one-to-one relationship between input and output. At the same time, however,

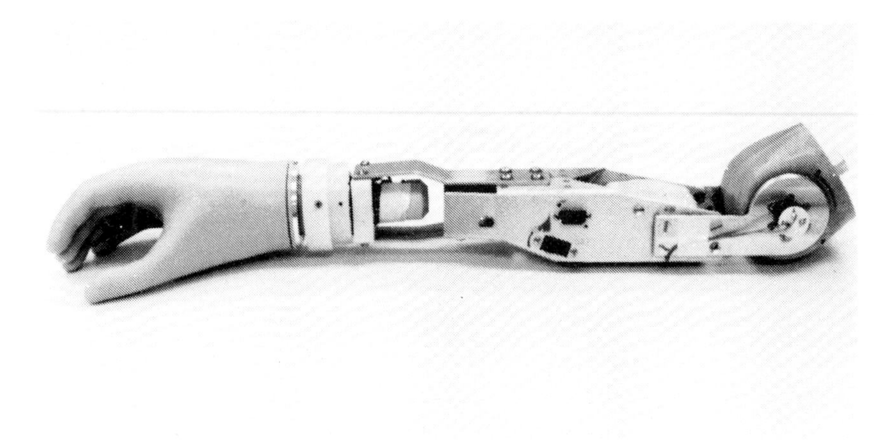

Figure 1 Above-elbow prosthesis.

this limited the use of the prosthesis to executing only a small number of motions. In an attempt to enhance the characteristics of the prosthetic arm and to provide a variety of different possible motions, we put the arm under microcomputer control. The basic input to the microcomputer is the shoulder position. The microcomputer uses the shoulder position to compute the elbow and wrist positions and to drive them accordingly. Because of the on-board memory space provided by the microcomputer, many different input–output relationships (linkages) can be programmed to suit the needs of different users and for the performance of different tasks. We use the term "linkages" to refer to the different input–output relationships.

3 DESCRIPTION OF THE NEW PROSTHESIS

The basic prosthesis is illustrated in Figure 1. It is made up of a modified Boston elbow, a modified Otto Bock wrist rotator and an Otto Bock hand. We have incorporated into the Boston elbow and into the Otto Bock wrist rotator, an absolute position encoder necessary for the type of control we are using. The controlling electronics and batteries are fitted inside the forearm. Figure 2 shows the goniometer that is used to measure shoulder flexion/extension (which is the primary input to the controlling system). This input is fed to the microcomputer which computes the corresponding elbow and wrist positions. It should be noted that shoulder abduction/adduction is not used in our present version. A major innovation in this prosthesis is that several input–output relationships (linkages) can be stored in the microcomputer memory for selection by the user depending on the task being performed. This is a major break with the original e.p.p. concept as

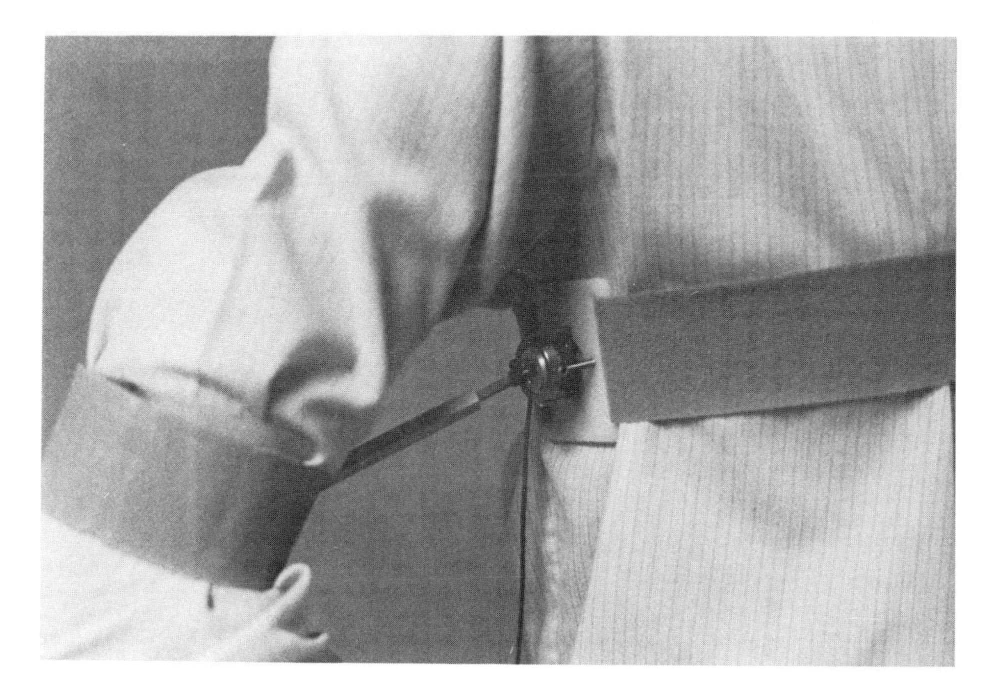

Figure 2 Shoulder goniometer.

introduced by Simpson, which had just a single input–output relationship. Allowing the users to select from a repertoire of input–output relationships, depending on the task being performed, will increase the potential flexibility and usefulness of the prosthesis. However, this strategy will inevitably degrade some aspects of the performance of the prosthesis, through the subject becoming confused between the different linkages available to him. This question is the subject of continuing investigation, as will be discussed further in this paper.

4 TESTS ON THE PROSTHESIS

We have carried out a variety of performance tests on the bench prototype of the prosthesis using normal subjects. The purpose of these tests was to verify the ability of the prosthesis to provide closed-loop control of the terminal device without using visual feedback. The performance of the prosthesis under both e.p.p. (closed-loop) and open-loop conditions was measured, and comparisons were made between these two operating modes and also with an intact arm.

For the purposes of the tests, the prosthesis was programmed with a simple linkage where the elbow extends as the shoulder is extended. Two series of tests were carried out on a group of six normal subjects. One series compared the ability of the intact elbow and the e.p.p. controlled elbow to set an angle. The second series compared the performance of the e.p.p. prosthesis and the prosthesis with open-loop control, to perform a spatial positioning task. The results, which are summarized below, are given in greater detail in an earlier paper [7].

Test of Angle Setting

For all six subjects tested, the e.p.p. system was found not to be significantly inferior to the intact arm in setting angles.

Positioning Test

In this test, five of the six subjects found a marked improvement in the e.p.p. prosthesis over the open-loop model. The last subject did not find any difference. He was able to control the e.p.p. and open-loop prosthesis equally well.

In the performance tests made with the bench prototype, our main goal was to determine how a group of subjects could adapt to the use of an e.p.p. controlled prosthesis. In doing so, we were also able to draw two important conclusions. Firstly, we determined that the e.p.p. prosthesis was not significantly inferior to the intact arm in angle setting. Secondly, we determined that the e.p.p. prosthesis was significantly superior to an open-loop prosthesis in executing a spatial positioning task. The natural type of feedback present in the e.p.p. control, enhanced the operation of the device.

5 COMPUTER SIMULATION

Having established that the e.p.p. control strategy is a useful one, it was necessary to see how many different linkages a group of test subjects could learn without too much confusion. This would fix the upper level of the repertoire of linkages that could be made available to an amputee. The other interesting question that arises relates to the ability of a subject to pick, from a set of different linkages, the one that would best fit the task he is about to undertake.

To answer the above questions it was decided that, for several reasons, a computer simulation of the above-elbow prosthesis would be more appropriate than the bench model. For the purpose of the present work, a pure software simulation (as opposed to hybrid software/hardward simulation) possesses many advantages. These include:

a) Safety. When working with volunteers who are usually unconnected with the research, this is a most important factor. There is no risk of accident since the only hardware involved is the digital computer and a small passive goniometer used to measure shoulder angle.

b) Flexibility. This is also an important factor since this allows us to build a general model for the prosthesis and refine it as needed. For instance, the operation of the Boston elbow, or any other elbow, is very easily modeled. There is also a great ease in simulating various conditions:

- the prosthesis operating in the presence or absence of gravity.
- the prosthesis operating loaded or unloaded.
- different types of actuators such as ones not yet developed or even "ideal" ones.
- the geometry of the prosthetic device can be altered as desired and various lengths of members can be tested.
- different prosthetic components (e.g., terminal devices) can be tried.

c) Practical aspects and realism. When using this simulation we found many important advantages over the bench model, especially in the case of reaching a target in space. True positioning tests can be performed without the subjects getting cues when they touch a target, since it is only an image. Also, motor sounds, that could be used as an avenue of feedback, are totally absent. Furthermore, it is simple to implement the unbeatable servo feature that prevents the input from exceeding the capabilities of the output. Finally, the data gathering process is reduced to a minimum, since the precise measurement of the distance between the terminal device and the target is done automatically without disturbing the subject. This was a problem with the bench model. Time elapsed between starting and ending of each test is also monitored by the system, thus reducing the risk of timing errors.

In implementing this computer simulation, our goals were to be able to display on a computer monitor, the complete real-time motion of an above-elbow prosthesis worn by an amputee. The image had to be as realistic as possible such that the test subject seated in front of the screen had the impression that he was actually wearing the prosthesis. The achievement of this goal required three major steps. The first one was to write the software such that a comprehensive model of the prosthetic arm could be defined for use by the simulation. The second one was to implement a life-like image of the complete arm prosthesis that would move just like the real system. The third step was to implement various tasks incorporating preprogrammed linkages.

5.1 *The Model Definition*

Since the computer simulation of the arm is a pure software implementation, it is very flexible. It can simulate the operation of almost any kind of prosthesis, even ideal ones. Thus, given a particular prosthesis defined by a set of parameters such as actuator friction, inertia and stiffness, and given a specific kind of controller such as proportional error (PE), proportional derivative (PD) or proportional integral (PI), we can simulate the behavior of the prosthesis. One of the simple models that we are using is that of a prosthetic device driven by a DC motor. The method of modelling is given in the following paragraph.

The primary loads on a DC motor are: friction, inertia and constant or varying torque loads [8]. Considering only friction and inertia designated respectively by F and J, a physical system in rotation without application of outside forces can be modelled as follows:

$$T = J\ddot{\Theta} + F\dot{\Theta}, \tag{1}$$

where T is the torque, $\dot{\Theta}$ is the angular velocity and $\ddot{\Theta}$ the angular acceleration. From control theory, we know that to avoid excessive overshoot and to keep the steady state error at a relatively low level, we can make the drive torque proportional to the derivative (rate of change) of position with respect to time [8].

$$T = K_e(\Theta_d - \Theta) - K_d\dot{\Theta}, \tag{2}$$

where Θ_d is the desired angular position in radians, Θ is the present joint angular position in radians.

This equation (2) is that of a proportional derivative controller (PD controller). K_e and K_d are the gains of the controller. K_e is a proportionality constant whereas K_d is a constant which when increased adds up to the friction F of the system. This can be shown by equating (1) and (2). This gives us:

$$J\ddot{\Theta} + F\dot{\Theta} = K_e(\Theta_d - \Theta) - K_d\dot{\Theta}. \tag{3}$$

Rearranging terms with $\Theta_d = 0$ gives:

$$-K_e\Theta = J\ddot{\Theta} + (F + K_d)\dot{\Theta}. \tag{4}$$

This does not in any case alter the generality of the problem. Thus K_d adds up with F to increase the system friction. From Eq. (2), it can be seen that, if the error is large (the present position is far from the desired one) and the velocity is small, the applied drive is large. Also, if the error is small (we are almost at the desired position) and the velocity is high, the applied drive is negative so as to slow down the system. Hence a braking effect is achieved. With this proportional derivative (PD) controller we obtained good performance from the control of our bench and simulation models. The common problem of drooping was not encountered since the Boston elbow provides a reverse-locking clutch which locks the arm unless the motor is being driven. This prevents the elbow from being driven by external torques. Note also that since we are using a computer to control the system, the drive torque equation (2) was expressed in discrete form and was used as such:

$$T(i) = K_e[\Theta_d - \Theta(i)] - K_d\dot{\Theta}(i), \tag{5}$$

where $\Theta(i)$ and $\dot{\Theta}(i)$ are the respective values at the i-th sampling instant.

The prosthetic arm in the simulation was purposely programmed to behave like the Boston elbow. It can flex 135 degrees in 1 s and is capable of lifting 4.5 lb at

1 ft. The Boston elbow goes faster down than up when operating in a vertical plane. However, since this was corrected in the bench model, it was decided to equalize the up and down velocities of the elbow in the simulation. An interesting aspect of the system was that we were able to model, to some extent, Simpson's unbeatable servo, whereby the input is never allowed to go faster or further than the output can. If at any instant the input were to exceed the capabilities of the output an alarm would warn the user. The alarm was equipped with two tones in order to differentiate between flexion and extension of the shoulder.

5.2 *The Computer-generated Image of the Prosthesis*

Our aim was to make a computer image of the end effector of the prosthesis. Figure 3 shows the image on the screen together with a simulated background which enhances the spatial perception of the hand. No attempt was made to display the forearm and upperarm as these images were confusing and detracted from the image of the end effector. The visual impression we wanted to give to the test subject was that of watching his hand performing a spatial positioning task. For the simulation to be realistic, real-time motion of the picture had to be ensured and proper visual perspective had to be incorporated. These requirements were hard to meet using commercial computer graphics packages. For this reason we decided to create our own display software. The high speed motion display was achieved by using the principle of look-up tables. The position of the prosthesis is precalculated for every possible shoulder angle (the input control signal) and the corresponding image is stored in memory. Hence when needed, these images can be displayed in the required sequence.

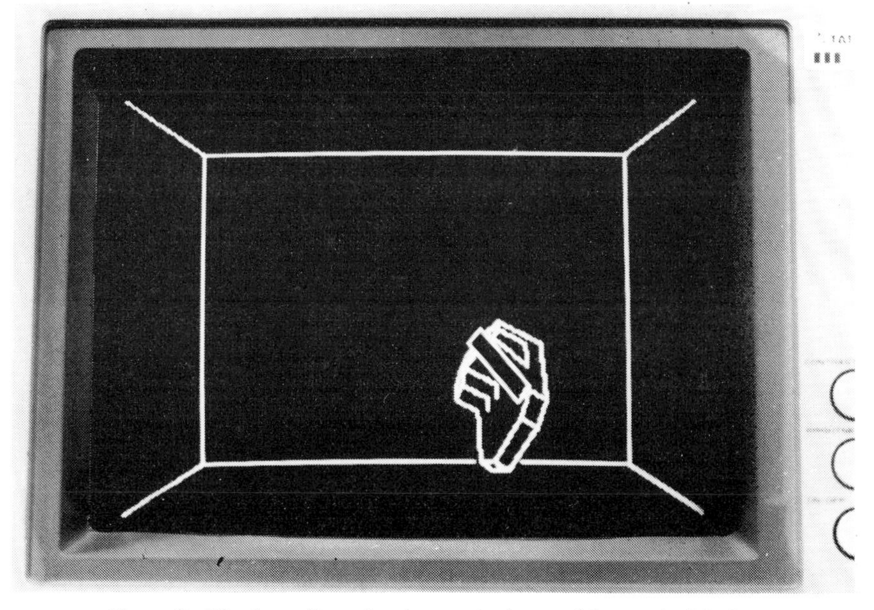

Figure 3 The three-dimensional computer image of the terminal device.

It is assumed at this stage that when the simulation is being executed, the patient is seated in front of the screen. There are three basic inputs to the computer which are sketched in the following line drawings:

- Shoulder flexion/extension with 90 degrees of range of motion measured with a goniometer, Figure 4.
- Chair rotation with 180 degrees range of motion (from extreme left to extreme right) measured with a single turn potentiometer, Figures 4 and 5.
- Trunk forward bending with a 30 degree range also measured with a single turn potentiometer, Figure 4.

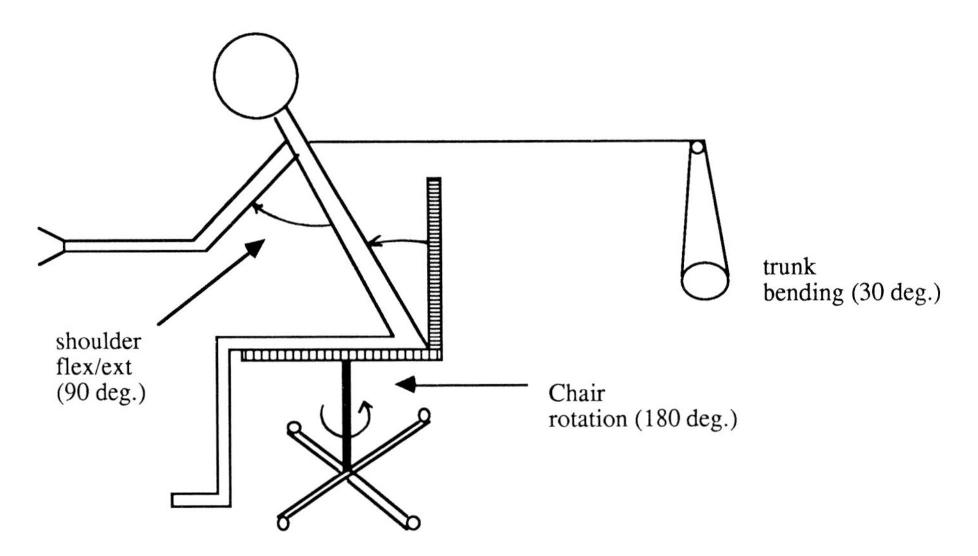

Figure 4 Schematic of the inputs to the simulation program.

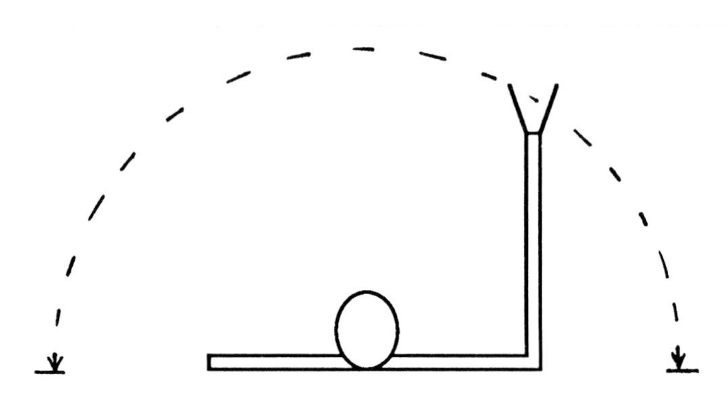

Figure 5 Top view of subject. Surface swept by using chair rotation.

These inputs were selected because of the relative ease with which one can measure their corresponding angles of rotation, and also because they allow the patient to reach almost any target within the space in front of him from the waist up. Trunk bending is required for fine positioning, or for cases where the selected linkages are not exactly appropriate for the tasks being performed. All of these inputs are fed to the computer via an analogue multiplexer. They are then digitized and used to alter the position of the image on the screen as shown on Figure 6.

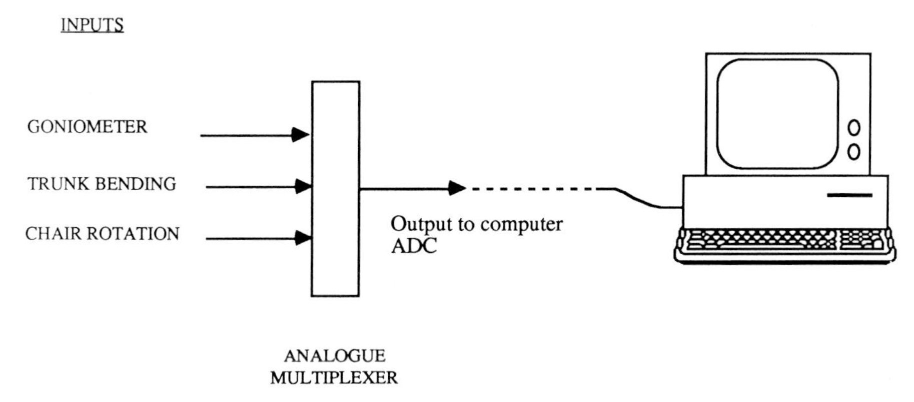

Figure 6 Schematic of the simulation equipment.

5.3 *Preprogrammed Linkages*

We have incorporated into the simulation some preprogrammed linkages that can be selected via the computer keyboard. All of them involve a coordinated motion of shoulder joint and elbow joint or shoulder joint and wrist joint. In all cases, however, the shoulder joint has to participate in the motion since this is a basic requirement of our e.p.p. control strategy. A prime example of the programmed linkages in our repertoire is a motion where the elbow extends as the shoulder is extended. The hand also rotates as the shoulder is extended. The elbow and wrist angles vary at twice the shoulder angle, see Figure 7.

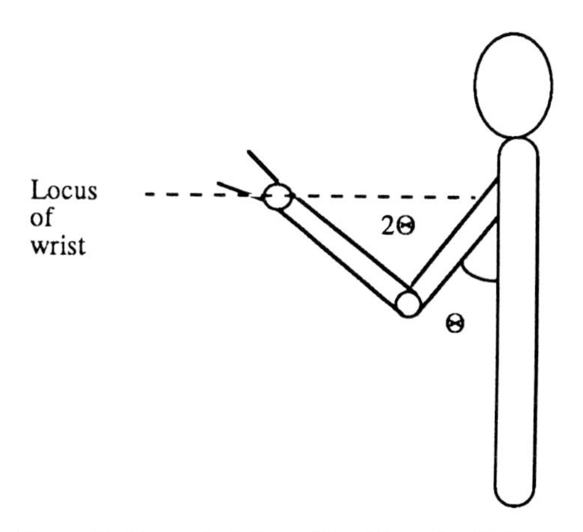

Figure 7 Example linkage: Shoulder extend/elbow extend.

Many other predefined linkages, can be incorporated as a function of the user's requirements and the task to be performed. It should be noted that these linkages or motions appear as files in the simulation system. When one of them is needed, the file is opened and the linkage is read. Hence, there is only one active linkage at any time in the simulation system. This approach is convenient to use since there can be virtually an unlimited number of linkages occupying disk space. The computer RAM can then be reserved for more time-critical applications.

5.4 *System Operation*

The operation of the system was designed to be very general so as to accommodate different types of tests. The basic operation can be described as follows:

1) The simulation is initiated by choosing one among a variety of tests, for example a positioning task.

2) A linkage is picked and the system automatically selects a target which is randomly positioned. The target can always be reached by the selected linkage.

3) Some time is given to accommodate to system operation.

4) Once ready to be tested the user depresses a start key. The subject then tries to reach the target with the screen blanked (to remove visual feedback). Once the maximum trial time has elapsed or the subject feels that he is close enough to the target, the trial is stopped and an absolute distance and test time is recorded and written on to the subject's own file.

5) Successive trials can be done and results are automatically appended to the given file until completion of the experiment.

The simulation system is totally user friendly and at different stages throughout execution the patient or researcher is prompted for pertinent information. At the same time, the patient is informed of his performance. Data processing can be done when enough information has been gathered. Again, the researchers do not have to record the results manually. The simulation system can provide, on request, statistical information on tests completed such as: mean, standard deviation, "*t*" test, etc.

6 CONCLUSION

A new prosthesis incorporating the control strategy known as extended physiological proprioception with the addition of programmable input/output relationships (linkages) has been designed. Tests performed on this prosthesis have demonstrated the validity of the e.p.p. control technique. In order to precisely evaluate the system, a three-dimensional computer simulation has been implemented. This strategy has many advantages over the use of a bench-prototype, since it enables almost any type of prosthesis to be simulated. The effects of changes to certain parameters can be studied while the other system parameters remain unchanged. This allows for parameter optimization.

ACKNOWLEDGEMENTS

This work is supported by the Natural Science and Engineering Research Council of Canada.

REFERENCES

1. R. W. Wirta, D. R. Taylor and F. R. Finlay, Pattern-recognition arm prosthesis: A historical perspective—a final report, *Bull. of Prosthetics Research*, **BPR 10:31,** 8–35 (1978).
2. J. H. Lyman, A. Freedy, R. Prior and M. Solomonov, Studies toward a practical computer-aided arm prosthesis system, *Bull. of Prosthetics Research*, 213–225, Fall (1974).
3. D. Graupe and W. K. Cline, Functional separation of EMG signals via ARMA identification methods for prosthesis control purposes. *IEEE Trans. System, Man and Cybernetics*, **SMC-5,** 252–259, March (1975).
4. S. C. Jacobsen, D. F. Knutti, R. T. Johnson and H. H. Sears, Development of the Utah artificial arm, *IEEE Trans. on Biomed. Eng.*, **29:4,** 249–269 (1982).
5. T. B. Sheridan and R. W. Mann, Design of control devices for people with severe motor impairment, *Human Factors*, **20,** 321–338 (1978).
6. D. C. Simpson, The choice of control system for the multimovement prosthesis: extended physiological proprioception (e.p.p.), in *The Control of Upper-Extremity Prostheses and Orthoses* (Eds. P. Heberts, *et al.*), C. Thomas (1974), ch. 15.
7. D. T. Gibbons, M. D. O'Riain and J. S. Philippe-Auguste, An above-elbow prosthesis employing programmed linkages, *IEEE Trans. on Biomed. Eng.*, **34(7)** (1987).
8. W. E. Snyder, *Industrial Robots: Computer Interfacing and Control, Englewood Cliffs, New Jersey, Prentice-Hall, Inc.* (1985).

Automedica, 1989, Vol. 11, pp. 111–122
Reprints available directly from the publisher only
Photocopying permitted by license only

A MATHEMATICAL MODEL TO ANALYZE CONTROL OF UPRIGHT POSTURE AND BALANCE

JOHN R. BUHRMAN and CHANDLER A. PHILLIPS

Department of Biomedical Engineering, Wright State University, Dayton, Ohio 45435

(Received September, 1988)

A mathematical model was designed to simulate balancing in the paraplegic individual as a closed-loop feedback system. The model was described by a second-order linear differential equation and the system transfer function for a step input perturbation was obtained. A physical model of the system using an inverted pendulum was constructed. The model was tested for step inputs of 2.5, 3.5, and 4.5 degrees by placing the physical model on a platform which was dropped from each of these inclinations. The constant values of the damping ratio ζ and the undamped frequency ω_n, and the similarity of the measured curves to the computed curve of θ_0 vs. t/T, indicated a correct model. The model was then tested with a human subject using the same procedure. The subject was a 30-year-old male who had been previously diagnosed as a level T-5 paraplegic, complete lesion. The same parameters were calculated and observed to be fairly constant but with more variability than in the simulated runs. The measured curves were similar in shape and periodicity to the computed curve although the magnitude of the peaks was somewhat lower. It was concluded that the mathematical model could be useful in analyzing the effectiveness of balancing strategies and feedback control system in lower extremity orthoses. It was also concluded that the platform drop is an effective method of creating a step perturbation which can be used to study the system response.

Keywords: Balance control, inverted pendulum, F.E.S.-orthosis, paraplegia, closed-loop feedback.

INTRODUCTION

In order to maintain a state of balance or to initiate a posture change, various types of feedback are employed to tell a person when the desired position has been attained. This feedback is needed continuously since even in a seemingly still position the person needs to make slight adjustments in muscle activity to avoid falling [1]. Since the feedback information is used by the central nervous system to initiate movement, with skeletal muscle as the actuator, this network can be viewed as a closed-loop feedback control system [2].

The feedback for the human system occurs both from impulses generated reflexly by internal stretch receptors in the muscles and joints, and by more externally apparent "sensory" feedback, consisting of vision, touch, and audition, with vision being the most important. In this type of feedback the eyes, ears, and skin act as transducers which convert outside information into impulses recognizable by the central nervous system [3].

In a paraplegic individual, some of this information flow has been interrupted, since this person no longer has the link between the spinal cord and the lower alpha motor neurons. The muscle tissue itself remains healthy but begins to atrophy with disuse [4]. Various functional electrical stimulation (FES) orthoses, including

Address for correspondence: Chandler A. Phillips, Department of Biomedical Engineering, Wright State University, Dayton, Ohio 45435.

those at Wright State University, have succeeded in providing a means for the paralyzed individual to generate movement and thus prevent further deterioration of the limbs [4, 5].

For a standing or walking orthosis to be continuous and automatic, a closed-loop feedback system is desirable [5] and should provide some other type of feedback, such as vibrotactile, in addition to the individual's own all important visual type [6, 7]. A model of this control system could prove very useful in choosing optimal balancing strategies, assuming certain important parameters could be determined and analyzed.

A model is employed which views the paraplegic in the standing position as the familiar inverted pendulum. Feedback control is introduced and the model is described by a second-order, linear differential equation. The transfer function of the model is obtained and a physical simulation capable of being perturbed by a step input is designed to test the model. The results of the simulation are used to obtain values of the damping ratio ζ and the undamped frequency ω_n. The mean values of these parameters, along with a transient scaling constant G, are used to plot a curve of θ_0 vs. t/T by solving the transfer function over time. A similar technique is used to test a human paraplegic subject and the results are analyzed and compared to the experimental curves.

METHODS AND PROCEDURES

Simulation

A theoretical model of an inverted pendulum perturbed by a step input along with its differential equation is shown in Figure 1. The system transfer function is derived in Appendix A and is shown below:

$$\frac{\theta_0}{\theta_i} = \left[1 + \frac{1}{\sqrt{1-\zeta^2}} e^{-\zeta\omega_n t} \sin(\omega_n \sqrt{1-\zeta^2}\, t + \phi) \right].$$

A physical model of the inverted pendulum to simulate erect posture and balance was constructed and is shown in Figure 2. An eyebolt with a stiff shaft was spring mounted to a platform adjustable from 0–5 degrees and a position transducer (Spectron L-210, Uniondale, NY) with a time constant of 24 ms was

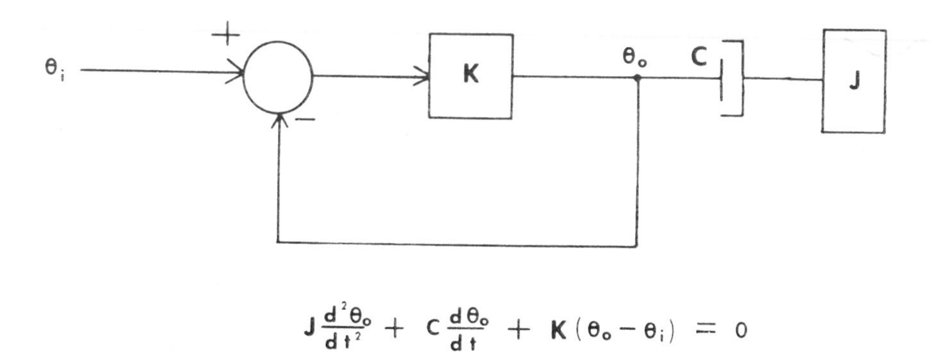

$$J\frac{d^2\theta_o}{dt^2} + C\frac{d\theta_o}{dt} + K(\theta_o - \theta_i) = 0$$

Figure 1 Model of feedback control system with differential equation.

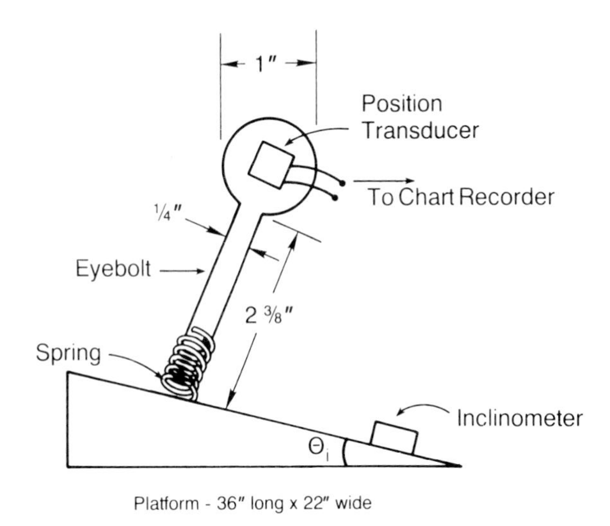

Figure 2 Physical model of inverted pendulum used to simulate balance response.

attached to the eyebolt head. The transducer consisted of a metal ball pivoted inside a housing filled with a viscous fluid. An AC voltage was applied to the leads connected to opposite ends of the fluid. As the ball changed position in response to anterior-posterior sway, the voltage at the conducting lead of the pivot varied proportionally. The voltage variations were recorded on a strip-chart recorder and later converted to degrees by calibrating platform tilt in degrees/mV. A string was run on either side of the eyebolt shaft in the anterior-posterior direction in order to prevent any appreciable lateral sway of the model. A switch was mounted to record the time when the platform reached ground level.

The platform was dropped from inclinations of 2.5, 3.5, and 4.5 degrees to the horizontal. The oscillations in response to these step perturbations were plotted over time by the chart recorder. Three trials for each of the three inclinations were run.

The natural logs of the amplitude peaks of the recorded curves were plotted over time for each trial. An exponential decrease in the magnitude of these points would be characteristic of the curves normally generated in a second-order control system [8].

The parameters ζ and ω_n for the theoretical model were determined by calculating their mean values from the experimental data (See Appendix B). After comparing the amplitude peaks of the experimental data to the predicted values, it was found that a transient scaling constant G was needed in the equation. This was obtained by substituting the measured peak values of θ_0 into the following equation and solving for G:

$$\theta_0 = \theta_i \left[1 + \frac{G}{\sqrt{1-\zeta^2}} e^{-\zeta\omega_n t} \sin(\omega_n \sqrt{1-\zeta^2}\, t + \phi) \right].$$

The above equation was solved for a step perturbation of 4.5 degrees using the previously calculated values of ζ and ω_n, along with G, and plotted over two cycles of t/T, then superimposed on the recorded experimental data (see Figure 3).

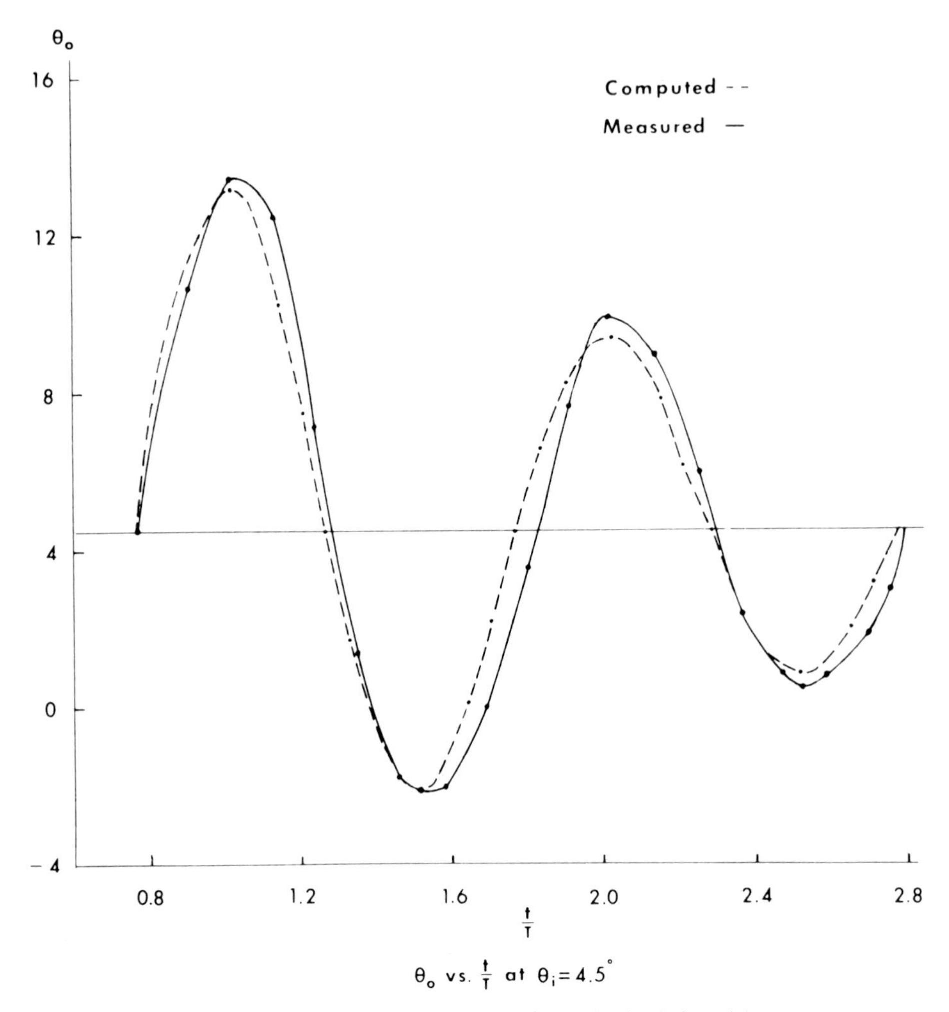

Figure 3 Computed vs. measured curves in simulation trial.

Subject Balancing Trials

A 30-year-old male subject who had been previously diagnosed as a T-5 level paraplegic (insensate and completely paralyzed below the lower chest but otherwise healthy) participated in the balancing trials. The subject stood 5 ft 9 in. in height and weighed 155 lb. The same adjustable platform used in the simulation was set to an inclination of 2.5 degrees. The subject performed all trials while wearing a "Reciprocating Gait Orthosis" (RGO) brace which kept his body rigid below the waist. The RGO is essentially a long leg brace made out of lightweight aluminium and plastics which can be locked at the knees [9]. To eliminate high frequency noise, a Spectron L-211 position transducer with a time constant of 125 ms was employed. The transducer was mounted on the RGO at the back of the left thigh.

Several weeks of preliminary balancing was done where the subject attempted to balance with his arms and legs in various positions and with the platform at different settings. The final experimental procedure involved having the subject stand on the platform with hands on hips and feet spread 13.5 in. apart. He was instructed to remain as still as possible when balancing both before and after the perturbations. To begin the trial, the subject removed his hands from a walker upon which he had been holding onto and let the experimenter know when he was completely balanced. The chart recorder was then started and the platform was dropped. The chart recorder was allowed to run until after the subject was observed to have completely regained his balance.

A total of 35 trials were run in five different sessions over a period of five weeks. The number of trials per session varied from 5 to 9, with the sessions lasting from 20 to 30 min. The sessions were terminated if the subject felt that he was becoming overly stiff or tired. Four trials in which the subject completely lost his balance were discarded, along with four others in which the subject could not remain steady after he had regained his balance. Of the remaining trials, 16 had at least three distinguishable peaks, the minimum number necessary to demonstrate an exponential curve characteristic of a second-order control system.

As in the simulation trials, the natural logs of the amplitude peaks of the recorded curves were plotted over time, and the deviation from a perfect exponential decrease was measured in degrees. Of the 16 plots with at least three distinguishable peaks, 11 were found to be linear with an error of 18 degrees or less. These 11 curves were then used to determine the parameters ζ, ω_n, and G by the same methods used in the simulation. The system equation was solved over time using these parameters and the computed curve was compared to the experimental curves (see Figure 4).

Although the inverted pendulum is potentially an unstable system [10], a stability analysis was conducted to determine the proportional band of stability (See Appendix C). The proportional band of stability was found to lie between $0 < K < 6.22$.

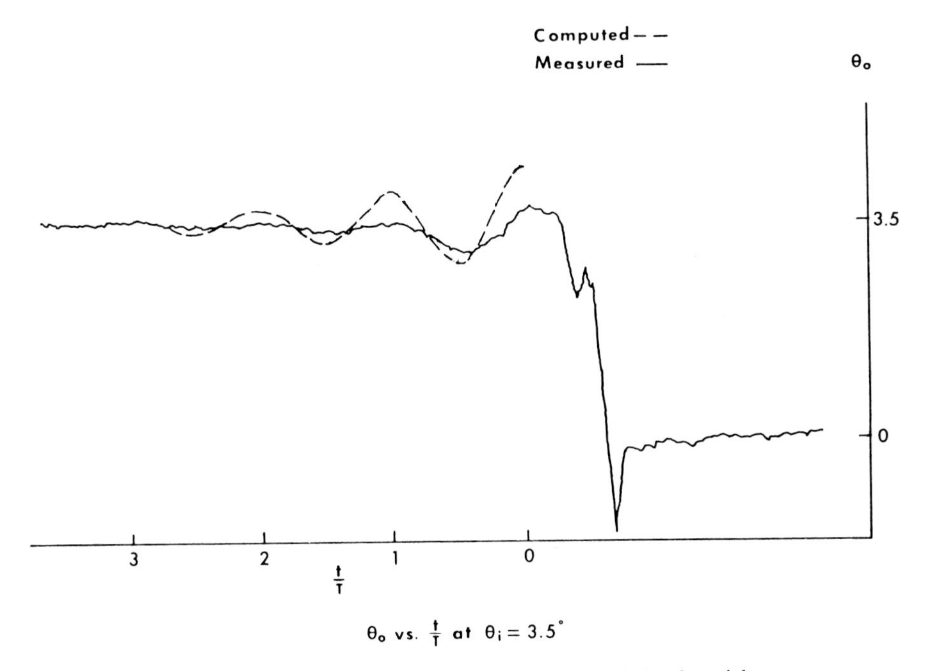

Figure 4 Computed vs. measured curves in subject balancing trial.

RESULTS

Simulation

The results of the simulation trials are shown in Table 1. The values of the parameters ζ, ω_n, and G each remained relatively constant over the three different angles of input perturbation θ_i. The average values of ζ were found to be 0.0885, 0.0969, and 0.0858 for 2.5, 3.5, and 4.5 degrees of perturbation, respectively, giving a mean of 0.0904 ± 0.006. The average values of ω_n were found to be 18.64, 17.92, and 17.47 rads/s for 2.5, 3.5, and 4.5 degrees, respectively, giving a mean of 18.01 ± 0.59 RADS/s. The values of the transient scaling constant G were calculated at 3.22, 3.73, and 3.44 for 2.5, 3.5, and 4.5 degrees, respectively, giving a mean of 3.46 ± 0.26. A comparison of the values of the first measured amplitude peaks within each group of trials yielded an experimental error of $\pm 8.0\%$ for amplitude measurements and $\pm 4.4\%$ for time measurements.

Figure 3 shows the computed values of θ_0 for a step perturbation of 4.5 degrees plotted over two cycles of t/T and superimposed on a plot of the corresponding measured values. The two curves are almost identical with the measured curve slightly greater than the computed curve in magnitude and slightly lagging in phase.

Subject Balancing Trials

The results of the subject balancing trials are shown in Table 2. The average parameter values are $\zeta = 0.112 \pm 0.032$, $\Omega_n = 2.57 \pm 0.16$ rads/s, and $G = 0.262 \pm 0.09$. Although some variation in ζ occurred over the 11 trials, the overall values were fairly consistent, with none of the values varying more than ± 0.05 from the mean. The transient scaling constant G was also fairly uniform with 6 out of the 11 values occurring between 0.2 and 0.3, and only 2 of the values varying more than ± 0.1 from the mean. The values of G became more consistent in the final 5 trials, varying only between 0.201 and 0.242. The values of ω_n remained very close over all the trials, varying only between 2.38 and 2.93 rads/s.

Although the initial angle of the platform itself was always set at a uniform inclination of 2.5 degrees, the actual angle of perturbation varied somewhat between trials. While the platform angle was read directly from an inclinometer, the angle of perturbation was considered as the difference between the initial and final tilt of the subject, as measured by the transducer attached to the back of the subject's leg. This angle was slightly higher than the platform angle and varied between 2.98 and 5.25 degrees with a mean of 4.02 ± 0.66 degrees.

Figure 4 shows the computed values of θ_0 for $\theta_i = 3.5$ degrees plotted over 3 cycles of t/T, along with the corresponding measured values. The measured curve shows a steady-state amplitude of $\theta_0 = 0$ degrees before the platform is dropped. Upon dropping of the platform, θ_0 rises rapidly to a slight overshoot and then decreases over three oscillations toward a steady-state value of $\theta_0 = \theta_i = 3.5$ degrees. The periodicity of the measured curve is similar to that of the computed curve but the magnitude of the measured peaks is somewhat smaller.

DISCUSSION

Model

The theoretical model shown in Figure 1 was based on a type 0 servomechanism system using simple position feedback. The use of position feedback in

Table 1 Simulation results for ζ, ω_n, and G

$\theta_i = 2.5°$			
Trial	ζ	ω_n(rads/s)	G
A	0.0924	18.78	3.09
B	0.0869	18.60	3.20
C	0.0861	18.55	3.37
Ave.	0.0885	18.64	3.22
$\theta_i = 3.5°$			
Trial	ζ	ω_n(rads/s)	G
A	0.101	17.74	3.73
B	0.0968	17.73	3.85
C	0.0930	18.29	3.59
Ave.	0.0969	17.92	3.73
$\theta_i = 4.5°$			
Trial	ζ	ω_n(rads/s)	G
A	0.0846	17.61	3.50
B	0.0836	17.66	3.48
C	0.0892	17.14	3.36
Ave.	0.0858	17.47	3.44
Mean (all trials)	0.0904 ± 0.006	18.01 ± 0.59	3.46 ± 0.26

Table 2 Subject balancing results for θ_i, ζ, ω_n, and G

Test	θ_i(degrees)	ζ	ω_n(rads/s)	G
1	4.03	0.125	2.75	0.155
2	4.20	0.0911	2.93	0.286
3	2.98	0.123	2.48	0.452
4	5.25	0.0721	2.43	0.195
5	4.20	0.157	2.65	0.362
6	4.38	0.141	2.48	0.334
7	4.03	0.0754	2.47	0.201
8	2.98	0.0635	2.57	0.201
9	3.48	0.137	2.48	0.216
10	4.45	0.110	2.38	0.242
11	4.25	0.135	2.64	0.235
Mean	4.02 ± 0.66	0.112 ± 0.032	2.57 ± 0.16	0.262 ± 0.09

biomathematical modeling is in agreement with Jaeger [11], Hamami and Stokes [12], and Petrofsky and Phillips, [2]. As in any standard second-order control system, it was assumed that a frictional force proportional to the output angular velocity was present which opposed the motion of the output, along with torque (moment of inertia × angular acceleration) and controller gain K proportional to the error signal [3].

The physical model (see Figure 2) employed biomathematical modeling assumptions in order to simulate human balancing. The stiff eyebolt shaft correlated with the RGO encased lower extremities of the paraplegic subject. The combination of the transducer, which was firmly mounted to the eyebolt head, and the rigid shaft of the eyebolt, is a single link inverted pendulum. Employing the inverted pendulum as a model for a balancing human is in agreement with studies by Jaeger [11] and Ishida and Miyazaki [14]. The steel spring allowed flexion only at the junction of the eyebolt and the platform, which was similar to a balancing human subject with movement occurring only at the ankle joints [1]. Also, the direct connection of the spring to the platform corresponds to the assumption of a massless foot in the human subject, as put forth by Ishida and Imai [15]. Standard simplifying assumptions for biomathematical modeling by Hemami and Stokes [12] were also employed.

The model was tested in two ways. The first test was to obtain constant values of ζ, ω_n, and G. If the model is correct, these values should remain constant for all values of θ_i. As can be seen from the results in Table 1, they were reasonably close over the three trials, allowing for experimental error. The second test was to solve the system equation for θ_0 over time and compare these values to the actual measure of values of θ_0. Time was normalized since the peaks of both the computed and measured values of θ_0 could be more accurately obtained for periodic values of t/T. The close similarity of the curves again indicates a correct model.

Subject Balancing

When determining the step input θ_i, the subject could not be instructed to lean forward a specific distance since his initial loss of balance was due to the dropping of the platform. Therefore, θ_i could not be maintained at exactly the same value over all trials. However, the platform method of creating a step perturbation appears to have some distinct advantages. Since the amount of time required for the platform to reach the ground in the subject trials was measured at less than 0.08 s, the initialization time for the step perturbation can be considered negligible. Also, the loss of balance is involuntary so the subject cannot consciously vary his velocity or acceleration at the onset of the step perturbation. It appears, therefore, that the platform drop is a superior method of creating a step perturbation since the loss of balance in the subject was both instantaneous and natural.

Variations in the values of ζ and G were probably due in part to the subject making slight adjustments in his balancing strategies by varying the amount of shift of his upper torso. This is very apparent in observing the values of G, as they become very consistent over the last five trials, apparently reflecting the subject's ability to better control the magnitude of his sway with additional practice. The variations may also be due to the initial stiffness of the subject prior to the first run of a test day and also to the stiffness and fatigue sometimes encountered towards the end of a group of trials.

The fact that ω_n remained nearly constant over all trials as well as within each trial points to the subject's uniform periodicity of motion while attempting to balance. The mean value of 2.57 rads/s translates to a value of 0.41 Hz, which is in agreement with previous studies by Ishida and Miyazaki [14] and Jaeger [11]. The mean value of $\zeta = 0.112$ was within the range expected since the human pendulum feedback system would predictably be more heavily damped than the spring-controlled simulation where ζ was calculated at 0.09.

Using the mean values of ζ, ω_n, and G, θ_0 was plotted vs. t/T and superimposed on a set of experimental values as shown in Figure 4. Although the general shape of the curves is reasonably close, the magnitude of the computed curve is noticably greater, especially at the positive peaks. This is due primarily to the fact that the computations of G were based on the negative peaks of the measured curves, as well as to the variability of ζ and G among the trials. Since the values for the computed curves were compiled using the mean values of ζ, ω_n, and G, none of the measured curves were expected to fit the computed curve precisely.

As predicted by the results of the simulation, the balancing subject appears to conform to the theoretical model of an oscillating inverted pendulum with closed-loop feedback. The subject was able to maintain his balance through a swaying motion consisting of oscillations similar to those in a second-order (or higher) control system.

The system parameters ζ and ω_n, along with G, can theoretically be used to analyze and compare various feedback control systems for use in conjunction with FES orthoses in paralyzed individuals. Before this could be accomplished, however, further testing with human subjects would be necessary in order to determine the normal, stable, and unstable ranges of each parameter. Once these ranges are established, feedback control systems such as the one developed by Phillips and Petrofsky [6, 7] employing vibrotactile feedback, could be more objectively analyzed. Their system could be tested by recording the mean parameter values over several trials with and without visual and/or vibrotactile feedback. The system could then be evaluated by a comparison of the parameters under the different test conditions to determine the improvement in stability using the sensory feedback system.

REFERENCES

1. D. P. Thomas and R. J. Whitney, Postural movements during normal standing in man, *Journal of Anatomy*, **93**, 524–539 (1959).
2. J. S. Petrofsky and C. A. Phillips, Closed-loop control of movement of skeletal muscle, *CRC Critical Reviews in Biomedical Engineering*, **13**, 35–96 (1985).
3. A. J. Vander, J. H. Sherman and D. S. Luciano, *Human Physiology—Mechanisms of Body Function* (*2nd Ed.*), McGraw-Hill Book Co., New York (1975), pp. 549–550.
4. J. S. Petrofsky, H. H. Heaton III and C. A. Phillips, Leg exerciser for training of paralysed muscle by closed-loop control, *Medical & Biological Engineering & Computing*, **22**, 298–303 (1984).
5. J. S. Petrofsky, C. A. Phillips and H. H. Heaton III, Feedback control system for walking in man, *Computers in Biology and Medicine*, **14**, 135–149 (1984).
6. C. A. Phillips and J. S. Petrofsky, Cognitive feedback as a sensory adjunct to functional electrical stimulation (FES) neural prosthesis, *J. Neurological & Orthopaedic Medicine & Surgery*, **6**, 231–238 (1985).
7. C. A. Phillips and J. S. Petrofsky, A total neural prosthesis for spinal cord injury rehabilitation: the cognitive feedback system with a functional electric stimulation (F.E.S.) orthosis, *J. Neurological & Orthopaedic Medicine & Surgery*, **7**, 225–234 (1986).
8. B. C. Kuo, *Automatic Control Systems* (5th Ed.), Prentice-Hall, Inc., Englewood Cliffs, NJ (1987).

9. J. S. Petrofsky, C. A. Phillips, R. Douglas and P. Larson, A computer-controlled walking system: the combination of an orthosis with functional electrical stimulation, *J. Clinical Engineering*, **11**, 121–133 (1986).
10. H. Hemami, F. C. Weimer, C. S. Robinson, C. W. Stockwell and V. S. Cvetkovic, Biped stability considerations with vestibular models, *IEEE Trans. Automatic Control*, **AC-23**, 1074–1079 (1978).
11. R. J. Jaeger, Design and simulation of closed-loop electrical stimulation orthoses for restoration of quiet standing in paraplegia, *J. Biomechanics*, **19**, 825–835 (1986).
12. H. Hemami and B. T. Stokes, A qualitative discussion of mechanisms of feedback and feedforward in the control of locomotion, *IEEE Trans. Biomed. Eng.*, **BME-30**, 681–688 (1983).
13. F. L. Westwater and W. A. Waddell, *An Introduction to Servomechanisms*, English Universities Press, London (1967).
14. A. Ishida and S. Miyazaki, Maximum likelihood of a posture control system, *IEEE Trans. Biomed. Eng.* **BME-34**, 1–5 (1987).
15. A. Ishida and S. Imai, Responses of the posture-control system to pseudorandom acceleration disturbances, *Med. &Biol. Eng. &Comput.*, **18**, 433–438 (1980).
16. J. J. DiStefano III, A. R. Stubberud and I. J. Williams, *Theory and Problems of Feedback and Control Systems*, Schaum Publishing Co, New York (1967), p. 362.

APPENDIX A

Derivation of the System Transfer Function

From Figure 1,

$$J\frac{d^2\theta_0}{dt^2} + C\frac{d\theta_0}{dt} = K(\theta_i - \theta_0), \quad [13] \tag{1}$$

$$\frac{d^2\theta_0}{dt^2} + \frac{C}{J}\frac{d\theta_0}{dt} + \frac{K}{J}\theta_0 = \frac{K}{J}\theta_i,$$

where

$$\omega_n = \sqrt{K/J} \text{ and } 2\zeta\omega_n = \frac{C}{J} \quad [13].$$

Substituting into Eq. (1),

$$\frac{d^2\theta_0}{dt^2} + 2\zeta\omega_n\frac{d\theta_0}{dt} + \omega_n^2\theta_0 = \omega_n^2\theta_i .$$

For a step input,

$$[S^2 + 2\zeta\omega_n S + \omega_n^2]\theta_0 = \frac{\omega_\eta^2\theta_i}{S},$$

$$\theta_0 = \omega_\eta^2\theta_i\left[\frac{1}{S(S^2 + 2\zeta\omega_n S + \omega_n^2)}\right].$$

Using Laplace transforms,

$$\frac{\theta_0}{\theta_i} = \left[1 + \frac{1}{\sqrt{1-\zeta^2}}e^{-\zeta\omega_n t}\sin(\omega_n\sqrt{1-\zeta^2}\,t + \phi)\right], \tag{2}$$

where $\phi = \cos^{-1}\zeta$ [16].

APPENDIX B

Derivation of ζ and ω_n

From Eq.(2),

$$\frac{\theta_0}{\theta_i}=\left[1+\frac{1}{\sqrt{1-\zeta^2}}\,e^{-\zeta\omega_n t}\sin(\omega_n\sqrt{1-\zeta^2}\,t+\phi)\right],$$

where the damping constant

$$\alpha=\zeta\omega_n \quad [8]. \tag{3}$$

Since

$$\omega_n=\frac{\omega}{\sqrt{1-\zeta^2}} \quad [8]. \tag{4}$$

Substituting Eq.(3) into Eq.(4),

$$\alpha=\frac{\zeta\omega}{\sqrt{1-\zeta^2}},$$

$$\zeta=\sqrt{\frac{\alpha^2}{\sqrt{\omega^2+\alpha^2}}}. \tag{5}$$

α and ω can be obtained from the empirical curves where

$$\alpha=\frac{ln\,(\text{peak A})-ln\,(\text{peak B})}{t_A-t_B} \text{ and } \omega=\frac{2\pi}{T}.$$

APPENDIX C

System Stability Test

From Eq.(1),

$$J\frac{d^2\theta_0}{dt^2}+C\frac{d\theta_0}{dt}=K(\theta_i-\theta_0),$$

$$\frac{\theta_0}{\theta_i}=\frac{K}{JS^2+CS+K},$$

$$2\zeta\omega_n=\frac{C}{J} \quad [13],$$

where $J=75\ \text{kg}\cdot\text{m}^2$ for the human inverted pendulum [11].

Substituting the mean experimental values of $\zeta = 0.112$ and $\omega_n = 2.57$ rads/s along with $J = 75$ kg·m² into the above equation gives $C = 43.2$ kg·m² (in limit where θ approaches 0, C can be approximated by kg·m²),

$$JS^2 + CS + K = 0,$$

$$75S^2 + 43.2S + K = 0,$$

$$S = \frac{-43.2 \pm \sqrt{1866.24 - 300K}}{150},$$

$0 < K < 6.2$, for a stable system.

NOMENCLATURE

C	frictional force
G	transient scaling constant
J	moment of inertia (kg·m²)
K	controller gain
t	time (s)
T	period of sinusoid (s)
t/T	normalized time

Greek Symbols

α	damping factor
ζ	damping ratio
θ_i	input angle (degrees, rads)
θ_0	output angle (degrees, rads)
ϕ	phase shift of sinusoid (rads)
ω	conditional or reference frequency (rads/s)
ω_n	natural undamped frequency (rads/s)

Automedica, 1989, Vol. 11, pp. 123–143
Reprints available directly from the publisher only
Photocopying permitted by license only

CONTROL STRATEGIES WITH APPLICATION TO LIMB PROSTHESES

D. W. REPPERGER and J. W. FRAZIER

Armstrong Aerospace Medical Research Laboratory, Wright Patterson Air Force Base, Dayton, Ohio 45433

and

C. GOODYEAR

Systems Research Laboratory, Dayton, Ohio 45440

(Received September 1988)

This study investigates control strategies through the analysis of surface electromyographic (EMG) signals from the bicep and tricep arm muscles of healthy subjects. Two different types of metrics used on data from the surface EMGs were compared to determine the best type of a decision rule to ascertain whether a particular muscle EMG pattern is representative of one of 13 possible primitive arm motions. The integrated threshold detector is shown to be the best metric to assist in the discrimination of the EMG signals resulting from the different primitive motions for one subject. This procedure has great application in the use of limb prostheses control because of the ability of this procedure to discern these EMG signals into as many as 13 different categories.

Keywords: EMG signal analysis, prosthesis aid, control strategies.

1 INTRODUCTION

An area of important research related to the effective upper and lower extremity prostheses involves the control strategies necessary to operate a prosthesis. In this study we investigated the manner in which control action can be delivered to a forearm prosthesis from an intact bicep and tricep muscle of the upper arm. Thus the implicit assumption is made that the subject can make muscle movements using only intact bicep and tricep muscles but the subject needs to make sufficiently distinct muscle movements using only bicep and tricep muscles to control a prosthesis. The objective is to control at least 13 different forearm motions using only EMG signals from the bicep and tricep arm muscles.

It is a difficult task to discern 13 different responses from only two muscles and it is further complicated due to the fact that surface EMG electrodes process electrical signals which are actually a summation of different motor signals which appear on the skin. This summation consists of motor unit potentials staggered both in time and space. In addition, it is required to synthesize and discriminate at least 13 distinctly different muscle responses (using only the two muscles, biceps and triceps) to operate a prosthesis. In this paper a comparison is made between two metrics of these data. The commonly known procedure termed the IEMG (Integrated EMG) is studied as well as a second method introduced here which is called the integrated threshold detector (ITD) method. The second method provides the best data analyzing procedure of discerning different muscle response signals for one subject. Data from 10 different subjects are averaged to examine this approach when studied across subjects.

2 REVIEW OF EXISTING METHODS

An extensive amount of research has been accomplished in studying muscle electrical signals and modelling [1]. The surface electromyographic signal (EMG) provides a summation of a great deal of electrical activity whenever a voluntary or involuntary muscle contraction occurs. This myoelectric activity is measured by a differential voltage at the surface electrodes and it is a complex interference pattern representing a linear, spatial, and temporal summation of motor unit action potentials [2]. The surface electrodes must be properly placed [3] on large muscles to attempt to eliminate extraneous signals appearing in the received voltage from the surface EMG electrodes.

An alternative to the surface EMG electrodes are needle electrodes. Needle electrodes are a more accurate method to obtain a specific muscle signal. The disadvantages of using needle electrodes are that they require pinpointing the exact position of the muscle fiber and they are invasive and somewhat painful. It would be difficult to use these signals for prosthesis control because these electrodes would have to be continuously applied every day, hence, creating other types of problems. Ideally to implement prosthesis control, the less cumbersome the electrode, the easier the application. Conversely, the noninvasive surface electrode gives information from electrical signals which are complicated summations of signals of both a spatial and temporal nature. These signals are extremely difficult to interpret. Thus, from an implementation point of view, it is desired to have the surface electrodes because of their simplicity. However, the tradeoffs associated with this choice may not be desirable because of the poor quality of information obtained from them. However, it may be possible to use such a surface EMG signal if an algorithm can be developed to analyze the data obtained from them in such a manner as to carefully discern muscle movements.

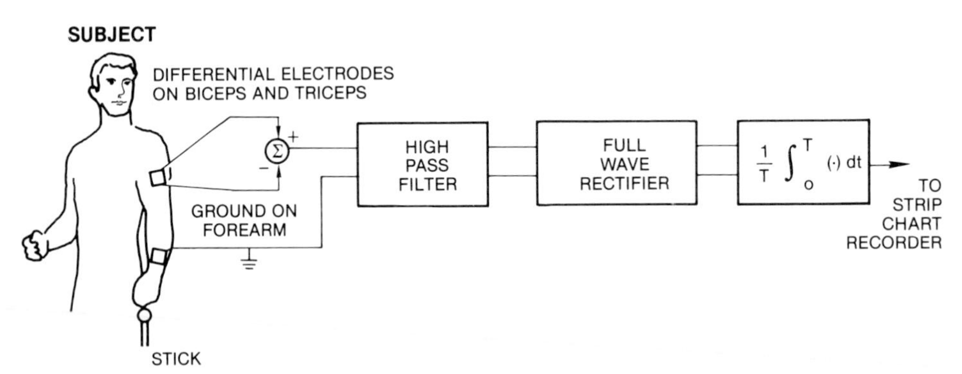

Figure 1 The IEMG signal processing procedure.

3 USE OF THE IEMG

One of the most useful methods of processing information of muscle activity through a measured electrical signal (including the surface EMG signal) is termed the IEMG. The IEMG signal is obtained (Figure 1) by first high-pass filtering the raw EMG signal to remove any artifact of low frequency arm movement, and then taking the absolute value of the raw, filtered EMG signal (full-wave rectification).

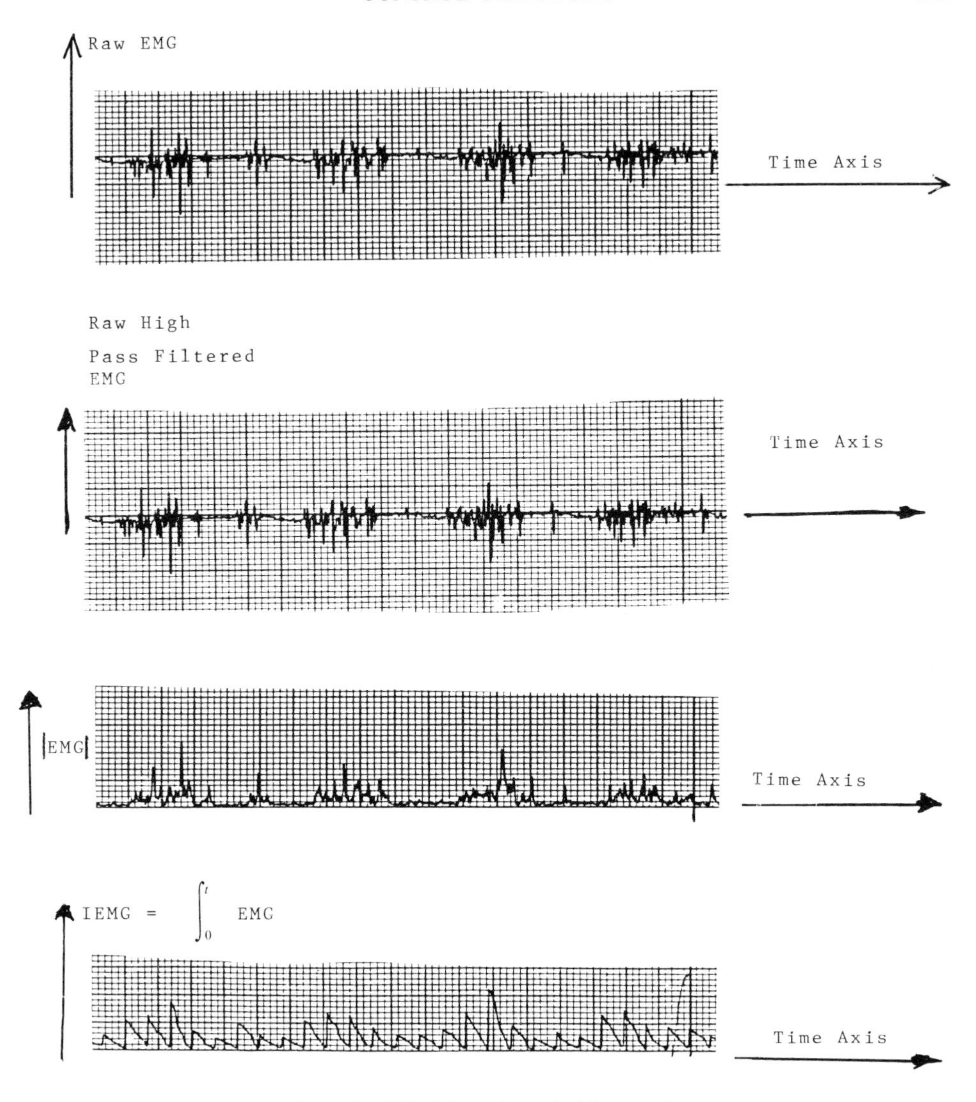

Figure 2 IEMG from data of subject no. 1.

These positive signals are then integrated with respect to time and the integral function is reset to zero every T = 160 milliseconds (ms). Thus by measuring the height of this ramp signal at the end of the 160 ms. period, a measure of the intensity of the electrical signal from the surface EMG electrode pair is obtained. Figure 2 illustrates the raw signal, the high pass filtered raw signal (full-wave rectified with a break point of 100 rad/sec), the full-wave rectified signal, and the IEMG for a subject's bicep muscle used in this experiment. The use of this signal, however, is somewhat distorted because the full-wave rectifier adds a non-linear operation on the signal waveform. If we do not rectify the signal, however, then when it is integrated (because it is a zero mean, changing amplitude and frequency time series), it would only yield a meaningless result of an approximately zero-mean process. It loses a great deal of information in this regard.

Much research has been accomplished with the IEMG signal. For example, Person and Libkind [4] showed that the absolute value of the EMG is proportional to the square root of the number of active motor units. This would imply that the force developed would be proportional to the IEMG [1]. From experimental data, however, Lippold [5] showed this was not the case. Using mathematical analysis, Lipkind [6] showed that the IEMG would be proportional to the square root of the number of active motor units if the motor units fired asynchronously and the firing was of the form of a biphasic pulse.

Empirically, there are some interesting results concerning the use of the IEMG. If one were to relate IEMG to muscle forces, there exists interactions between tension, velocity, and whether or not the IEMG was maintained in steady-state conditions. For example Inman *et al.* [7] (as did Lippold [5]) saw that for isometric conditions, the strength of the IEMG signal was strongly correlated with muscle force. However, if the muscle length changed, this relationship did not hold. Bigland and Lippold [8] found that if the muscle fiber were kept at constant tension, then when it lengthened, this occurred independent of the velocity of movement. However, when the muscle length decreased, the tension was linear with the velocity of movement. Moller [9] used both needle and surface electrodes and he observed that the IEMG versus tension relationship depended on interelectrode distance. Liberson *et al.* [10] found that isometric IEMG depends on the joint angle and Lam *et al.* [11] showed, by using 5 pairs of EMG electrodes on cat muscle, that by averaging the 5 separate electrode pair recordings, there existed a much more linear relationship of IEMG with tension. Thus the use of the IEMG is somewhat viewed with controversial results in the literature. In this paper we would like to consider the threshold IEMG as an alternative metric to study muscle activity using as our data base surface electrode information from the experiment described herein.

4 THE INTEGRATED THRESHOLD DETECTOR-MOTIVATION

In the analysis of the raw (high pass filtered) EMG signal, several characteristics about it need to be discussed. This electrical signal is an approximately zero-mean signal, nonstationary, somewhat random process, which varies in both frequency and amplitude. In resting muscle there is a certain minimal level of activity due to just coactivation forces in the antagonist–agonist pair (tricep–bicep). As a muscle goes into fatigue both the amplitude of the surface EMG increases and the mean frequency of the measured electrical signal from the muscle decreases all the way to fatigue. These results [12] are for static (isometric) contractions. When dynamic motions occur, the spurious arm activity can cause other extraneous electrical signals as well as the muscle length change which adds further complications.

Figures 3a–c illustrates the concept of a threshold detector. If one looks at the EMG activity in Figure 3a, during the rest activity, this signal level is not being utilized to perform outside work. This part of signal activity may bias IEMG measurements or other metrics used to describe muscle activity. It is desired here to have the rest activity portion of the measured electrical signal have no effect on the metric used to evaluate muscle output during the performance of actual work. The threshold detector (Figure 3b) is a comparator circuit. The output of this circuit is zero if the absolute value of the raw, filtered EMG is less (in absolute value) than the threshold level $\delta 1$. When the raw, filtered EMG exceeds the value $\pm \delta 1$, then the output of the threshold detector is equal to that part of the EMG greater than $\delta 1$. From Figure 3c it is seen that the output of the ITD (Integrated Threshold Detector) is zero when the EMG activity falls within normal resting levels.

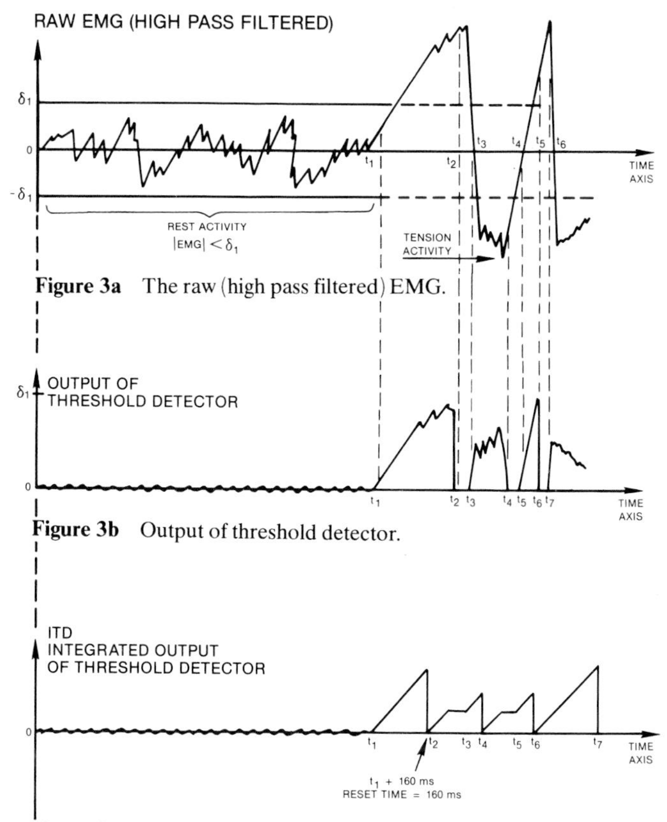

Figure 3a The raw (high pass filtered) EMG.

Figure 3b Output of threshold detector.

Figure 3c Integrated output of threshold detector.

When the muscle is forced, the electrical activity increases dramatically and in Figure 3c it is seen that the ITD output is equal to the integral of that part of the EMG signal greater than δ_1. Thus, by observing the output of ITD, this provides a method of observing the level of work associated with a muscle action. Note that it is necessary to develop a reasonable estimate of δ_1 and to make good use of this metric. Figure 4 illustrates the electronic circuit required to implement the ITD.

There are also some heuristic arguments associated with the ITH as a useful measure of EMG activity. Consider the case that N motor units are firing to result in a given muscle tension. Let *ia* be the electric current associated from the firing of one motor unit consisting of a number of muscle fibers. If N of these motor units fire, then the electric current between two electrode points would be *Nia* (or a constant). The electrode potential difference (motor unit action potential) between the two points (differential voltage) would be the measured EMG voltage = VEMG at that instant of time. The instantaneous power between the points separated by electrodes could be written:

$$P \text{ instantaneous} = N\, ia\, \text{VEMG}\,(t) \tag{1}$$

and the instantaneous energy used by the muscle is simply the instantaneous power times time. Thus electric energy from muscles is given by:

$$\text{Electric Energy from Muscles} = \int_0^t P \text{ instantaneous} \tag{2}$$

or

$$\text{Electric Energy from Muscles} = \int_0^t N\, ia(\tau)\, \text{VEMG}(\tau) d\tau. \tag{3}$$

Thus the energy delivered by the muscles would be proportional to the integral of VEMG(t) if $N\, ia(t)$ were constant, i.e.

$$\text{Energy} = N\, ia(t) \int_0^t \text{VEMG}(\tau) d\tau. \tag{4}$$

But if VEMG(t) does not produce work unless it is above a threshold level δ_1, then:

$$\text{Energy} = N\, ia(t) \int_0^t |\text{VEMG}(\tau) - \delta_1|\, d\tau, \tag{5}$$

which is proportional to the ITD measure adopted in this paper.

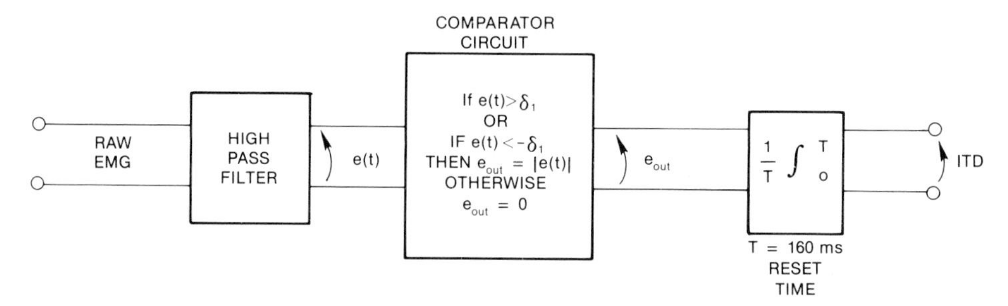

Figure 4 The electronic circuit to implement the ITD.

Thus, heuristically, one would expect that the ITD measure may be a more sensitive indicator of work or effort accomplished by the muscle. This paper examines the ITD metric, as well as the IEMG to study the feasibility of using these postprocessing signal analysis procedures in discriminating different arm motions to move a prosthesis using simple, primitive muscle responses.

It should also be noted that there are some interesting theoretical characteristics of the ITD which need to be discussed. In signal processing [13], the detection part of the ITD may be considered a "Level Crossing Detector" and it has been a subject of some interesting theoretical research. It has found some application, however, it has not found extensive use in the area of signal processing. In [14] a review paper on several diverse applications of this procedure is given and one successful implementation of the level crossing detector occurred when data compression for an image recognition problem [15] was conducted quite efficiently using this device. In general, it has a nonlinear characteristic which precludes its extensive use in signal analysis, where linearity is the desired characteristic when it is necessary to try to reconstruct information from a time series signal.

5 EXPERIMENTAL DESIGN

Ten healthy male subjects between the ages of 24 and 51 years (both active duty Air Force military personnel and civilians) participated in this experiment. They were involved in a preliminary training period of the primitive arm motions and data were then collected. Each subject had four silver/silver chloride surface electrodes (a differential pair) placed on the biceps and triceps and a reference electrode located on the forearm. These electrodes were all placed on the subject's right arm. Three of the ten subjects were left handed. The subject's skin resistance was required to be less than 5,000 ohms before the signal could be recorded. This was accomplished by first rubbing with an alcohol swab the skin surface to remove oil. Secondly, a piece of very light sandpaper was used to remove dead skin and then the skin surface was cleaned again with the alcohol to clear the surface. This procedure was repeated until the skin resistance was reduced to 5,000 ohms or less.

The EMG signals that were obtained were first high-pass filtered to remove limb artifact motion (less than 2 Hz, see Neilson [16]) and then the IEMG and ITD were calculated. The electrodes were placed 2 cm apart on each subject to ensure the same bandwidth of the bipolar signal [17]. The subjects were then required to perform ten trials each of the following list of 13 basic primitive motions (Table 1).

A small weight of approximately 1500 g was used for the bicep and tricep curls. In the wrist pronation, supination, and in the hands open and hand closed primitives, the forearm was rested on a table and the upper arm was maintained at the same height above the ground and essentially in a stationary position.

Table 1 Primitive movements used for baselines

(1)	3 s MVC for Biceps—Rest 1 min
(2)	3 s MVC for Triceps—Rest 1 min
(3)	10 s recording of resting muscle activity
(4)	10 Bicep Curls—Rest 1 min
(5)	10 Tricep Curls—Rest 1 min
(6)	10 Wrist Pronations alternating with
(7)	10 Wrist Supinations—Rest 1 min
(8)	10 Wrist Flexions alternating with
(9)	10 Wrist Extensions—Rest 1 min
(10)	10 Wrist Abductions alternating with
(11)	10 Wrist Adductions—Rest 1 min
(12)	10 Open Hand Motions alternating with
(13)	10 Close Hand Motions

6 METHODS

During the training period, each subject rehearsed the 13 different types of experimental conditions considered here. During the data collection, each subject was prepared (as described in the Experimental Design section) by properly scraping skin and applying the EMG surface electrodes. The 3-s Maximum Voluntary Contraction (MVC) for the bicep muscle was performed first. The forearm pressed against a stationary object such that the arm was truly in an isometric condition. After 1 min of rest, the tricep MVC was then performed against the same stationary object. It was difficult to get a true tricep MVC but it was accomplished by keeping the upper arm stationary (and braced) and the lower arm would press against a stationary object using extensor motion. There was no arm movement so the muscle action was truly isometric. After 1 min rest, the rest values of the bicep and tricep EMGs were recorded. At this point in time the choice was made for the value of δ_1. The δ_1 value was chosen to be approximately twice the root mean square value of the rest value of the EMG. Hence it was subject dependent.

The next primitive was bicep curls using a small weight of 1500 g and ten of these motions were performed using a time duration of about 3 s for one curl. After about a minute of rest, the tricep curls were performed. In this case the upper arm was stationary and the lower arm curled the 1500 g weight using mainly tricep force. It was difficult to get a pure tricep signal because of the strong influence of the bicep muscle in any motion of this type. After resting 1 min, 10 wrist pronation and supination movements were made in succession with the upper arm and the forearm fixed in relatively stationary positions.

After another minute of rest, the 10 wrist flexions and extensions were conducted in an alternating manner. Again, the upper and lower arm were supported and stationary. Again resting 1 min, the 10 wrist abductions and adductions were conducted with the upper and lower arm supported as before. Finally, after resting 1 min, the 10 open hand and 10 close hand movements were conducted in alternating sequence with both the upper arm and the lower arm supported.

The EMG signals were processed from analog computers to strip chart data. Ten replications of each condition were read off the strip chart recorder. The recorder was run at a fast speed (100 mm/s) for the first 5 primitive motions and a slow speed (10 mm/s) for the last 5 primitive motions.

7 RESULTS (ANALYSIS)

The two metrics of interest were the IEMG and the ITD. For each of the ten subjects and thirteen experimental conditions, 10 measurements were made of both the bicep and tricep EMG signals for each metric.

The purpose of the analysis was to determine how significantly different the experimental conditions were (no comparisons were made with the MVCs). The reason why no comparisons were made with the MVCs was because these were exclusively static muscle exercises. The remaining 12 conditions were considered dynamic motions. When one compares dynamic motions to a static MVC measurement, the EMG metric may yield larger values for the dynamic motions. This is contrary to standard work in fatigue studies where all EMG signals are normalized with respect to the maximum static signal (the MVC). Thus, these are entirely different types of muscle responses. To accomplish the analysis, a multivariate analysis of variance was used with bicep and tricep EMG as the dependent variables. Assumptions of equal variance for the dependent variables within the remaining 12 experimental conditions could not be made. Therefore, pairwise comparisons were made for the thirteen experimental conditions using only data from those conditions.

7.1 *Analysis of Data Within One Subject*

It is desired to study the data from one subject for 3 reasons:

1) Using data from one subject significantly reduces the variability. This gives us an opportunity to investigate if we can discern any differences in these experimental conditions less prone to the influence of other variables.

2) In designing prostheses, usually EMG data are extremely subject dependent (e.g., in measurements of skin resistance, muscle site location, etc.), and it is reasonable in the design of a prostheses for a subject to consider the design very subject specific. Across subject variability is not an important issue here.

3) These data were all taken on one day, so using only one subject significantly reduced the variability in these data.

Table 2 IEMG data—subject no. 1
Means, standard deviations and coefficients of variation for each condition

Code	Condition	Mean bicep	STD bicep	CV bicep	Mean tricep	STD tricep	CV tricep
BC	Bicep curl	12.2	1.2	9.5	3.7	0.7	18.9
BM	Bicep MVC	14.5	2.9	19.9	12.0	1.5	12.3
HC	Hand closed	1.1	0.1	6.3	3.4	0.3	9.1
HO	Hand open	1.1	0.1	5.0	3.1	0.1	4.4
R	Rest	1.0	0.0	4.7	3.0	0.1	2.1
TC	Tricep curl	4.2	0.3	6.9	7.6	1.2	16.5
TM	Tricep MVC	8.9	1.1	11.9	21.4	3.3	15.6
WAB	Wrist abduction	1.9	0.2	10.7	6.8	1.4	20.0
WAD	Wrist adduction	3.4	0.5	13.9	4.4	0.6	14.3
WE	Wrist extension	4.4	0.4	9.4	12.4	1.8	14.9
WF	Wrist flexion	3.8	0.4	11.1	8.7	0.9	10.6
WP	Wrist pronation	3.3	0.4	11.1	3.0	0.2	6.4
WS	Wrist supination	3.4	0.4	11.8	2.5	0.2	8.0

Table 2 illustrates the means, standard deviations, and coefficients of variation (ratio of s.d./mean) for these 13 conditions for subject No. 1 for IEMG data. Table 3 illustrates these same variables for the metric ITD. Figure 5 illustrates a plot of IEMG data on bicep and tricep axis for subject No. 1. Figure 6 is the corresponding plot on the ITD data axis. In Figures 5 and 6 if no lines are drawn between variables, this indicates that they are significantly different ($p<0.05$). Thus in Figure 5, the only variables that could not be discriminated are the rest condition and hand open. In Figure 6, all 13 conditions could be distinguished from each other, illustrating the clear advantage of the ITD metric when data is analyzed within a subject.

Using a multiple comparison between the variables, in Table 4, * indicates the variables are different at $p<0.0001$. The numerical values are differences $p<0.01$, but in all cases $p<0.05$. Note that for the individual subjects, the multivariate analysis of variance is equivalent to using the Mahalanobis distance between the two conditions as the test statistic. Comparing Tables 4 and 5 it is seen that ITD outperforms IEMG except for the one comparison of Hands Open (HO) versus Hands Closed (HC). Of course, Table 4 is implicitly dependent on our choice of δ_1.

Table 3 ITD data—subject no. 1
Means, standard deviations and coefficients of variation for each condition

Code	Condition	Mean bicep	STD bicep	CV bicep	Mean tricep	STD tricep	CV tricep
BC	Bicep curl	5.2	0.4	8.4	1.0	0.2	21.5
BM	Bicep MVC	6.1	1.2	19.2	4.3	0.6	14.9
HC	Hand closed	1.0	0.0	4.7	1.0	0.1	9.2
HO	Hand open	1.0	0.0	3.7	0.9	0.1	8.6
R	Rest	0.5	0.1	10.1	0.7	0.1	13.1
TC	Tricep curl	2.0	0.1	4.4	2.5	0.6	22.4
TM	Tricep MVC	3.9	0.4	10.1	7.9	1.3	16.4
WAB	Wrist abduction	1.0	0.1	7.7	2.3	0.5	22.5
WAD	Wrist adduction	1.7	0.3	15.7	1.3	0.3	24.4
WE	Wrist extension	2.1	0.1	5.0	4.6	0.6	13.8
WF	Wrist flexion	1.9	0.1	6.9	3.1	0.3	9.4
WP	Wrist pronation	1.7	0.2	12.5	0.9	0.1	10.5
WS	Wrist supination	1.7	0.3	17.9	0.6	0.1	19.6

Table 4 IEMG data—subject no. 1
P-values for pairwise comparisons

	R	BC	TC	WP	WS	WF	WE	WAB	WAD	HO
BC	*									
TC	*	*								
WP	*	*	*							
WS	*	*	*	*						
WF	*	*	0.0349	*	*					
WE	*	*	*	*	*	*				
WAB	*	*	*	*	*	*	*			
WAD	*	*	*	*	*	*	*	*		
HO	0.1402	*	*	*	*	*	*	*	*	
HC	0.0027	*	*	*	*	*	*	*	*	0.0195

*$P < 0.0001$.

Table 5 ITD data—subject no. 1
P-values for pairwise comparisons

	R	BC	TC	WP	WS	WF	WE	WAB	WAD	HO
BC	*									
TC	*	*								
WP	*	*	*							
WS	*	*	*	*						
WF	*	*	0.0269	*	*					
WE	*	*	*	*	*	*				
WAB	*	*	*	*	*	*	*			
WAD	*	*	*	0.0006	*	*	*	*		
HO	*	*	*	*	*	*	*	*	*	
HC	*	*	*	*	*	*	*	*	*	0.0388

*$P < 0.0001$.

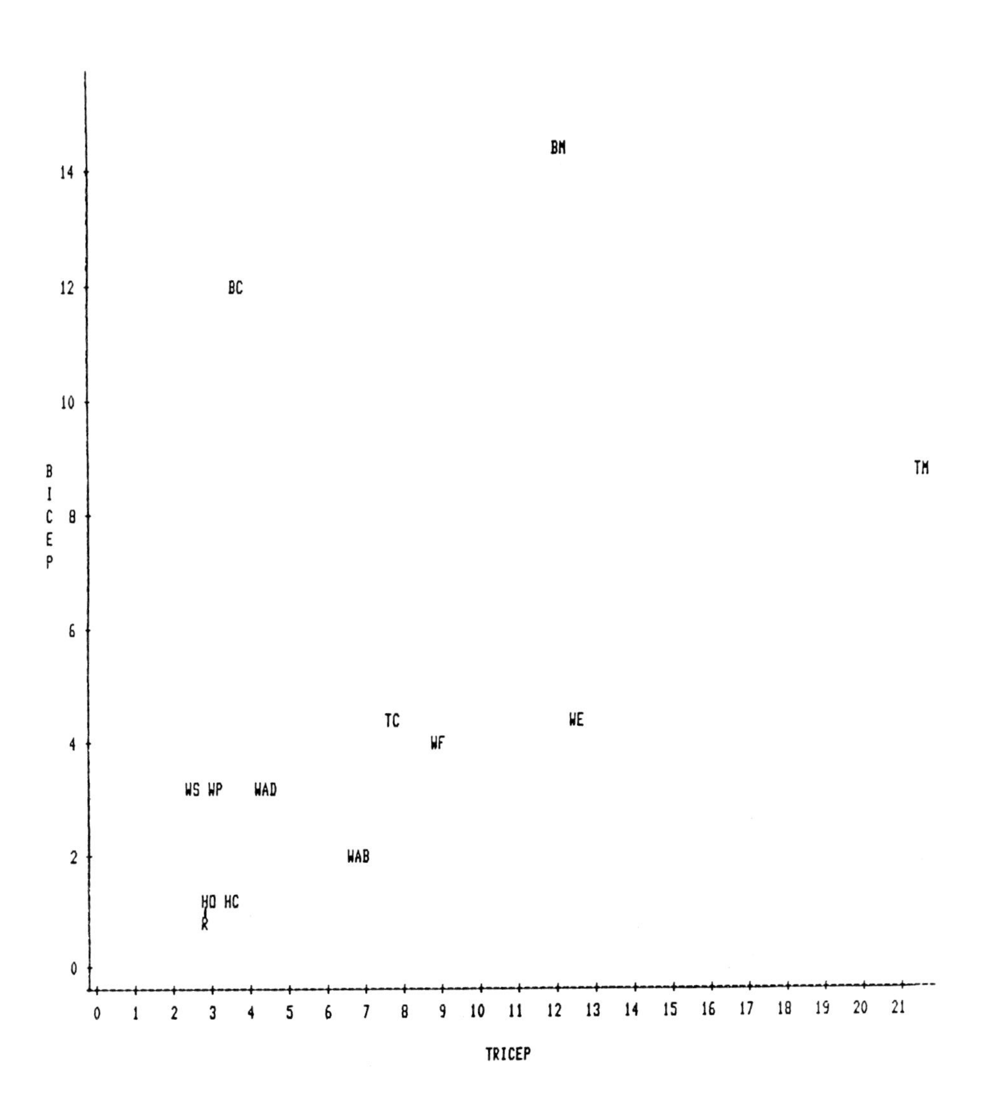

Figure 5 IEMG data—subject no. 1.

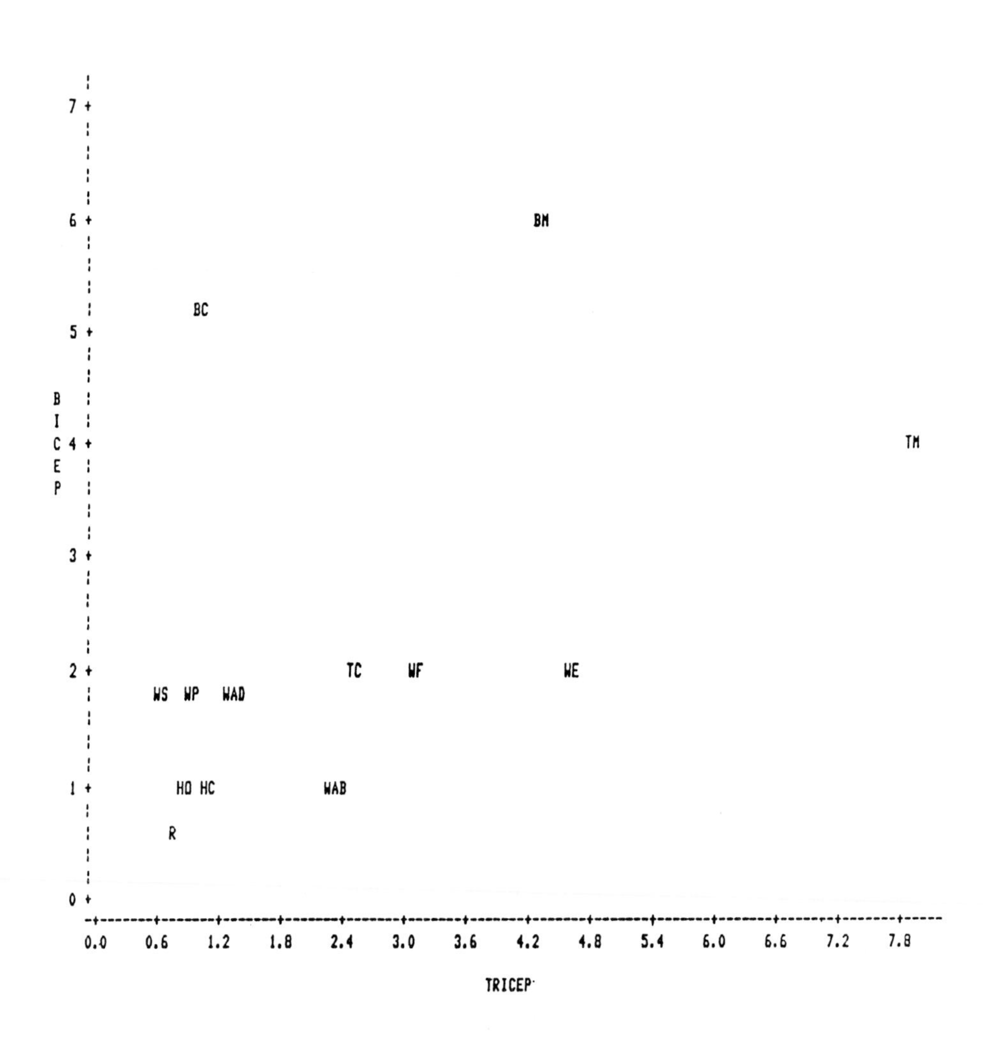

Figure 6 ITD data—subject no. 1.

7.2 *Across Subjects*

Across subjects we can say the following about the comparison of the metrics IEMG versus ITD:

To determine the maximum dynamic EMG value obtained from a metric, only data involving dynamic motions were considered. For each subject individually, using the 11 dynamic motions, the maximum measurement for both triceps and biceps was first obtained. All the remaining dynamic values (and also the MVC responses) were normalized with respect to this maximum dynamic measurement. The remaining responses were then expressed as a percent of this measurement for each individual. Then, using a percent of the maximum value within each subject, the percentages could be averaged across subjects. Table 6 illustrates these data when averaged across subjects in terms of percent of each individual subject's maximum dynamic EMG. Table 7(a,b) illustrates the F ratios and P values for the two metrics. Figure 7 shows the plots of these data for IEMG when averaged across subjects using percent of the respective maximum dynamic measurement. Figure 8 is the corresponding plot of the ITD. Figures 9 and 10 are blown up versions of Figures 7 and 8 using only the comparisons of the dynamic motions not equal to the curls. It also notes the only case where the experimental conditions are significantly different. In observing Figures 9 and 10, it is obvious that averaging across subjects is not a useful method to analyze data or to be used in the implementation of a prosthesis aid.

8 DISCUSSION

There is great interest in the use of the surface EMG signal with an appropriate classification algorithm to measure muscle activity and to apply this research to investigate the implementation of prosthesis aides. In [18] Morris and Repperger studied the change of muscle function when assistive devices are used to improve tracking performance using an aircraft control stick. This concept has been explored before using basic hand motions [19], nearest neighbor pattern classifier techniques [20], using both pattern recognition and regression techniques [21], time series analysis [22], and recently using parameters from an autoregressive time series [23].

9 SUMMARY AND CONCLUSIONS

A comparison of the two metrics used to evaluate data on primitive arm movements demonstrates that the ITD has great promise as an analysis tool to assist in moving a lower arm prosthesis. The important characteristic of such a metric is that it must clearly distinguish different arm movements using empirical data from only the bicep and tricep EMG arm muscle signals. The metric must require clean bicep and tricep muscle signals and a good signal processing algorithm to help distinguish different muscle actions and provide sharp comparisons. These results suggest (with the ITD metric) a procedure with great potential to control a lower arm prosthesis.

Table 6 Across subjects

Mean and standard deviation of subjects, data is percent of maximum dynamic EMG

Type	Condition	Code	Mean % bicep	STD % bicep	Minimum % bicep	Maximum % bicep	Mean % tricep	STD % tricep	Minimum % tricep	Maximum % tricep
	Bicep curl	BC	87.8	10.3	60.5	94.8	28.8	18.9	8.3	63.6
	Bicep MVC	BM	250.7	168.7	33.1	629.5	79.9	48.5	10.5	181.5
	Hand closed	HC	15.3	18.6	0.9	59.5	22.9	23.6	4.4	70.2
	Hand open	HO	11.0	8.8	1.5	34.6	18.1	13.2	3.5	42.8
	Rest	R	4.8	3.6	0.4	12.5	4.2	5.8	0.5	19.9
	Tricep curl	TC	35.0	18.6	15.0	80.7	84.2	14.3	50.5	94.4
IEMG	Tricep MVC	TM	83.0	49.5	10.7	155.8	267.7	171.1	52.8	612.4
	Wrist abduction	WAB	11.8	5.4	2.9	19.4	22.9	12.4	6.8	45.5
	Wrist adduction	WAD	18.7	11.9	2.5	36.0	31.9	20.9	12.6	70.5
	Wrist extension	WE	19.8	13.2	1.9	37.3	36.4	23.3	7.0	82.9
	Wrist flexion	WF	13.1	10.1	2.0	33.3	22.0	15.1	7.5	58.0
	Wrist pronation	WP	9.8	6.1	2.1	23.6	23.3	16.6	2.0	58.5
	Wrist supination	WS	11.8	7.0	1.9	23.9	20.4	15.9	2.4	55.5
	Bicep curl	BC	90.1	4.9	80.2	98.0	29.5	18.6	8.0	61.0
	Bicep MVC	BM	272.5	190.4	26.4	714.5	78.4	49.0	12.5	175.5
	Hand closed	HC	18.6	22.7	0.1	63.2	25.5	24.0	4.6	69.0
	Hand open	HO	12.0	12.6	0.2	45.0	20.1	13.5	0.6	46.7
	Rest	R	3.7	4.0	0.1	10.4	2.9	4.1	0.1	13.5
	Tricep curl	TC	37.5	22.1	12.5	90.0	82.7	15.7	45.6	95.3
ITD	Tricep MVC	TM	90.0	64.1	8.2	228.2	262.2	160.3	50.1	523.8
	Wrist abduction	WAB	13.8	12.8	0.8	45.5	24.4	12.0	11.7	43.9
	Wrist adduction	WAD	18.7	15.9	0.7	50.0	30.5	16.0	11.4	55.8
	Wrist extension	WE	20.8	15.0	0.1	44.5	38.4	22.9	8.7	83.8
	Wrist flexion	WF	16.0	13.4	0.1	40.5	26.0	14.9	8.6	56.4
	Wrist pronation	WP	9.4	8.5	0.8	28.8	22.1	16.1	2.7	49.9
	Wrist supination	WS	10.1	10.2	0.2	28.7	20.9	21.3	1.7	70.5

Table 7a Across subjects
F-values (2 and 8 degrees of freedom) and P-values for pairwise differences between conditions
Type = IEMG

	R	BC	TC	WP	WS	WF	WE	WAB	WAD	HO
BC	233.3									
	0.0001									
TC	82.7	36.1								
	0.0001	0.0001								
WP	9.22	172.3	45.4							
	0.0084	0.0001	0.0001							
WS	17.6	248.8	55.5	2.15						
	0.0012	0.0001	0.0001	0.1791						
WF	13.8	66.6	22.8	1.15	0.12					
	0.0025	0.0001	0.0005	0.3629	0.8921					
WE	11.5	62.9	10.2	4.95	3.29	3.17				
	0.0045	0.0001	0.0062	0.0398	0.0904	0.0968				
WAB	20.3	166.2	50.9	0.50	0.07	0.23	3.57			
	0.0007	0.0001	0.0001	0.6244	0.9356	0.8031	0.0782			
WAD	12.0	56.4	19.4	4.88	4.38	2.30	0.09	1.91		
	0.0039	0.0001	0.0008	0.0411	0.0518	0.1627	0.9124	0.2101		
HO	4.76	78.7	36.6	0.64	0.08	0.42	2.97	0.31	4.37	
	0.0435	0.0001	0.0001	0.5532	0.9231	0.6683	0.1087	0.7415	0.5021	
HC	2.78	67.7	26.7	0.63	0.13	0.14	1.30	0.37	0.32	0.36
	0.1209	0.0001	0.0003	0.5581	0.8779	0.8677	0.3233	0.7051	0.7339	0.7100

Legend: F-value, P-value.

Table 7b Across subjects
F-values (2 and 8 degrees of freedom) and *P*-values for pairwise differences between conditions
Type = ITD

	R	BC	TC	WP	WS	WF	WE	WAB	WAD	HO
BC	1288.5 0.0001									
TC	89.9 0.0001	42.8 0.0001								
WP	9.16 0.0085	331.7 0.0001	48.3 0.0001							
WS	5.79 0.0279	247.2 0.0001	38.3 0.0001	0.20 0.8209						
WF	18.2 0.0011	117.5 0.0001	20.2 0.0007	1.10 0.3770	0.74 0.5065					
WE	14.4 0.0022	97.7 0.0001	9.51 0.0077	3.51 0.0803	2.33 0.1592	2.71 0.1263				
WAB	25.4 0.0003	163.8 0.0001	35.4 0.0001	0.41 0.6774	0.28 0.7604	0.42 0.6698	4.12 0.0589			
WAD	12.2 0.0037	77.8 0.0001	25.5 0.0003	2.96 0.1089	1.55 0.2688	0.44 0.6585	0.38 0.6932	1.22 0.3452		
HO	6.86 0.0184	150.0 0.0001	38.5 0.0001	0.60 0.5713	0.14 0.8685	1.53 0.2734	4.02 0.0618	1.05 0.3947	3.31 0.0897	
HC	4.11 0.0593	63.7 0.0001	23.4 0.0005	0.81 0.4798	0.54 0.6050	0.51 0.6200	0.90 0.4446	0.36 0.7107	0.32 0.7345	0.73 0.5135

Legend: *F*-value, *P*-value.

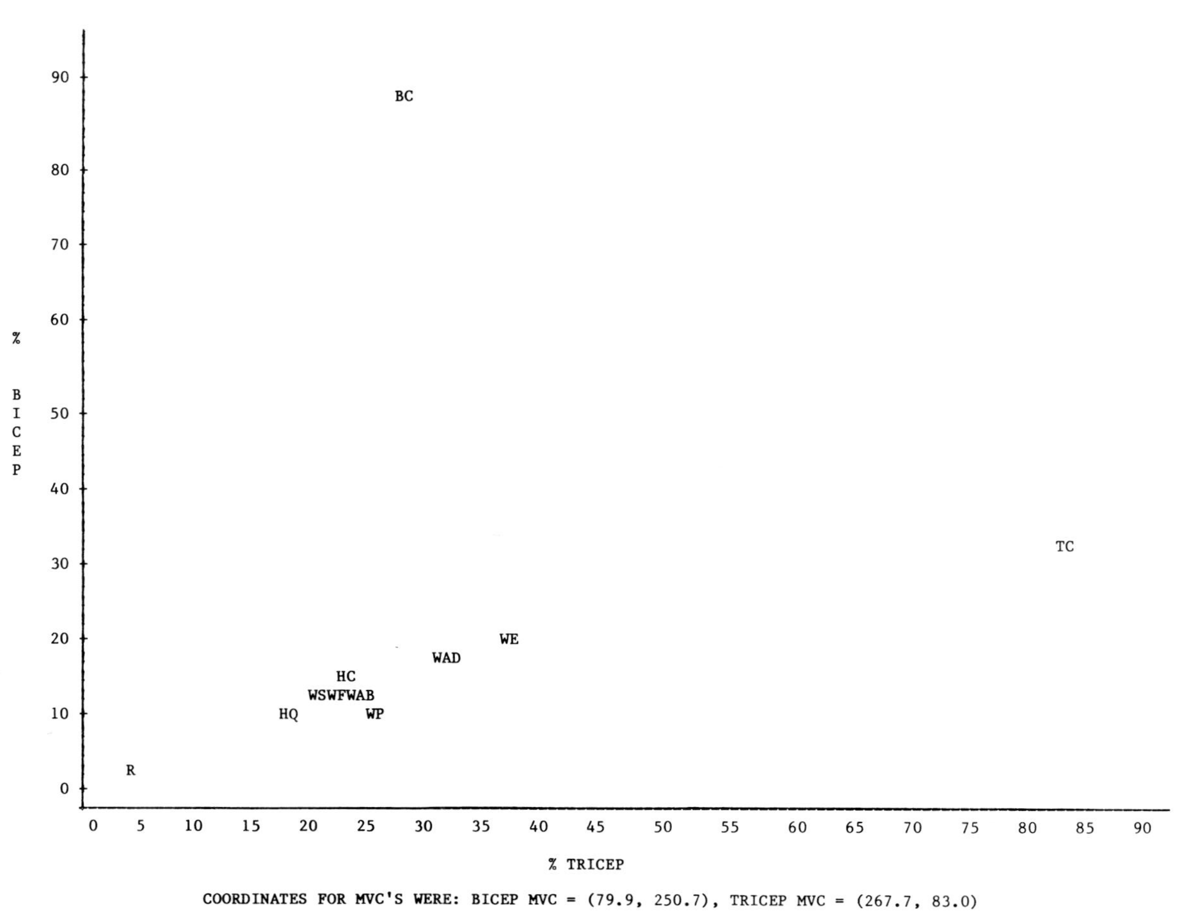

Figure 7 Across subjects—mean percent of maximum dynamic EMG for all dynamic conditions— type = IEMG.

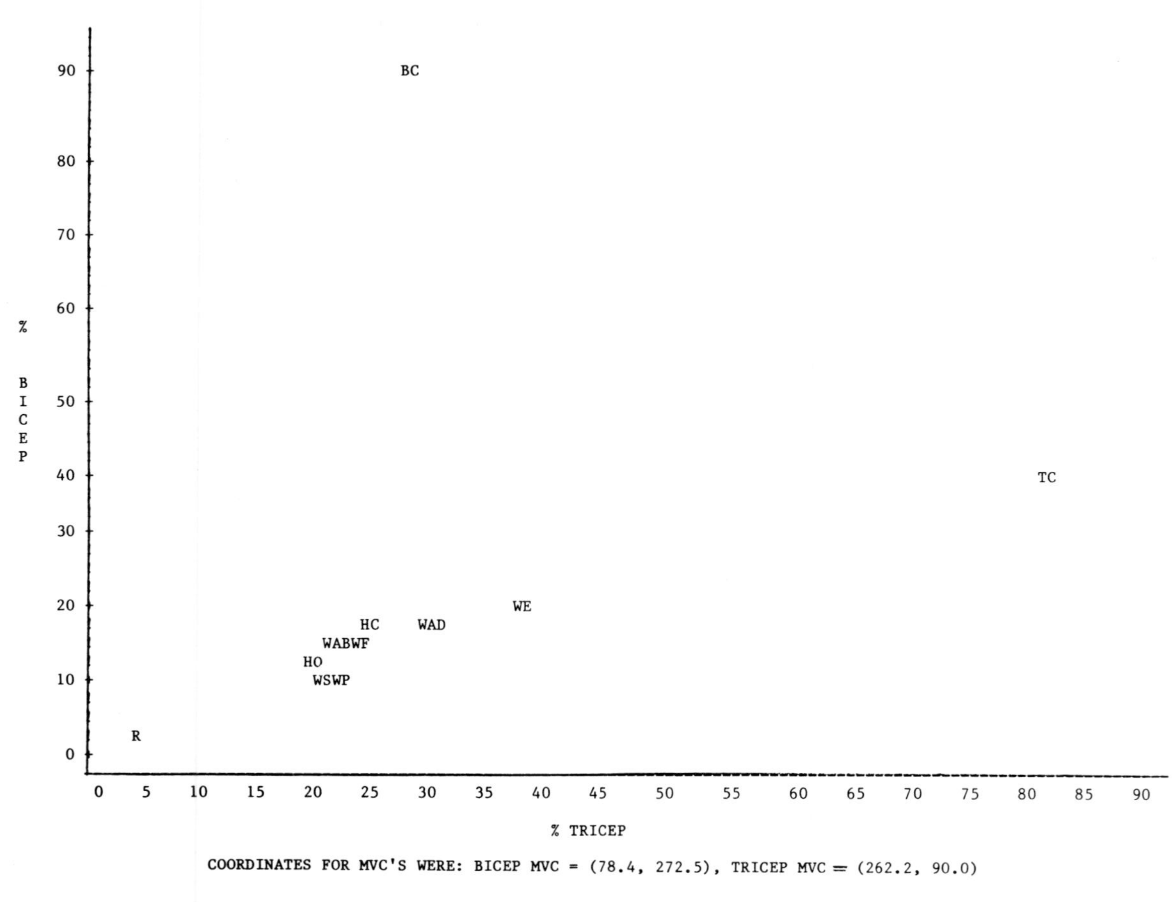

Figure 8 Across subjects—mean percent of maximum dynamic EMG for all dynamic conditions— type=ITD.

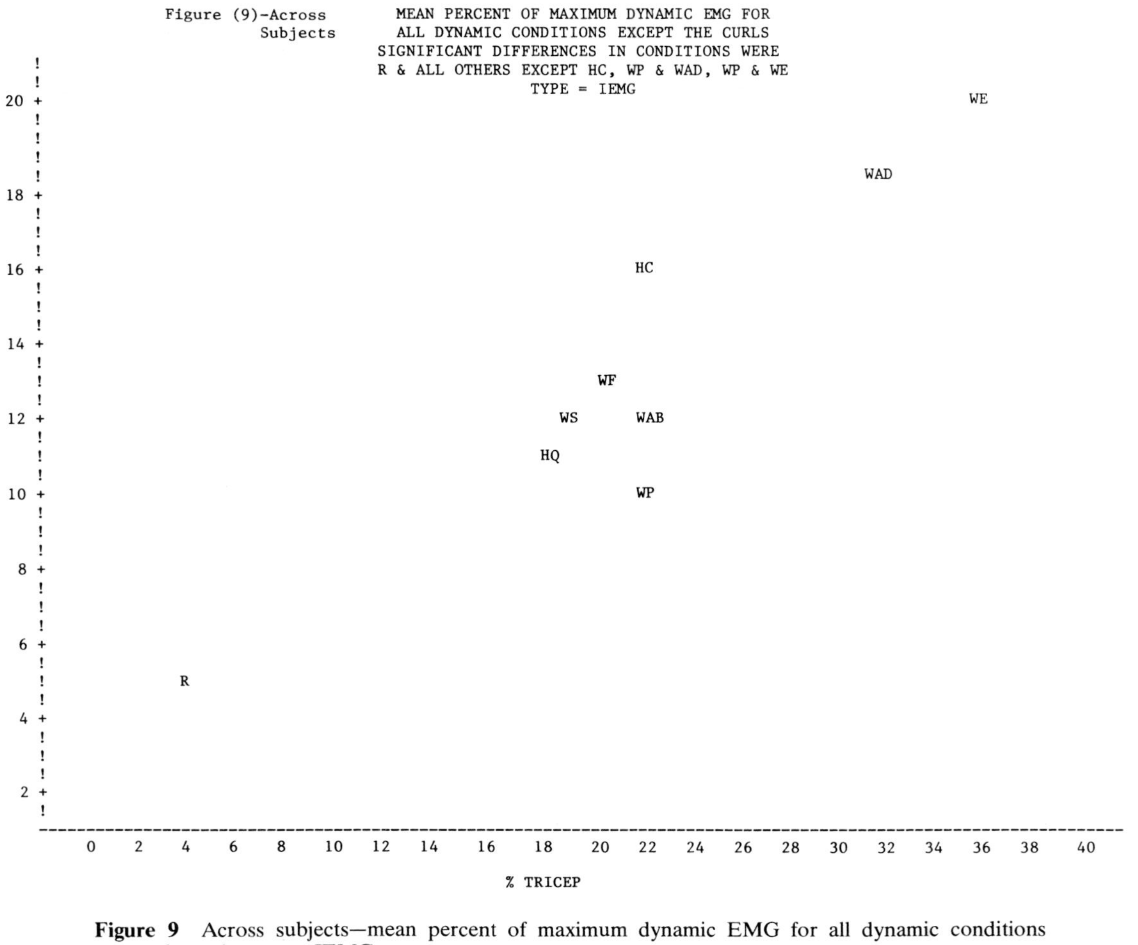

Figure 9 Across subjects—mean percent of maximum dynamic EMG for all dynamic conditions except the curls—type = IEMG.

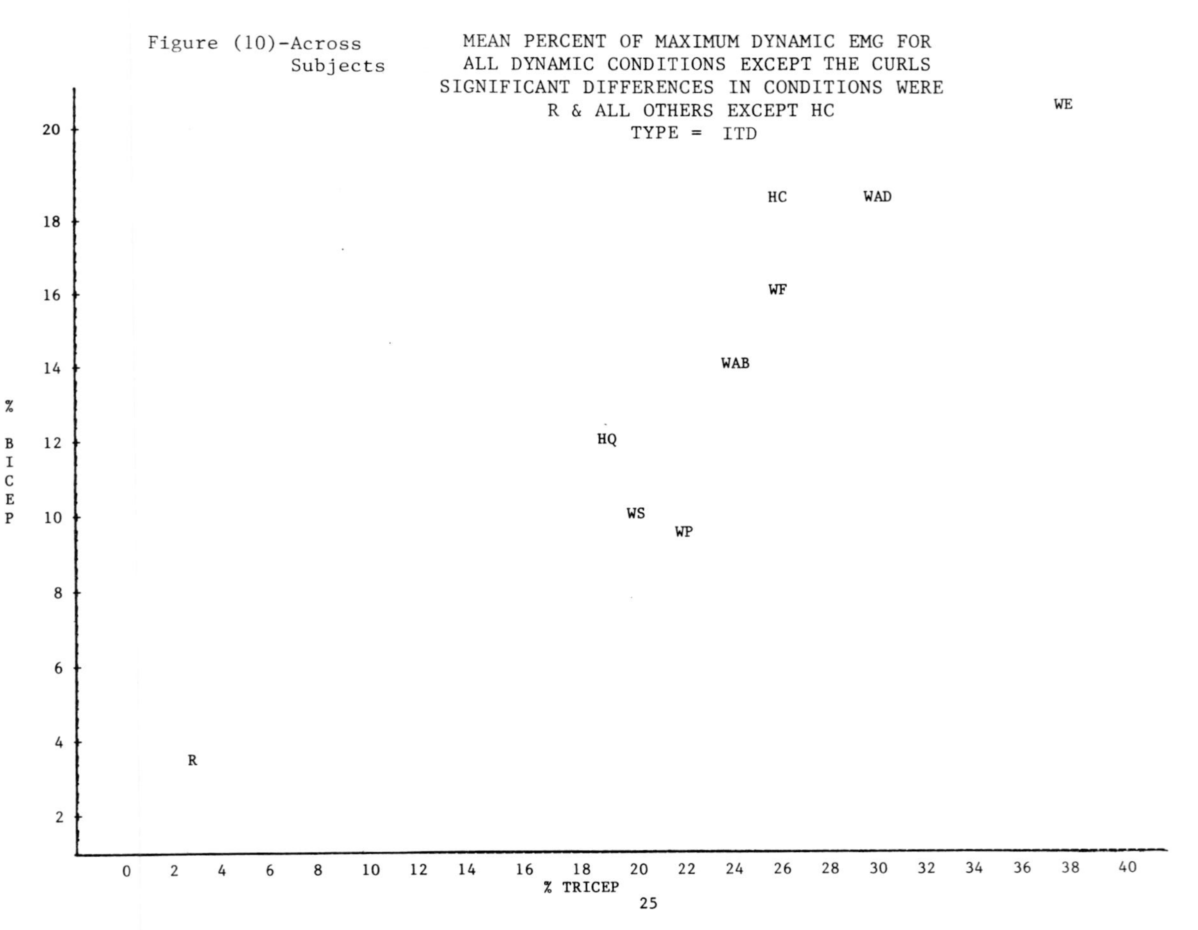

Figure 10 Across subjects—mean percent of maximum dynamic EMG for all dynamic conditions except the curls—type = ITD.

REFERENCES

1. G. C. Agarwal and G. L. Gottlieb, Mathematical modeling and simulation of the postural control loop: Parts I and II, *CRC Critical Reviews in Biomedical Engineering*, **8,** 93–134.
2. C. J. Deluca, Physiology and Mathematics of Myoelectric Signals, *IEEE Trans. Biomed. Eng.*, **26,** 313-325 (1979).
3. K. T. Cavanaugh, E. A. Clancy, J. A. Natrillo, R. J. Paquette and F. J. Looft, Optimal site selection for prosthetic control, *IEEE Frontiers of Engineering and Computing in Health Care*, 565–569, (1983).
4. R. S. Person and M. S. Libkind, Modelling of interference bioelectrical activity, *Biofizika*, **12,** 127 (1967) (English Translation-Pergamon Press, 145).
5. C. J. Lippold, The relation between integrated action potentials in a human muscle and its isometric tension, *J. Physiol.* (London), **117,** 492 (1952).
6. M. S. Lipkind, Modelling of interference bioelectrical activity II, III, and IV, in *Biofizika*, vols 13, 14, and 17, 1968, 1969, 1972. English Translation Pergamon Press.
7. V. T. Inman, H. J. Ralston, J. B. Saunders, B. Feinstein and E. W. Wright Jr., Relation of human electromyogram to muscle tension, *Electroenceph. Clin. Neurophysiol.*, **4,** 187 (1952).
8. B. Bigland, O. C. J. Lippold, The relation between force, velocity, and integrated electrical activity in human muscles, *J. Physiol.*, **123,** 214–224 (1954).
9. E. Moller, The chewing apparatus: an electromyographic study of the action of the muscles of mastication and its correlation to facial morphology, *Acta Physiol. Scand.*, **69** (Suppl.), 280 (1966).
10. W. T. Liberson, M. Dondey and M. M. Asa, Brief repeated isometric maximal exercises: an evaluation by integrative electromyography, *Am. J. Phys. Med.*, **41,** 3 (1962).
11. H. S. Lam, D. L. Morgan and D. G. Lampard, Derivation of reliable electromyograms and their relation to tension in mammalian skeletal muscles during synchronous stimulation, *Electroencephal. Clin. Neurophysiol.*, **46,** 72 (1979).
12. J. S. Petrofsky and A. R. Lind, Frequency analysis of the surface EMG during sustained isometric contractions, *Eur. J. Appl. Physiol.*, **41,** 1–15 (1980).
13. F. A. Marvasti, *A Unified Approach To Zero-Crossings and Nonuniform Sampling*, Illinois Institute of Technology (1987).
14. I. F. Blake and W. C. Lindsey, Level crossings problems for random processes, *IEEE Trans. Inf. Theory*, **IT-19,** 295–315 (1973).
15. J. W. Mark and T. D. Todd, A nonuniform sampling approach to data compression, *IEEE Trans. Commun.*, **COM-29,** 24–32 (1981).
16. P. D. Neilson, Interaction between voluntary contraction and tonic stretch reflex transmission in normal and spastic patients, *J. Neurol. Neurosur. Psych.*, **35,** 853 (1972).
17. P. Zipp, Effect of electrode parameters on the bandwidth of the surface EMG power-density spectrum, *Med. Biol. Eng. Comp.*, **16,** 537 (1978).
18. A. M. Morris and D. W. Repperger, Discriminant analysis of changes in human muscle function when interacting with an assistive aid, *IEEE Trans. Biomed. Eng*, **35,** 316–322 (1988).
19. P. C. Herberts, C. Alstrom, R. Kadefors and P. D. Lawrence, Hand prosthesis control via myoelectric patterns, *Acta Orthop. Scand.*, **44,** 389–409 (1973).
20. D. C. Dening, F. G. Gray and R. M. Haralick, Prosthesis control using a nearest neighbor electromyographic pattern classifier, *IEEE Trans. Biomed. Eng.*, BME-30, 356–360 (1983).
21. P. D. Lawrence and W. C. Lin, Statistical decision making in the real-time control of an arm aid for the disabled, *IEEE Trans. Systems, Man, and Cybernetics*, **SMC-2,** 35–42 (1972).
22. P. C. Doershuk, D. Gustafson and A. S. Willsky, Upper extremity limb function discrimination using EMG signal analysis, *IEEE Trans. Biomed. Eng.*, **BME-30,** 18–29 (1983).
23. G. Hefftner and G. G. Jaros, The electromyogram (EMG) as a control signal for functional neuromuscular stimulation-Part II: Practical demonstration of the EMG signature discrimination system, *IEEE Trans. Biomed. Eng.*, **35,** 238–242 (1988).

Automedica, 1989, Vol. 11, pp. 145–162
Reprints available directly from the publisher only
Photocopying permitted by license only

HUMAN PERFORMANCE WITH PROSTHETIC DEVICES AND SURGICALLY MODIFIED SKELETAL ELEMENTS

R. SELIKTAR

Department of Mechanical Engineering and Mechanics, Drexel University, Philadelphia, Pennsylvania

J. MIZRAHI

Department of Biomedical Engineering, Technion, Israel Institute of Technology, Haifa, Israel

T. VACHRANUKUNKIET

Moss Rehabilitation Hospital, Philadelphia, Pa.

and

M. BESSER and D. KUENZIG

Department of Mechanical Engineering and the Institute for Biomedical Engineering and Science, Drexel University

(Received September 1988)

The paper describes a general philosophy and some specific results which evolved from a continuing study of human performance under normal and pathological conditions. The results presented focus on the conclusions derived particularly from human locomotion studies performed with several groups of subjects. A control group of normal individuals was used to establish the reference data set. Motion studies were then performed with groups of above-knee amputees, below-knee amputees, patients with hip and knee joint replacements in the pre and post operative state and patients with various degrees of ankle joint fusion. The discussion is also based on some limited clinical and research experience with arm motion studies of normal subjects and amputees and other subjects who have undergone different surgical modifications of their musculoskeletal system. The discussion centers on some fundamental features of normal motion performance such as repeatability; learning and control of motion tasks; individual adaptations and motion under pathological conditions. Some results are presented on performance of subjects from the different groups listed above. Performance criteria and methods of evaluation developed during the course of the study are discussed. The general conclusions reached in the global study to date are summarized at the end of the paper.

Keywords: Motion, locomotion, gait, performance, prosthetics, rehabilitation.

1 INTRODUCTION

Human motion seems to us so obvious and unquestionable, that only when the natural system is impaired we begin to appreciate its complexity. Our appreciation is also aroused when an athlete or a dancer challenges his neuromuscular system, in an attempt to reach the upper limits of his performance capabilities.

Human physical performance has traditionally been modeled and treated with such basic functional criteria as: minimum energy principles; stability considerations; symmetry; and a variety of optimization criteria considering primarily physical parameters. Very little weight (if any at all) has been assigned to psychological factors such as: motivation; courage; concern and response to pain. It is obvious that many of us are unable to perform complex athletic activities because we do not dare try and not because we cannot do it. It is also apparent that a locomotor irregularity such as a limp is, most often, a precautionary measure aimed at reducing pain, rather than a faulty function of the neuromuscular system. Therefore, modeling pathological gait by using models composed of functional elements only, may be found occasionally misrepresenting.

Surgical procedures such as joint replacement, joint fusion and osteotomy are aimed at the correction of a functional problem, usually associated with joint degeneration, skeletal trauma and pain. The procedure usually results in realignment of the skeletal system and alteration of its mechanical response to external loading. In turn, this may lead to the development of new problems which may be unrelated to the original ones. In order to plan appropriately a corrective surgical procedure, one should be able to anticipate the functional changes that will occur as a result of the surgery. Therefore, the variables involved in the control of motion should be clearly understood and accounted for, in any modeling procedure.

Traumatic injury and skeletal deformity produce similar biomechanical effects. In such events, even if there is no immediate indication for surgery, one should be able to determine the long term effects and intervene if necessary, as a preventive measure.

Amputation or congenital malformation of a limb, is usually treated by providing a prosthetic substitute. The artificial limb is attached to the residue (stump) in a non-invasive fashion through a "soft" interface, which is formed by fitting a relatively hard socket over the stump soft tissue. The socket conforms to the shape of the stump, with certain modifications made to reduce freedom and facilitate appropriate load transmission with minimal discomfort. In lower limb prostheses the mechanical features of the interface play a critical role in the successful delivery of the limb to the amputee. This is due to the substantial load that is transmitted through the soft tissue. In the case of the upper limbs the stump socket interface is subjected to considerably smaller loads and, therefore, its biomechanics is, in most cases, of a lesser importance.

From the functional and control points of view, feedback information on the position of the joints is critical to the sound operation of the prosthesis. This information is of particular importance to the control of the upper limbs due to the considerably broader range of tasks performed by them.

2 SOME FUNDAMENTAL FEATURES OF HUMAN MOTION

2.1 *Repeatability of Motion*

For several decades scientists have been investigating and modeling human motion with the goals of: understanding how normal motion is performed; creation of a reference data-set for the design of prosthetic devices; understanding pathological gait for diagnostic, therapeutic and follow-up purposes and; improving athletic performance. Due to the relative complexity of the phenomenon, studies focused primarily on specific task performance rather than the general control of motion. In that respect, locomotion was found most attractive. On the one hand, from the control point of view, it is a rather complex weight bearing activity, while on the other hand it constitutes a well defined task with fairly consistent characteristics and good resemblance between individuals. A so-called "Normal Gait" could therefore be defined, despite the fact that variability in task performance by the same individual and differences between individuals may be substantial at times, compared with the conventional definition of normality. The variablitiy depends on the choice of the measured parameters and the methods of their processing. In a most general way we can state that resemblance and consistency exist primarily with respect to the pattern of the measured or calculated variables such as: ground reaction forces, joint forces and joint torques etc. However, comparison of performance within or between individuals through quantified values, whether they constitute fragmentation of the information or an integrated form, often yield differences of the order of ten to fifty percent.

The key issue is: should we expect repeatability in the performance of a certain task by the same individual. In that respect one is tempted to suggest the analysis of an athletic activity and a search for repeatability in its performance, since it involves fine tuning and higher standards of accuracy. We know that athletic performance can be improved through "hard work" but, even then, repeatability is not perfect.

Locomotion is more attractive for comparative purposes and a search for repeatability since it is treated as a cyclic phenomenon and therefore the task is being repeated every cycle. However, despite the general resemblance of gait cycles one can easily observe differences of 10 to 15 percent between the measured quantities in two consecutive cycles of the same individual. Figure 1 depicts the antero-posterior (A-P) ground reaction force characteristics as measured during two different runs of the same individual. It is noticeable that the general features of the curve look alike, however the impulse values as obtained by integration of the individual phases differ by more than 10%.

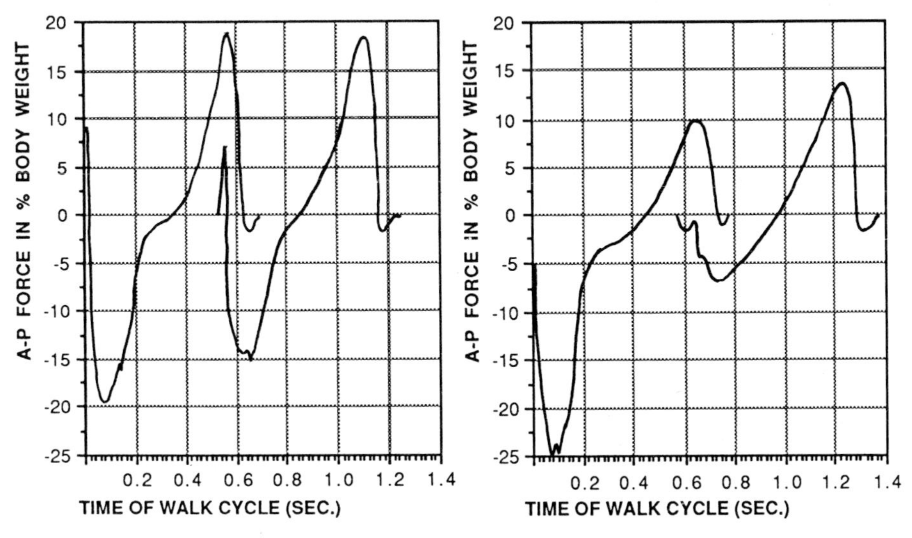

Figure 1 The (A-P) ground reaction force characteristics as measured during two different runs in the same individual.

Certain other values such as peak forces, speed of walking, cadence and length of stride would yield similar differences.

Based on our studies, we can state with a limited reservation, that human locomotion can be considered to consist of repeatable cycles. However, other tasks, which are not so routinely performed, are less so.

2.2 *Learning and Control of Motion Tasks*

The process of learning how to perform motor tasks appears to be a rather crude method of iteration. The so called primitive patterns exhibited by CVA patients and similarly by newborn children, suggest that the control of motion is performed in an open loop control mode. Skills seem to be acquired by trial and error, as if the body environment is being mapped in the memory of the central nervous system. There may even be a certain hierarchical order in the control of motion tasks.

The above hypothesis, if correct, raises some questions about the importance of the proprioceptive feedback in the continuous control of motion. Our general observations suggest that the sensory-motor system can operate in a closed-loop mode, particularly when the motion is slow, and in an open-loop mode, when the motion is fast. The need of proprioception however, is crucial for the achievement of good functional performance even when the system operates in an open-loop mode. Then, proprioception provides the position awareness for information only and if critical conditions such as danger develop, the nervous system may intervene and send a corrective command. This is clearly evident from the substantial ballistic component contained in every "rapid" motion. In one of our earlier studies, we attempted to model human locomotion with a purely ballistic model, utilizing the inverted pendulum concept[1]. Figure 2 depicts a comparison between a synthetic *vertical* ground reaction force characteristic which was derived with the aid of an inverted pendulum model (2a) and a curve representing experimental results as obtained on a normal subject (2b). The resemblance between the two

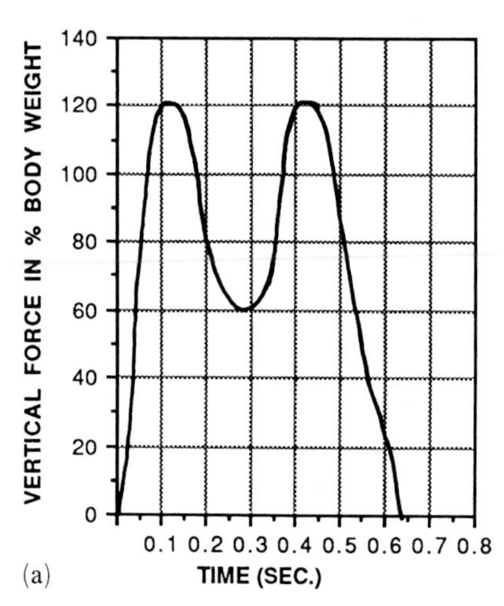

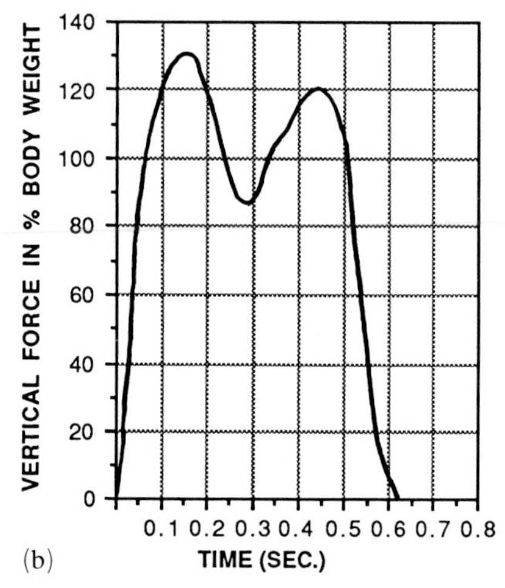

Figure 2 A comparison between a synthetic vertical ground reaction force characteristic which was derived with the aid of an inverted pendulum model (a) and experimental results as obtained in a normal subject (b).

curves clearly suggests that a ballistic motion of the head-arms-trunk (HAT), supported by a straight leg represents fairly well, the physical phenomena of *normal gait.* The gait characteristics, in such an event, are determined primarily by the initial conditions of the single leg support.

In pathologocial gait, in which there is an involvement of either pain or risk of falling, the picture is entirely different and only parts of the gait cycle can be modeled with a ballistic model. For example, Figure 3 illustrates the ground reaction force characteristics as obtained from below-knee amputee gait and it is clearly noticeable that the more "ballistic" curve (left) is the one that represents the prosthetic side. This is typical and represents the "mistrust" of the amputee with respect to his artificial limb. All the stabilizing and control work is therefore done by the normal leg while the leg on the prosthesis side acts as the arm of the inverted pendulum. The perturbations superimposed on the curve are indicative of stabilization activity of the calf muscles in particular.

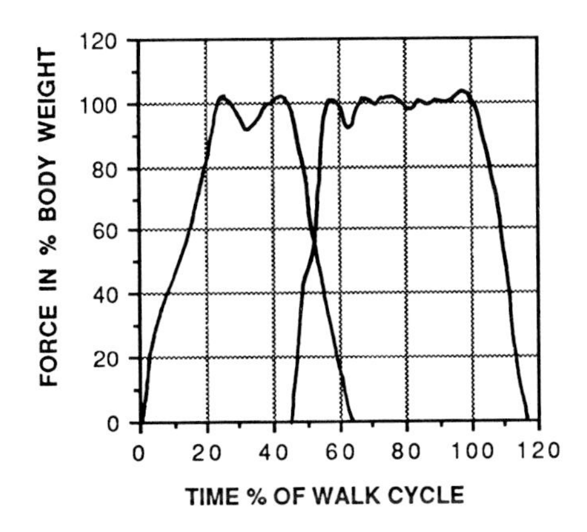

Figure 3 Vertical component of the ground reaction force characteristics as obtained from below-knee amputee gait.

In the case of arm motion a greater dexterity is required since there is no specific task of preference. There are obviously tasks that are repeated more frequently and the arm becomes more specialized in their performance. Training is an important factor in the refinement of motion skills, but every individual also has his own upper limit of achievement. For instance to draw a circle or to follow with a pencil over a pre-drawn path, one can reach much better results in an open loop control mode. In other words to follow slowly over the path would probably yield worse results than to perform the task with a swing of the arm after a few trial runs. Regarding the ballistic nature of the motion of the arm, we are currently investigating arm motion in general and the method of selection of trajectories. However, except for general impressions we do not have evidence to support a hypothesis similar to the one suggested for the lower limb performance during locomotion.

2.3 *Individual Adaptations*

Every individual develops his own typical pattern of movements which make up his gait characteristics. Differences between individuals are evident and we easily recognize a person by his gait. Some of the more visible differences are observed in the lateral sway of the torso, foot plantar flexion past mid-stance accompanied by an excessive elevation of the whole body c.g., non-pathological asymmetry in the coronal plane, etc. As stated earlier, despite the differences, the general pattern of the characteristic of the measured variables (at least in the time domain) for all individuals remains similar. Certain differences in the motion performance between individuals suggest that everyone has his own optimal conditions for the performance of motion tasks at a minimum effort and inconvenience. Some other differences definitely indicate that motion task is not necessarily performed at a minimum energy cost. For instance, differences between individuals are observed in the combinations of cadence and stride-length, which may produce equal walking speeds. Also, different individuals have different walking speeds at their convenient paces, although their body build may be of the same general dimensions.

One important factor in the determination of the nature of performance is fatigue. Some investigators relate to fatigue as if it is synonymous with high energy expenditure. It is only logical to assume that person will perform in a manner which is the least fatiguing. However, since motion is produced by synergistic activity of different muscle groups, they cannot all work at their optimal conditions (e.g., length-force characteristics) and as a result, some muscles fatigue earlier than others. That type of localized fatigue, although not caused by excessive energy cost, is as undesirable to the individual as an overall state of exhaustion. The person will therefore try to avoid the development of such fatigue which is usually accompanied by pain.

3 PATHOLOGICAL MOTION

3.1 *Rehabilitation Objectives*

The two main objectives in pathological motion training are: to restore a sound biomechanical function and, to achieve good cosmetic (aesthetic) appearance in both the static and the functional sense.

The traditional attitude which is most commonly found in rehabilitation practice is to try to bring the patient's functional capabilities as close as possible to those of a normal individual. Since most of the neuro-muscular patients are severely limited, the above goal seems utopian. However, this attitude may prove dangerous or even fatal in certain cases, in which the subject is capable of reaching such a level of perfection. For instance, in the case of endo-prosthetic joint replacement (arthroplasty) one may be able to perform in a perfectly normal fashion, but the prosthesis will wear or fail as a result of extensive use and over-loading. On the other hand, one has to accept the fact that a person with a limb disability is limited by the different functional constraints and therefore his optimum is not necessarily the same as that of a normal individual. Every case should therefore be reviewed according to its own criteria and the rehabilitation goals should be set correspondingly.

3.2 *Motion Studies*

a. *Sites* Different parts of the study were performed at four different sites: The Biomechanics Laboratory of the Loewenstein Rehabilitation Hospital in Israel; The Prosthetics laboratory at the Department of Biomedical Engineering, Technion

Israel Institute of Technology; The Rehabilitation Engineering Center at Moss Hospital and the Biomechanics Laboratory of the Mechanical Engineering Department at Drexel University.

b. *Equipment* The two laboratories in Israel were equipped similarly to the Drexel and Moss laboratories. Special prosthetic arrangements were provided by prosthetic orthotic facilities available at both the Loewenstein and Moss hospitals. The principle laboratorial systems used in these human performance studies were: an instrumented walkway containing two force measuring platforms; a kinematic documentation system composed of a "Selspot" position monitoring system, Videotape recorder and cameras, connected to a special effect generator, Bolex H-16 cine-camera and projection facilities. The different kinematic systems were used as three optional alternatives. The force measuring system was favored over the others due to its simplicity, accuracy and ease of information acquisition and processing.

3.3 *Performance with Artificial Limbs*

a. *General* Artificial limbs for the upper and lower extremities are extensions of the residual limb (the stump). However, the functional requirements of these two systems are so different that they can in no way be treated with the same set of rules. From the point of view of operation and control of the prosthesis, an extension of the residual limb, which is not a direct skeletal attachment, forms a pseudo-joint with a very limited motion. That type of joint does not impair severely the position awareness of the amputee. On the other hand, if an artificial joint is incorporated in the prosthesis, then hardly any proprioception is transmitted across this joint. The wearer can then rely on the interpretation of inertial effects and his vision as a partial substitute for the lost proprioception. Since the functional expectations of a lower limb amputee are that his legs are used primarily for walking, the task is relatively simple and the prosthesis can be "pre-programed" to perform this limited function. Lower limb prostheses are therefore most commonly made to act as constrained compound pendulums during the swing phase of the gait cycle. During the stance phase the prosthetic joints are either mechanically or geometrically locked or severely constrained so as to avoid uncontrolled collapse of the prosthesis and fall of the amputee. Lower limb prostheses are driven by the stump and very seldom utilize energy from an external source. However, performance becomes severely impaired with the increase of the number of prosthetic joints.

Arm prostheses on the other hand, involve voluntary control of coordinated motion in several degees of freedom. Due to the versatile and dexterous functional requirmements of the arms, the sense of position awareness, e.g., proprioception, is critical to the successful control of an arm prosthesis. In the absence of proprioception, the control of the arm becomes an impossible mission.

b. *Lower limb prostheses* The main goal of our studies of amputee gait was to create the basic knowledge needed in order to improve the quality of prosthetic care delivery. To serve this purpose we focused on the following objectives:

1) Develop a quantitative method of evaluation of gait to be used clinically for the evaluation of any pathological condition.

2) Evaluate the quality of performance with the prosthesis.

3) Predict long term effects due to the use of the prosthesis.

4) Identify general instability of performance.

5) Identify stump-socket interfacial instability.

6) Identify inadequate interfacial force transmission.

7) Determine the effect of structural alignment conditions on the quality of performance and interfacial discomfort.

Of all the different laboratorial systems available, the force plate transducers were found the most attractive. The ground reaction forces represent the dynamic interaction of the walker with the environment. These forces are therefore responsible for the resulting gross motion of the body. However, most of the locomotor irregularities are reflected in these characteristics. Once a method of identification of these reflections is developed, one can enjoy the relative simplicity of the force characteristics in both data acquisition and processing compared with any alternative kinematic variable.

Two features of the force information were used as principal factors in the evaluation of the gait performance, the impulses of the ground reaction forces and their general patterns. The impulses and their applications have been discussed by us extensively in [2, 3]. The usefulness of the impulse is twofold: it provides information on the *effort* involved in the performance of a certain function in an integrated but non *work* form; it also facilitates an easy examination of the data and confirmation of its validity[4].

Figure 4 presents a typical ground reaction force characteristic in the antero-posterior direction (A-P) as obtained from gait records of a normal individual. The area enclosed between the curve and the timebase represents the impulse. The impulse value for the entire "gait cycle" should therefore yield zero for each one of the force variables if we assume that locomotion is a cyclic phenomenon (see reference above). We named this concept a "gait consistency test" because it

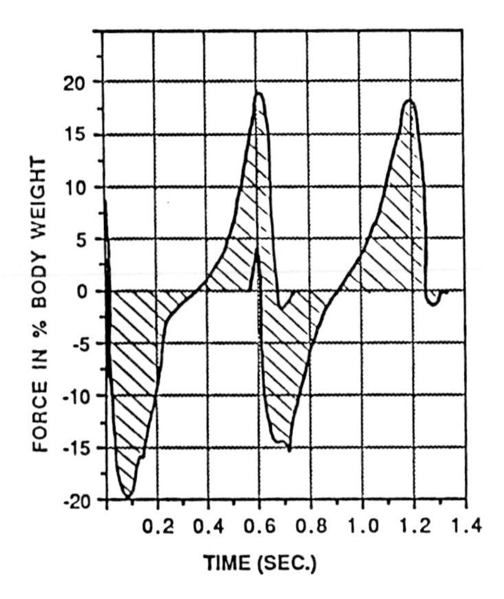

Figure 4 Typical ground reaction force characteristic in the (A-P) direction as obtained from gait records of a normal individual.

examines, based on impulse-momentum equality, whether the initial conditions of two consecutive cycles are the same. Such a test is extremely valuable in the laboratory because "natural gait" tends to be distorted considerably due to environmental and psychological influences. Since the development of this test we have been using it routinely in gait testing in order to discard any invalid information which may have been collected in the lab.

The choice of the A-P force component for this demonostration is made because this is the force responsible for the advancement in the direction of progression. It can be seen that each "stance phase" is composed of a "braking" action and a "pushing" action. These two activities are very important to the understanding of the different events of the gait cycle such as mechanical stability of the prosthetic joints and the interface, active and passive muscle function and more. Figure 5, for instance, presents a comparison between the impulse ratios of the prosthetic leg vs. the sound leg, in the different phases, as obtained on five below-knee (B-K) amputees. Although they do not all display the same behavior, it is evident beyond any doubt that all of them utilize the prosthesis considerably less than their sound leg. An exception to that is the braking impulse ratio of subjects 2 and 3. Particularly noticeable is the consistent inferiority of the pushing action of the prosthesis relative to this of the sound leg.

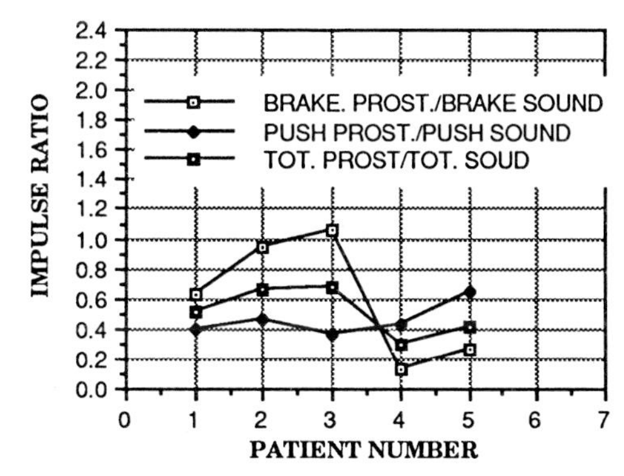

Figure 5 Impulse ratios of the prosthetic leg vs. the sound leg, in the different phases, as obtained on five below-knee (B-K) amputees.

Figure 6 presents similar results to the above, as obtained on 11 above-knee (A-K) amputees. A-K amputation is a considerably more crippling procedure and this is clearly visible from the results presented. Of particular consistency is the ratio between the total impulse (braking + pushing) on the prosthesis side and that on the sound leg side. This ratio is an index of overall activity of the leg and therefore every improvement of the prosthesis is directly measurable through this value. On the other hand, the pushing and braking of the prosthesis can be enhanced by re-structuring (realignment) of the prosthesis and these can also be measured with the corresponding impulse values. For instance, due to the obvious risk of knee-buckling during the braking phase of A-K amputees, these amputees tend to compensate by reducing the braking effort of the prosthetic leg. By mechanically stabilizing their knee, in a way that is noticeable by the amputee, the braking of the prosthesis can be enhanced and, again, this is directly readable in the braking impulse values.

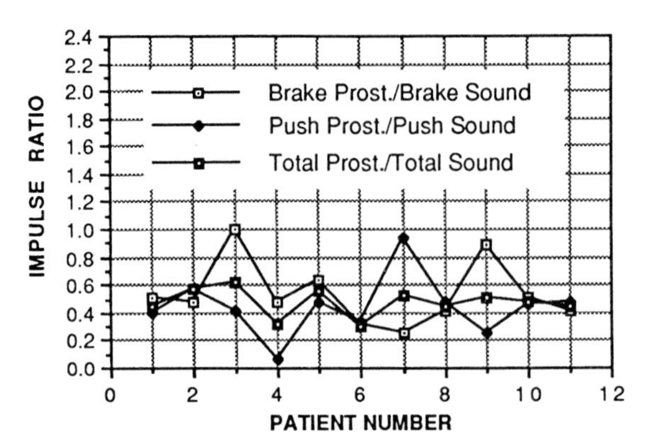

Figure 6 Impulse ratios of the prosthetic leg vs. the sound leg, in the different phases, as obtained on eleven (A-K) amputees.

As a reference we present in Figure 7 similar information to the one contained in Figure 6, which was obtained on four normal subjects (a greater number of tests were performed with b and d). It is evident from the results, b, c and d, in particular, that normal gait is very close to being symmetrical (within certain limits of variability).

Analysis of forces or kinematic variables by transformation into the frequency domain did not prove very rewarding in the past [5]. However, the presence of odd or repeatable perturbations superimposed on the force curves is evdient in A-K and B-K amputee results.

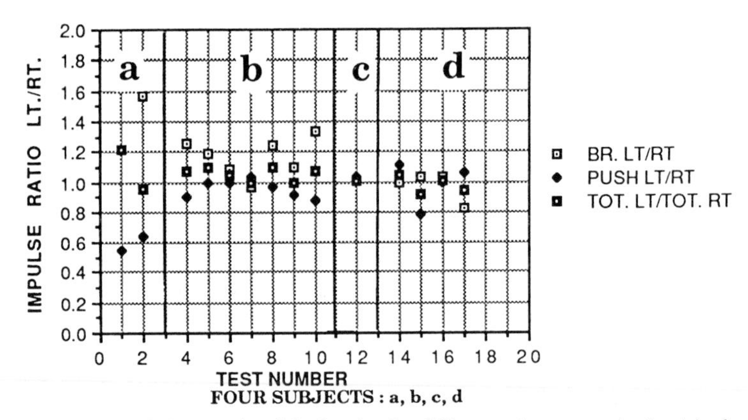

Figure 7 Impulse ratios of the left leg vs. the right leg, in the different phases, as obtained in four normal subjects.

Figure 8 depicts a typical antero-posterior (A-P) ground force characteristic of a B-K amputee. The single perturbation superimposed on the braking portion of the curve (left) was found to be typical and indicative of freedom at the stump socket interface. In a current study on patellar tendon bearing (PTB) prosthesis we have implemented means of augmentation of the patellar tendon portion and the end portion of the socket for the purpose of varying the weight bearing characteristics of the prosthesis. The above perturbation is considerably reduced with the "tightening" of the socket.

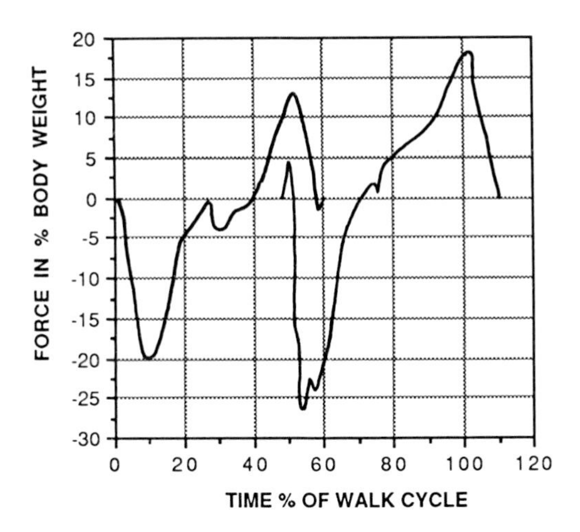

Figure 8 A typical Antero-Posterior ground force characteristic of a B-K amputee. The curve on the left represents the prosthesis.

We have shown in an earlier work [6] that this same effect of stabilization of the interface and reduction of the force perturbation above, can be achieved by re-alignment of the prosthesis. Through re-alignment one can phase-shift the moment of the ground reaction with respect to the stump centroid in the sagital plane and attenuate the toggle action created by the brief change in direction of that moment.

On the other hand, as seen earlier, from Figure 3, perturbations superimposed on the vertical force characteristic representing the sound leg is indicative of the presence of a stabilizing muscle activity. This activity is particularly the responsibility of the calf muscles which bear the burden of the prolonged weight bearing period of this leg, as suggested earlier.

3.4 *Performance after Total Joint Replacement*

a. *General* Total joint replacement does not necessarily create a functional problem. The anatomical joint is resurfaced by removal of the cartilage and a portion of the underlying bone and its replacement with an artificial component. The different techniques of joint replacement and their influence on the functional characteristics of the subjects is beyond the scope of this paper and will not be discussed here. However, since, in general, joint replacement constitutes a reconstruction procedure, the degree of reconstruction depends on the extent of damage to the original joint. If the joint surfaces alone are replaced and the ligaments are intact, then the potential for a complete functional rehabilitation is considerable. If however, the reconstruction involves soft tissue repair then the probability for a full functional recovery is rather low.

The mechanical limitations of the artificial joint and particularly its interface with the skeletal component are considerable. Wear is not repairable and therefore mechanical deterioration is inevitable. Naturally, the rate of deterioration, even if no catastrophic failure occurs, is related to the extent of use and exertion of the prosthetic joint. In any event it should be kept in mind that under no circumstances should the subject be encouraged to resume the full extent of normal activities.

Sports activities which may overload the reconstructed joint should be particularly discouraged.

b. *Knee joint replacement* Since the primary activity of the knee is in the sagital plain, its function and irregularities are reflected mainly in the A-P ground force characteristic. This is particularly true in the post surgical state, e.g., when the leg is equipped with the prosthetic knee. Some lateral instability may exist as a result of laxity of the collateral ligaments but this is less visible in the ground reaction force characteristics.

Figure 9 presents results as obtained on ten patients before undergoing a total knee joint replacement surgery. The results are presented in terms of the ratios of impulses of the ground reaction forces in the A-P aspect, between the affected side and the side of the healthy knee. Three values are presented per subject: the braking impulse ratio; the pushing impulse ratio and; the total impulse ratio. The braking impulse is indicative of the stability of the knee joint while the pushing impulse and the total impulse represent the extent of use of the leg in ambulation. It is generally expected that the affected leg should be functionally inferior to the sound leg and therefore cases such as 2, 3, 4, 5, 6, 7 and 9 are typical. Cases such as 8 and 10, in which all impulse ratios are consistently greater than 1 indicate that the affected leg is generally more active than the sound leg, are rather unusual in pathological gait. This may be a result of the use of pain relieving agents, taken by the subjects without our awareness or a result of the existence of problems in the "sound" leg which were not brought to our attention.

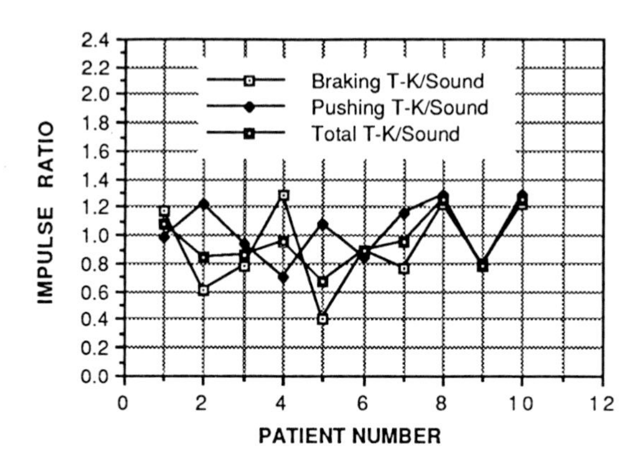

Figure 9 Results as obtained in ten patients before undergoing a total knee joint replacement surgery.

A spread of the individual results such as in cases 2, 4 and 5 is not unusual. Spread such as in case 4 on the other hand, could even occur in the gait of certain normal individuals, particularly in view of the fact that the total impulse ratio is very close to 1. This suggests that the general "effort" of both legs is the same but one leg is more active in propulsion while the other is more active in restraining, e.g., braking. Usually though, normals are expected to perform more symmetrically.

Figure 10 compares the values of the three impulse ratios before and after total-knee replacement surgery. Of particular importance is the total impulse ratio since it represents a more integrated form of overall activity. The post surgical improvement is notable in subjects 1, 2, 3, 5, 6 and 9. As to 4 there is not much change, 7 and 8 have definitely changed for the worst and 10, despite its decline, is

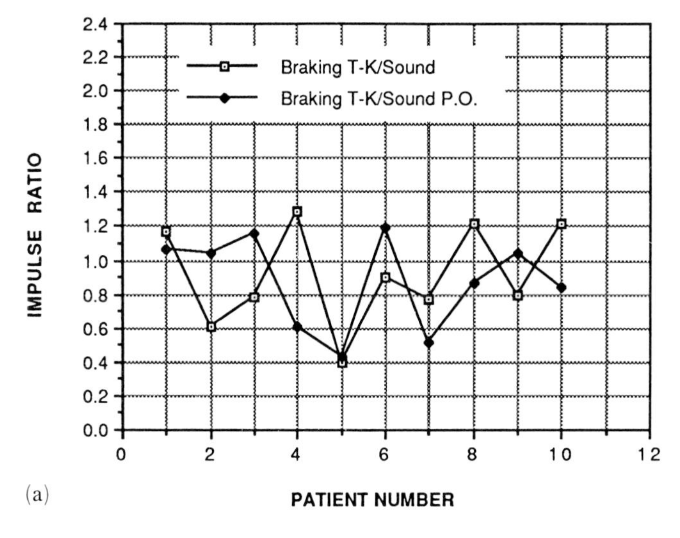

Figure 10 (a) Braking impulse ratio between the affected (total-knee replacement) and the sound leg, in the pre and post operative states.

still close to 1 which is the desirable value. As far as the braking impulse ratios are concerned (10a) 1, 2, 3, 6 and 9 show benefit from the surgery while 4, 7, 8 and 10 show a change in the undesirable direction. This may not necessarily represent a chronic problem but rather a result of a sense of post-surgical insecurity that may improve with time. The decline in pushing impulse ratios seen in Figure 10b is not necessarily a negative sign since subjects with lower limb disability often push excessively with their affected leg and brake more substantially with their sound leg. Reduction in pushing, therefore, if not excessive, may be an indication of improvement.

c. *Hip joint replacement* Our study of the performance of subjects who have undergone hip arthroplasty was not as informative as the study of the knee replacement group. With the joint being more proximal, the use of a lumped parameter model becomes more complicated since all the irregularities of the more distal joints (knee, ankle, foot) are also reflected in the same characteristics. However, instability of the hip is particularly disabling in the medio-lateral aspect (M-L). The effect of this instability is clearly visible in the corresponding ground force characteristic. A typical example of such instability is depicted in Figure 11. The M-L ground reaction forces as recorded on a subject who had undergone a bilateral hip arthroplasty are depicted in the figure. The substantial medial component of the force and the considerably reduced lateral component, are clearly abnormal and indicate a substantial atrophy of the abductor muscle group, particularly in the right leg. Kinematically, this results in a waddling gait in which the subject secures stability by leaning directly on his hip joint and hence minimizes the hip abductor moments [7].

3.5 *Performance after Joint Fusion (Arthrodesis)*

Fusion of the proximal joints of the lower limbs is becoming quite uncommon due to the relatively successful surgical replacement of the knee and the hip. The ankle

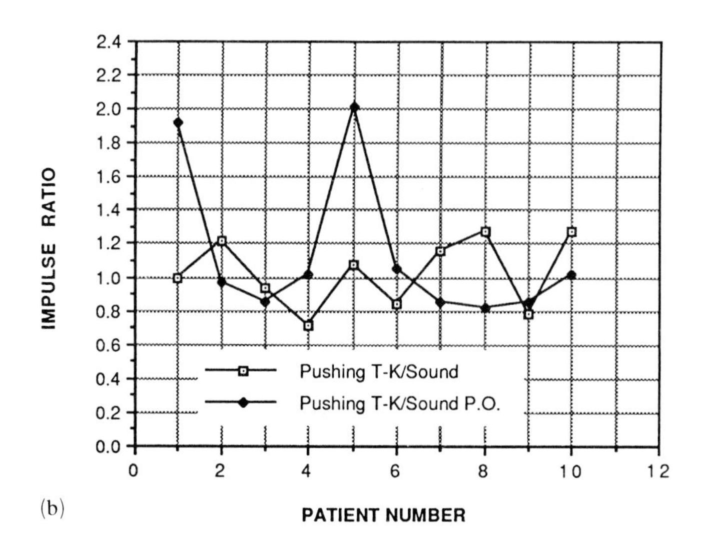

(b)

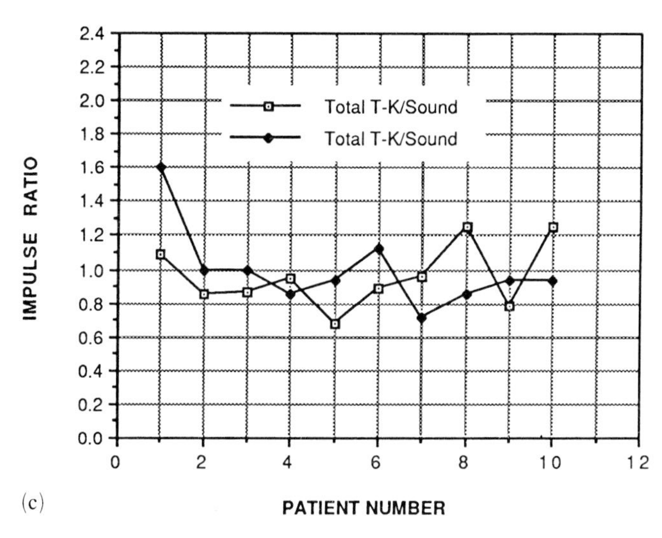

(c)

Figure 10 (continued) (b) Pushing and (c) total impulse ratios between the affected (total knee) and the sound leg, in the pre and post operative states.

joint complex, however, constitutes an elaborate bony and soft tissue structure and therefore prosthetic replacements of its components are presently by far less successful than those for the hip and the knee. Arthrodesis of the ankle therefore, either partial or full (triple), is much more common in orthopedic practice.

The triple arthrodesis is a rather radical procedure from the biomechanical point of view. The mechanical compliance of the joint is almost totally eliminated and the entire task of impulse attenuation is left to the heel pad, the metatarsals and the more proximal joints. The transfer of shock absorbing responsibilities to the more proximal joints, e.g., knee or hip or both of them jointly, requires some

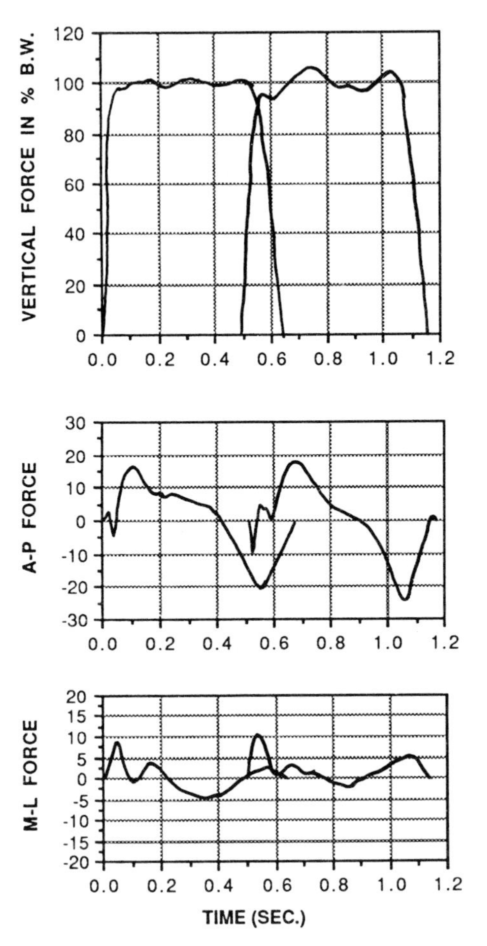

Figure 11 The ground reaction force components as recorded during gait of a subject with a bilateral hip arthroplasty.

modification of the locomotor kinematic characteristics. These modifications can be achieved through gait training or personal reaction of the subject. The kind of function modification which is quite common in triple arthrodesis is the adoption of a moderate knee and hip flexion at hell strike. This attenuates significantly the peak forces which are otherwise generated by the stiff ankle during the inital encounter of the foot with the ground.

In order for a modification of the gait characteristics to be acceptable to the subject he has to feel a certain inconvenience while walking in the "old" fashion with the fused ankle. This is not always the case with these patients since the restructuring of the joint often destroys some of the sensory receptors that would otherwise generate warning signals. Consequently, patients who have undergone that type of joint fusion will often subject their joints to substantial peak forces, containing a high frequency spectrum, without being congnitively aware of their existence. These cases are easily recognizable when the subject is examined in the gait laboratory with the aid of force-plates. Figure 12 depicts the ground reaction force characteristics as obtained on one subject following a triple arthrodesis of the left foot. The high frequency contents is evident from the sharp force spikes and since this patient had already undergone a full gait training this was considered to be his regular mode of walking. In such cases one has to rely primarily on footware for the attenuation of impulsive forces and Figure 12b demonstrates clearly the effect of special shoes. An appropriate design of footware produces a filtering action for high frequencies contained in the ground force characteristics. Further details of this study are obtainable from: Seliktar and Mizrahi [8].

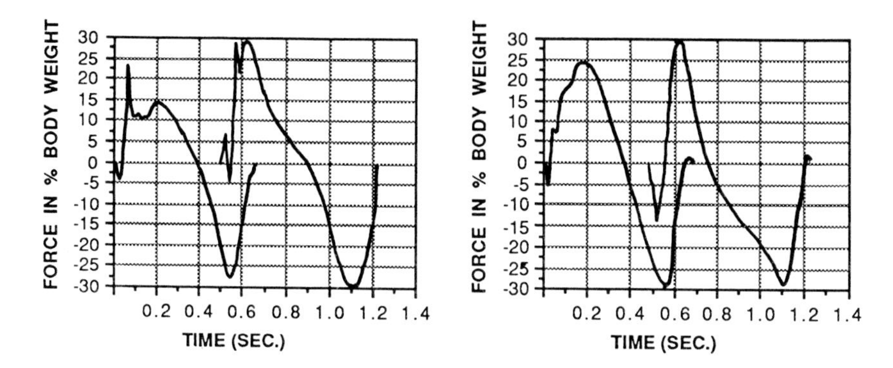

Figure 12 The A-P ground reaction force characteristics as obtained on one subject following a triple arthrodesis of the left foot (left figure: bare-footed, right figure: with shoes).

4 SUMMARY AND CONCLUSIONS

The work presented in this paper summarizes some of the findings of a continuing study of human performance with particular emphasis on pathological gait. A certain philosophy as to the way motion is controlled and what could the expectations be insofar as the repeatability of task performance, is discussed. Criteria for evaluation of gait performance, developed by the authors during the different stages of the research, have been used to demonstrate the reflection of various pathological conditions in the ground reaction force characteristics.

The general goal of this work was: *To understand how locomotor habits are acquired and how are they modified as a result of the introduction of artificial structural changes in the neuro-musculo-skeletal system.*

Since the thrust of our work was in the area of exo-prosthetics, some of the key subjects investigated were:

1) Investigations aimed at prediction of the long term effects due to the use of prostheses.

2) Identification of unstable locomotor performance.

3) Identification of stump-socket interfacial instability as reflected in the whole body kinetic characteristics.

4) Identification of inadequate interfacial force transmission.

5) Determination of the effect of structural alignment conditions on the quality of performance and interfacial discomfort.

Some of the findings of the different parts of the study are described in the text. The global conclusions derived to date are listed below:

a) The governing factors in the adoption of pathological motion habits are either psychological or precautionary, aimed at the prevention of pain and discomfort. They are very rarely a direct result of the mechanical constraints of the locomotor system.

b) The assumption that human performance is based on minimal energy principles has never been substantiated and, furthermore, appears to be invalid.

c) Modeling of the human locomotor system as a closed-loop control system is not always adequate. Proprioceptive feedback is present in the system, but is not always being utilized for control purposes. However, the absence of proprioception, severely impairs performance due to deficiency in the control as well as to the psychological effect of loss of confidence.

d) Performance with exoskeletal upper limb prostheses is still extremely poor due to major problems of man–machine interfacing, deficiency in the control strategies and inadequate hardware technology.

e) Performance with exoskeletal lower limb prostheses is reasonably good in below knee amputations and less so in above knee amputations. Higher levels of amputations are rare and our experience amounts to dealing with only a few cases of hip disarticulations and hemipelvectomies.

f) The overall performance of lower limb amputees is greatly affected by stump pain, stump-socket instability and a sense of insecurity. Performance can therefore be substantially enhanced by treating these factors directly or indirectly. For instance, all the above are influenced by the geometrical alignment of the prosthetic structure. Reformation of the socket is another option for the improvement of stability, reducing stump pain and enhancement of the sense of security through improvement of proprioception.

g) Anatomical joint replacements cannot be categorized from the functional point of view, by the replaced joint alone. The integrity of the soft tissue structure is extremely important to the sound operation of the joint. Elimination of pain is a key factor in the "success" of the surgery from the performance point of view. However, good post-surgical performance does not guarantee the biomechanical success of the procedure. Biomechanical consequences of an inadequate surgical methodology may become evident only years later in the life of the subject.

h) Ankle joint fusion is a radical procedure from the biomechanical point of view. If this surgery is strictly essential, a close followup should be maintained for a long time after surgery. Appropriate gait training scheme should be recommended for these patients, to teach them to attenuate impulsive forces by using alternative joints. Ground reaction forces are quite informative insofar as indication of hazardous conditions are concerned.

i) We have developed several methods of evaluation of performance, focusing primarily on pathological conditions of the locmotor system. Of particular importance was the subject of repeatability of motion execution. This led us to the development of the "Gait Consistency Test" which is currently being used routinely for laboratorial data screening.

j) Gait performance indices were developed, based on impulse momentum considerations and these served us in the evaluation of the different pathological conditions described in the paper. However, for a clearer understanding of these methods of evaluation, the reader is referred to the corresponding publications as referenced in the text.

ACKNOWLEDGEMENTS

The work presented constitutes a continuing study which was conducted at four sites. The early stage of the research was conducted at the Julius Silver Institute for Biomedical Engineering at the Technion, Israel Institute of technology in collaboration with the Bio-mechanics laboratory at the Loewenstein Rehabilitation Hospital. The latest part was conducted at Drexel University in collaboration with the Rehabilitation Engineering Center at Moss Rehabilitation Hospital in Philadelphia Pennsylvania.

The different phases of the research were supported by intramural research funds from all four institutions and by grants from: The National Institute for Handicapped Research, The Easter Seal Research Foundation and The National Science Foundation (Grant #ECE-841 9556).

REFERENCES

1. S. Siegler, R. Seliktar, and W. Hyman, Simulation of human gait with the aid of a simple mechanical model, *J. Biomechanics*, **15(6),** 415–425 (1982).
2. R. Seliktar, Integrated information approach to clinical analysis of gait, *Proceedings of the Third ASCE/Eng. Mech. Division* (Ed. C. Phill Johnson), Specialty conference, Austin, Texas (1979), pp. 586–589.
3. R. Seliktar, J. Mizrahi, and Z. Susak, Computer aided dynamic alignment of below-knee prostheses, in *Use of Computers in Aiding the Disabled* (Ed. J. Raviv), North Holland (1982), pp. 87–95.
4. R. Seliktar, M. Yekutiel, and A. Bar, Gait consistency test, based on the impulse momentum theorem, *Prosthetics and Orthotics International*, **3(2),** 91–98 (1979).
5. N. A. Jacobs, J. Skorecki, and J. Charnley, Analysis of the vertical component of force in normal and pathological gait, *J. Biomechanics*, **5,** 11–34 (1972).
6. R. Seliktar and J. Mizrahi, Some gait characteristics of below-knee amputees and their reflection in the ground reaction forces, *Engineering in Medicine*, **15(1),** 27–34 (1986).
7. Z. Susak, K. Katz, and R. Seliktar, Persistence of waddling gait after total hip replacement, for congenital dislocation of the hip, *Scand. J. Rehab. med.*, **12,** 113–114 (1980).
8. R. Seliktar and J. Mizrahi, Partial immobilization of the ankle and talar joints complex and its effect on the ground-foot force characteristics, *Engineering in Medicine*, **13(1),** 5–10 (1984).

Automedica, 1989, Vol. 11, pp. 163–174
Reprints available directly from the publisher only
Photocopying permitted by license only

CNS STRATEGIES IN HUMAN GAIT: IMPLICATIONS FOR FES CONTROL

DAVID A. WINTER

Department of Kinesiology, University of Waterloo, Waterloo, Ontario, Canada N2L 3G1

(Received September 1988)

Any FES control system for the gait of the paralyzed must mimic the intact motor patterns seen in normal walking. Therefore the designers of FES microprocessor controls must recognize the same strategies that the intact CNS takes into account. Details are presented to show that the CNS knows the characteristics of the "plant" it is controlling and its force environment: the transfer characteristics of each muscle, the segment inertias and inter-limb coupling, plus gravitational and ground reaction forces. Secondly, the designer must have an overall strategy to generate a total locomotor pattern and yet achieve safe transit. Based on the synergistic patterns seen in normal gait, information is presented to document how the CNS controls the balance of the upper body, prevents collapse during stance, achieves a safe toe clearance and prevents a damaging heel strike. Thus the stimulation profile to any given muscle must be generated by a controller that is programmed with all this information.

Keywords: FES control, human gait, inter limb coupling, muscle biomechanics, dynamic balance, support moment, swing trajectory.

1 THE PROBLEM

Considerable effort is now being directed to the FES of spinal cord injury patients in order to re-establish a safe walking pattern. Most of the research to-date has addressed the problems of electrode design and placement, stimulation patterns [1] and microprocessor design. Very little has been said regarding the overall control strategy and its essential ingredients, so that a safe walking pattern will result. The level of control has been limited to "on-off" stimulation patterns approximating average phasic EMG patterns that have been reported in the literature [2]. Such stimulation is open-loop and assumes that these slow-walking spinal injury patients require the same EMG patterns as normal adults walking their natural cadence.

What needs to be recognized is that any FES microprocessor system is replacing the CNS and therefore must mimic the same strategies that the intact CNS uses. Because these FES systems are being overlaid on an intact musculoskeletal system two things must be taken into account. Firstly, the characteristics of each muscle, joint and segment must be recognized along with gravitational and ground reaction forces and inter-limb coupling. These factors are referred to as the "plant" and the environment. Secondly, the FES system must have an overall priorized strategy to generate a total locomotor pattern and yet achieve safe transit. Safe transit means maintenance of balance, prevention of collapse, and achieving a safe toe clearance and gentle heel contact. Failure of any one of these four safety criteria will negate a satisfactory control system.

The total strategy must control the detailed stimulation patterns to a large number of individual muscles, such that any given muscle may be involved at any given time with more than one of the factors (e.g., simultaneous control of the balance of the head, arms and trunk (H.A.T.) and the collapse of the knee). Also, each muscle stimulation must be controlled from both peripheral feedback plus feedforward (anticipatory) control. Safe toe clearance during swing cannot be achieved by feedback alone because there is insufficient time to correct for toe trajectories after they have gone astray: the inertias and delays of the musculoskeletal system are such that swing trajectories must be monitored and corrected well in advance of the time of minimum toe clearance.

The purpose of this paper is to examine the body of knowledge of motor control of normal human gait and thereby get insight as to what the CNS recognizes about its plant and its environment. Also, such an examination provides strong suggestions to be made re overall strategies that the CNS is using which should be useful for FES researchers.

2 THE HUMAN MUSCULOSKELETAL PLANT AND ITS ENVIRONMENT

2.1 *Transfer Function of Muscles*

As actuators in the human system muscles have unique characteristics that any controller must account for, and the muscle twitch waveform gives us considerable insight into those transfer characteristics. Because the twitch is the tension response to a single stimulus it is an impulse response. The waveform has been shown to be close to a second-order critically-damped system [3] and has been modelled as such by many researchers [4–7]. Thus a linear envelope detector using a second-order low-pass filter models these characteristics with the cut-off frequency of the filter (1.2 Hz to 8 Hz) being related to the twitch times (20 ms to 133 ms). Inman *et. al.* [8] reported that the EMG processed through such a detector produced a waveform which rose and fell with the muscle force waveform. Olney and Winter [7] optimized the cut-off frequencies of five major muscles in a gait model and found frequencies as low as 1.2 Hz for the slow twitch soleus muscle and increasing to 1.8 to 2.8 Hz for the faster twitch quadriceps muscles.

Such electromechanical delays are recognized by the CNS, and result in both early activation and early deactivation of muscles. The raw EMG is a primary indicator of the total neural drive to a muscle, and Figure 1 shows such patterns for the tibialis anterior (TA) and soleus (S) muscles during the gait cycle. The solid and dashed lines plot the stride-to-stride ensemble average of the linear envelope waveform under two conditions for the TA and S. A 100 Hz low pass filter (solid line) provided some minimal smoothing of the sharp spikes to create a smoother stride-averaged activity waveform, yet it introduced a negligible phase delay (about 2 ms). The 3 Hz filter was chosen to model the twitch response of muscles with an average twitch time about 52 ms, and produced a waveform more closely in phase with the muscle tension or moment-of-force. TA reaches its peak activity just before heel contact (HC) in order that the dorsiflexor tension reaches its peak just after HC (in order to resist the expected ground reaction forces which will attempt to plantarflex the foot). Similarly, at the end of stance TA reaches its peak activity exactly at TO (60% of stride) but the tension does not increase to a reasonable level until about 5% after TO, when it dorsiflexes the foot for toe clearance. At the end of stance, about 200 ms before toe-off (TO) the soleus activity reaches its peak and suddenly drops off to near-zero 100 ms prior to TO. This sudden derecruitment of the plantarflexors recognizes the need to decrease the plantarflexor force from its peak at about 100 ms before TO to near-zero at TO. If this plantarflexor activation

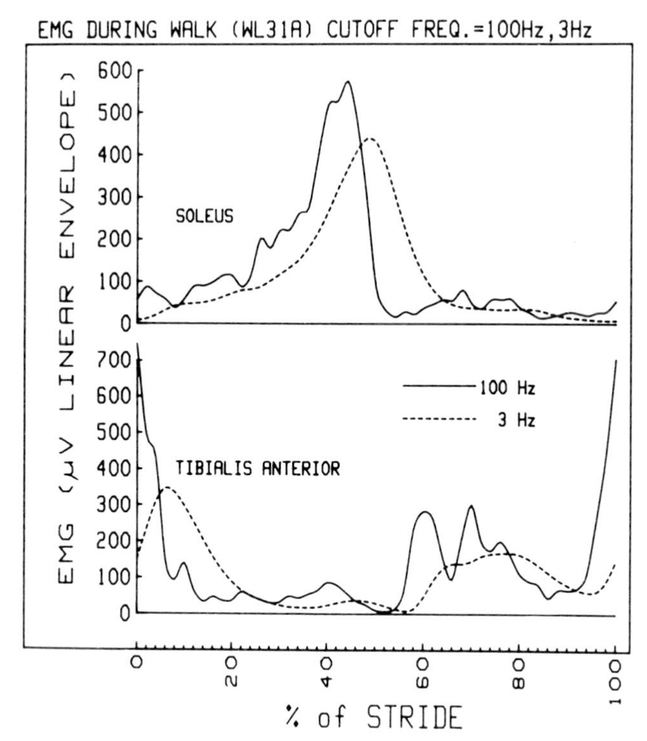

Figure 1 Computer averaged EMG waveforms over 10 successive strides for the soleus and tibialis anterior muscles. HC is 0% and TO is 60%. The solid line is the activation waveform filtered slightly (fc = 100 Hz) to remove sharp spikes but with negligible phase delay. The dashed line was filtered at 3 Hz to mimic the twitch response and yield a waveform closely following the muscle tension. See text for details.

had continued until TO there would be a large plantarflexor moment into early swing which would prevent the rapid dorsiflexion of the foot necessary for safe toe clearance [9].

As to how the CNS accounts for these muscle transfer characteristics is not known. However, modelling of this anticipatory control suggests that the CNS programs the inverse transfer function in order to cancel the low-pass filter effect of the muscle. Such modelling of the CNS anticipatory (feed-forward) control has been incorporated [10] in a swing phase model of human gait, more of which will be discussed in Section 3.3.

2.2 *Inertias of Limb Segments*

Moments of inertia of individual segments must be known in advance in order that appropriate acceleration profiles are generated. This knowledge is especially true during swing when the thigh and leg/foot segments are initially accelerated and then decelerated prior to HC. The moments-of-force cannot be too high or too low to achieve the correct trajectories, especially the foot segment which achieves a low but safe toe clearance and the heel is decelerated to near zero velocity prior to HC. The trajectory patterns during swing are extemely consistent as seen in intra subject repeat assessments done days apart or minutes apart [9]. The moment-of-force profiles associated with these intra-subject trials were also extremely consistent during swing showing that the CNS knew in advance the segment inertias and thereby generated the appropriate moments-of-force. Figures 2(a) and (b) are reproduced [9] to show these consistent swing phase kinematics and kinetics. Very few quantitative studies have documented gait patterns after the segment inertias were altered. Ralston and Lukin [11] added weights to the leg and reported increased kinetic energy changes of the leg and foot. How quickly this adaptation was made by the CNS is not known, but anecdotal evidence suggests that the adaptation is made within a few strides. FES patients have lost their sensory feedback to respond to such changes, thus a "gain" control would be necessary to mimic this adaptation when the inertial loads are changed (different footwear, bracing, etc.). One other factor that becomes evident from these consistent swing phase trajectories is the contribution of gravitational forces to the acceleration and deceleration of the swinging leg.

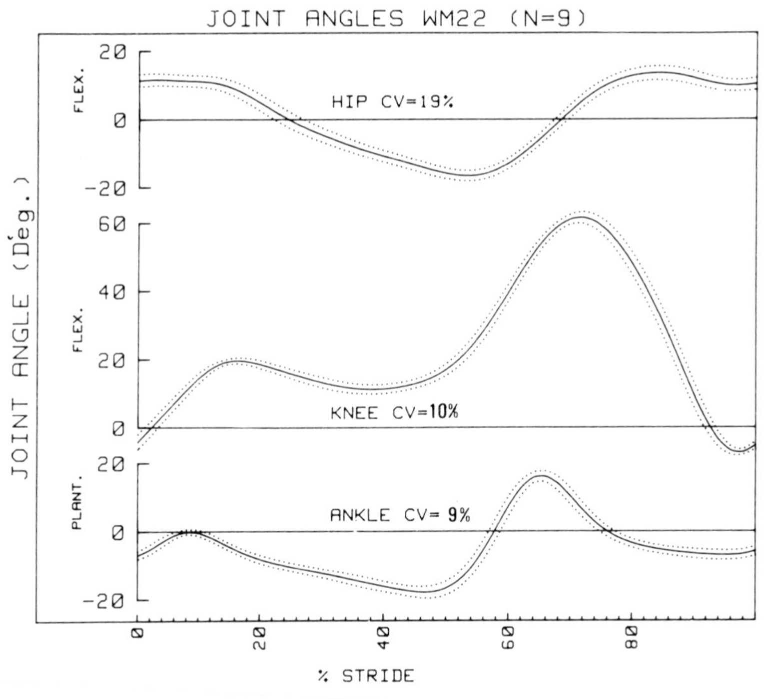

Figure 2(a) Averaged joint angle over the stride period for nine repeat trials done days apart on the same subject.

2.3 *Gravitational Forces*

During swing phase the thigh and leg/foot form a double pendulum system which is influenced at the hip and knee by gravitational moments. During early swing gravity assists the forward acceleration of the leg and foot and at the end of swing it assists in its decelerating prior to HC. Figures 3(a) and (b) are reproduced to demonstrate the importance of this gravity assist in a typical swing phase [12]. The knee's

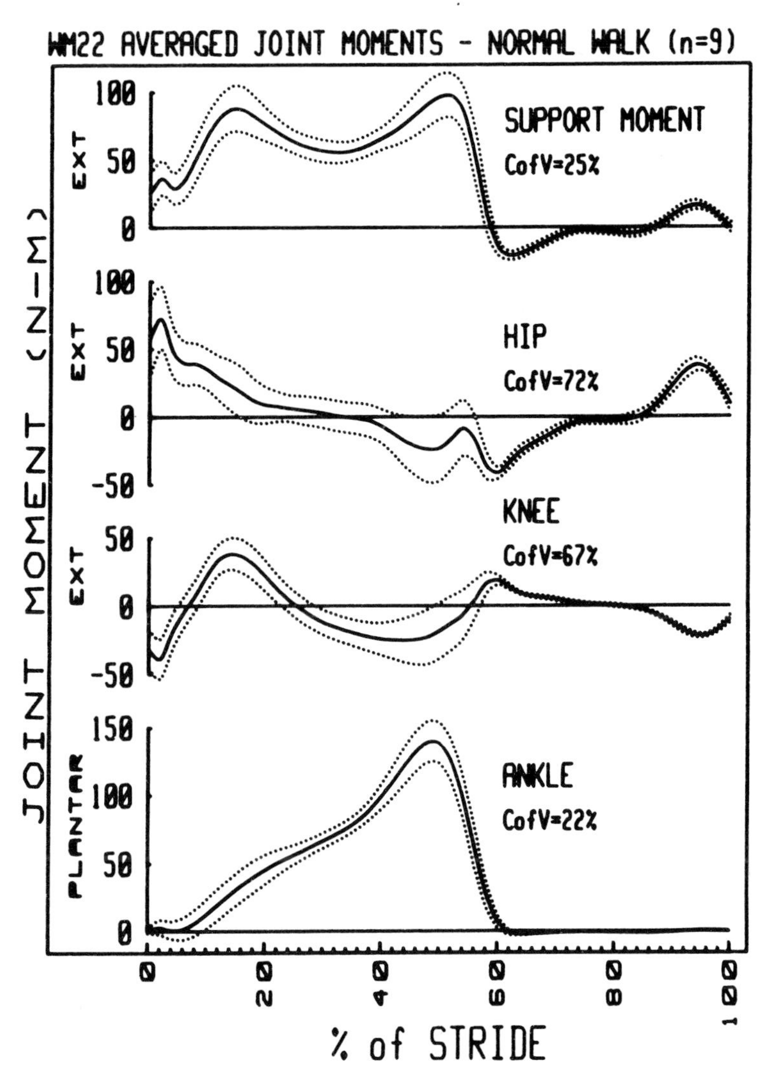

Figure 2(b) Ankle, knee and hip moments of force for the same subject reported in 2(a). Hip and knee movements are quite variable during stance but quite consistent during swing.

gravitational moment (Figure 3b) increases from TO to a maximum during early swing (at maximum knee flexion). Here gravity assists in decelerating the backward rotating leg and then is the major contributor to its forward acceleration. Then during late swing the leg is decelerated until heel contact, with the gravitational component being somewhat less but still important. The active contribution of the knee extensors was only about 20% to the leg's acceleration with gravity adding about 50% and the knee acceleration couple about 30%. However, during the deceleration in late swing, the knee flexors were responsible for about 80% of the joint moment-of-force. The CNS must recognize these non-muscular contributions and FES programs must also do so. When a subject walks more slowly the gravitational component remains about constant but the total inertia moment

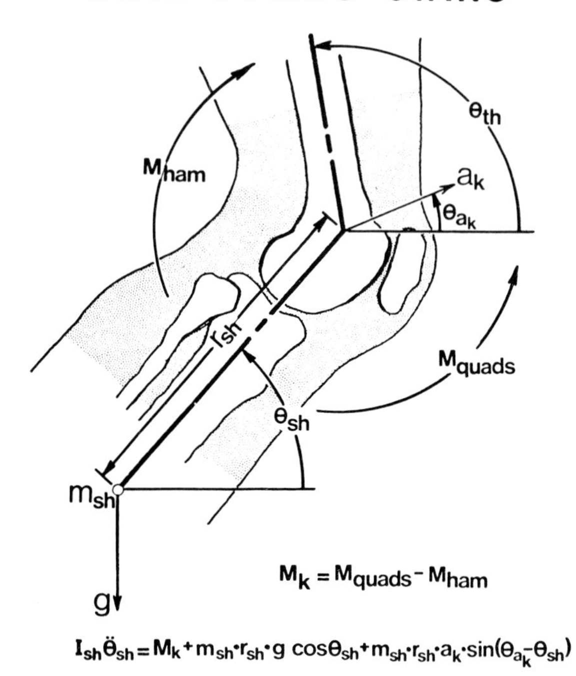

Figure 3(a) Knee kinetics during swing showing the knee acceleration, a_k, due to hip flexor/extensor activity.

decreases. Thus slow walking FES patients would require that the microprocessors reduce knee extensor stimulation during early swing to recognize the non-muscular contributions. Similarly, at the end of swing there would be less hamstring stimulation needed to decelerate the leg prior to HC. The third contributor to the leg's acceleration is inter-limb coupling which is now discussed.

2.4 *Inter-Limb Coupling*

a) Swing phase The angular acceleration of the thigh by the hip flexors/extensors causes a linear acceleration at the knee. In turn, this acceleration, a_k, results in a reaction force at the knee (Figure 3a). Thus a couple is created to assist in the leg's acceleration and deceleration (Figure 3b). As was noted previously, it was seen that this acceleration component is quite significant and is usually more important than the active muscle moment. Thus, we see evidence of active muscle activity at the hip (extensors) creating a reaction force at the knee to assist in the acceleration of the distal segments. Above-knee amputees recognize this coupling and use it to advantage to swing their prosthetic leg.

b) Control of knee flexion during stance Control of the knee joint during stance is essential for safe and efficient weight bearing. Collapse of the knee is really a collapse of all three joints, and therefore prevention of collapse of the ankle and hip will assist the quadriceps in controlling knee flexion. The total limb synergy which quantifies this three joint control is called the support moment [13], which is calculated to recognize inter-limb coupling such that above or below normal

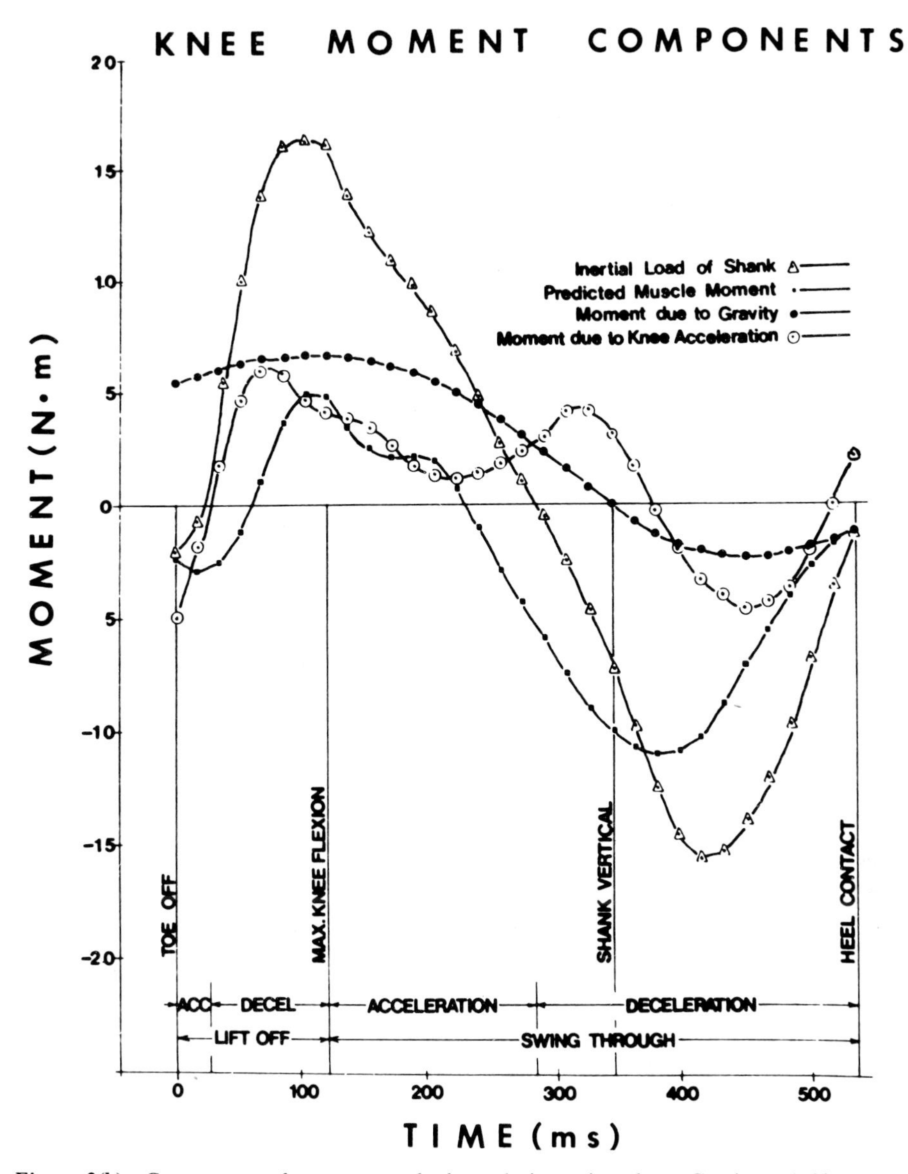

Figure 3(b) Components of moments at the knee during swing phase. Gravity and this a_k coupling from the hip muscles accounts for 80% of desired moment-of-force during early swing, but reduces to about 25% during late swing. (Reproduced with permission of *Biological Cybernetics.*)

extensor moments at the hip and ankle have a direct control of the knee joint. For example, hyperactivity of the plantar flexors during mid stance will slow down or stop the forward rotation of the leg over the foot, and thereby reduce knee flexion (or even cause the knee to hyperextend). Similarly, hyperactivity of the hip extensors early in stance will cause a backward rotation of the thigh and, again, knee flexion is reduced. Such inter-limb coupling gives greater flexibility to the CNS to accomplish the same knee control in more than one way. More will be said later about this total limb synergy and its role in walking.

2.5 *Concentric vs. Eccentric Contractions*

It is well known from muscle force–velocity curves that muscle force decreases as the velocity of shortening increases and increases as the velocity of lengthening increases. Thus, in order to generate the correct muscle force, the CNS activation must be altered depending on the lengthening or shortening velocity. Differences in force for the same level of activation can be drastically different [14, 15]. However, in slow walking this velocity effect is likely to be only of secondary importance.

3 AN OVERALL STRATEGY FOR FES CONTROL OF GAIT

In addition to these micro-level control factors that must be taken into account the FES researcher must overlay a total control strategy which integrates the activation profiles to all muscles being stimulated. This is not an easy task. However, an examination of normal human gait gives us insight into the sub-tasks which must be accomplished in order that safe and efficient progress can be made. Three sub-tasks have been identified [16] and any control strategy would be required to focus on stimulation control sequences that would independently control these tasks. The tasks are:

i) Support of the body by the total lower limb during stance (i.e., prevent collapse of the lower limb).

ii) Maintenance of upright posture and balance of H.A.T. in both the anterior–posterior and medial–lateral directions.

iii) Control of foot trajectory to achieve safe ground clearance and a gentle heel or toe landing.

The first two of these tasks are regulatory in nature while the trajectory task is cyclical. Also, support and balance are stance phase tasks involving quite high forces while foot trajectory control is primarily swing phase involving lower forces. Each of these individual tasks are necessary, but not sufficient, for safe walking. Only when all three are accomplished do we achieve a total gait pattern.

3.1 *Support Synergy*

Support of the H.A.T. by a total stance limb synergy was observed eight years ago [13] and is summarized in the support moment (Ms) profile, which is the algebraic summation of the moments at all three joints, with extensor being positive. Subsequent analysis of this pattern for slow, natural and fast walking subjects showed Ms to be positive during stance, negative during early swing and positive during late swing [17]. The variability in Ms was quite low in spite of high variability in the moment profiles at the knee and hip. Further analysis showed that the hip and knee profiles were not randomly variable but varied with respect to each other in a very deterministic way. This coupling between these motor patterns was documented several ways, possibly best by a covariance analysis, which was seen to be as high as 89% on repeat trials done days apart on the same subject. Figure 4 shows these patterns for this subject. The high hip/knee covariance results from a large cancellation between those moment patterns such that the sum of the hip + knee moment has very low variance. M_h has a mean coefficient of variation (CV) over stance of 68%, M_k has a CV of 60% while M_{k+h} has a CV of 21%. Thus

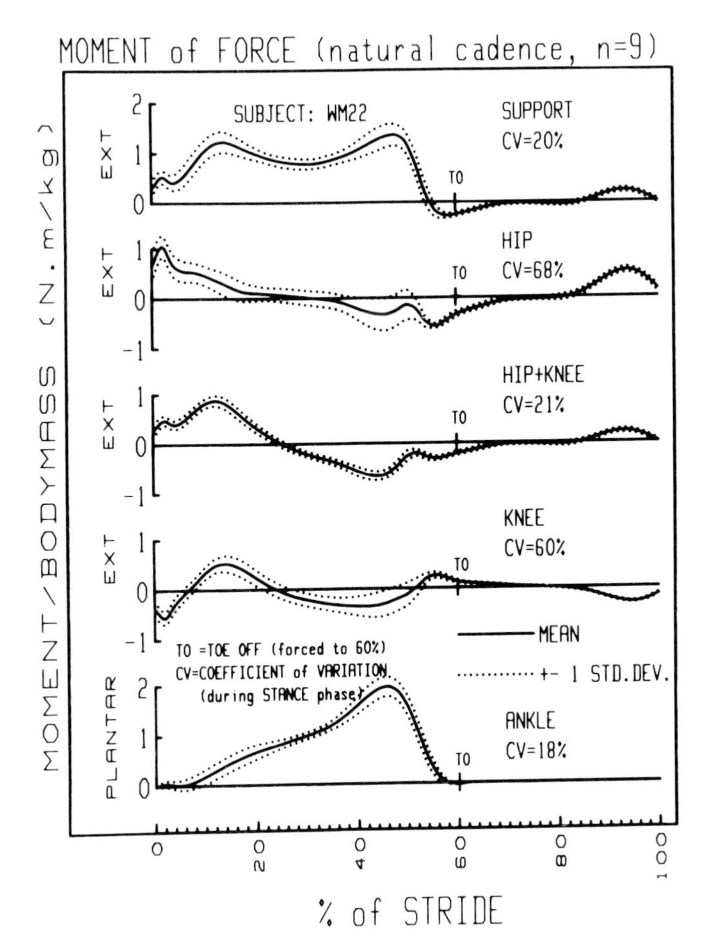

Figure 4 Ensemble averaged joint moment-of-force patterns for repeat trials done on the same subject. Support moment, which is the sum of ankle, knee and hip moments had a low variability ($CV = 20\%$) in spite of high variability at the hip and knee. A summation of hip and knee moments showed very decreased variability ($CV = 21\%$) during stance, which indicates a high degree of coupling between the hip and knee motor patterns. See text for further discussion related to support and balance sub-tasks.

the *CV* for M_s is low because M_s is the sum of $M_a + M_{k+h}$, each of which have low *CV*s. This total support synergy is partially due to the anatomical coupling of the biarticulate hamstrings and rectus femoris muscles. However, these biarticulate muscles represent only 33% of the physiological cross-section area of muscles crossing these two joints [18], thus most of the coupling must be neurological in origin. The reason for the low variability in M_{h+k} is an almost one-for-one trade-off between the moments at the hip and knee. Figure 5 presents the average hip and knee moments during stance for these 9 repeat trials, with positive moments being extensor. Thus strides in the second quadrant are controlled by hip extensors and knee flexors, while in quadrant four the control is knee extensors and hip flexors. Thus the trade-off is in the anterior/posterior direction and means that knee collapse can be equally well controlled by extensor moments at the hip as well as at the knee.

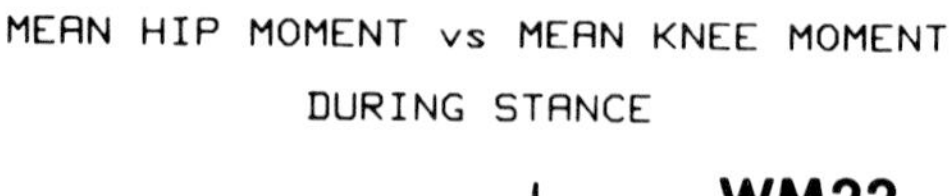

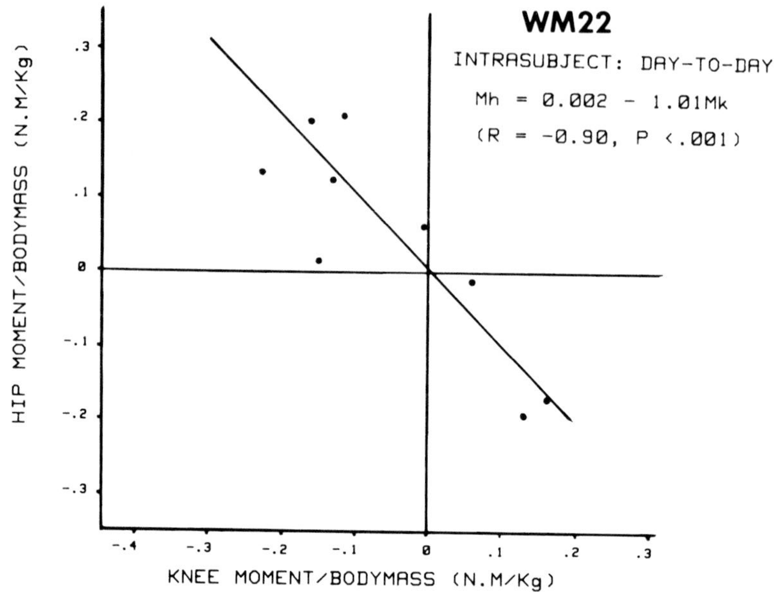

Figure 5 Average hip vs average knee moment during stance for the nine strides in Figure 4. The slope of the regression is -1 indicating a one-for-one trade-off for each stride and that the trade-off is in the anterior/posterior direction. See text for implications regarding independence of balance and support synergies.

3.2 *Balance Control*

The only reason why the CNS would allow these anterior/posterior changes on a stride-to-stride basis would be dynamic control of the balance of H.A.T. This has been shown to be true [16]. Thus on a stride-to-stride basis there is an anterior/posterior shift in the moment profiles at the hip and knee to control the angular acceleration of H.A.T. On one stride the posterior moments increase to accelerate H.A.T.'s position in the backwards direction, on a subsequent stride H.A.T. is accelerated forward by hip and knee moments biased in a forward direction. In the process M_{k+h} is kept almost completely constant, thus the support pattern remains essentially unchanged. The concept of two independent control systems is quite attractive, but with priority being given to balance control. Thus hip flexors and extensors would be controlled to maintain balance and posture of H.A.T. with the knee extensors or flexors adding an additional moment to control knee flexion. Therefore, the hip and knee flexor and extensor muscles have two primary tasks during stance and at any given point in time any one of these muscles may be equally involved in controlling both balance and support.

3.3 *Control of Foot Strategy*

The third and final task relates to the forward trajectory of the lower limbs, which starts at push-off and continues through swing until heel contact. The task is one of end-point control of the foot, requiring low but safe toe clearance and a gentle heel contact. Such control has been documented [9] and shows that toe clearance averages less than 1 cm as the foot reaches a forward velocity of 4–5 m/s. The heel velocity immediately prior to heel contact in both the vertical and horizontal directions was seen to be almost zero. This is a fine control task and is the integrated result of scores of muscles across a seven segment chain (from stance foot up, across pelvis and down to the swing foot) with at least 12 degrees of freedom that can influence end-point control. The moment-of-force patterns at all three joints of the swing leg are low but extremely consistent (see Figure 2b).

Classical feedback control is inadequate to achieve consistent and fine end-point control of the foot. It would be impossible for feedback sensors to respond using a motor system with so many inherent delays: electro-mechanical delays of the muscle (as indicated by the twitch waveform) and inertias of the lower limb. Thus the FES controller must incorporate a feedforward or anticipatory control [19] to take into account these delays, as discussed previously. The modelling of feedforward in a linear system requires that the inverse transfer function of the system be introduced at the controller input [10]. Thus the electromechanical delay of the muscle of the form $1/(\tau s+1)^2$ is cancelled out by a front-end controller with a transfer function $(\tau s+1)^2$. Also, an "inverse inertia" component, Js^2, cancels out the lag due to segment inertias.

4 CONCLUSIONS

Computer control of gait by means of FES of spinal cord injury patients is an extremely difficult problem. Not only is the researcher dealing with a complex multisegment system with scores of muscles but also the microprocessor is lacking any feedback from the periphery regarding the state of the joints and the muscles. Also, the microprocessor must replace the CNS and, as such, must take into account the delay characteristics of muscles, stimulation, the inertias of the segments, and inter-limb coupling, plus the influence of gravity. Overlaid on this information about the "plant" and its environment is the need for a total strategy. Based on a detailed examination of motor patterns in normal gait during slow, natural and fast cadences three fairly independent sub-tasks were identified, each of which must be satisfied to achieve safe and efficient locomotion. Two of the tasks are regulatory and are primarily stance phase: support against gravity and dynamic balance of the upper body. The final task begins late in stance and continues throughout swing: trajectory of the foot to achieve safe but minimal toe clearance plus a gentle foot landing. This latter motor control requires a strong anticipatory or feedforward component to account for the delays of the neuromuscular system. Within these three tasks priorities must be established to mimic what the CNS does: priority is given to dynamic balance, followed by support followed by swing trajectory.

The above compensations and strategies cannot be achieved with the current "on-off" type of stimulation patterns [2]. Graded analog contractions are required to overcome the extremely jerky type of movement that is now achieved. Because of the sharp and variable non-linearities in the stimulation–force curves more research is required into the factors that influence that relationship. Only through a graded turn-on and turn-off of each muscle can properly integrated motor strategies be implemented.

ACKNOWLEDGEMENTS

The author acknowledges the financial support of the Medical Research Council of Canada (Grant MT4343) and the National Science and Engineering Research Council (Grant 2917), and the technical assistance of Mr. Paul Guy.

REFERENCES

1. P. H. Peckham, Functional neuromuscular stimulation, *Phys. Technol.*, **12,** 114–121 (1981).
2. E. B. Marsolais and R. Kobetic, Functional electrical stimulation for walking in paraplegia, *J. Bone Jt. Surg.*, **69A,** 728–733 (1987).
3. H. S. Milner-Brown, R. B. Stein and R. Yemm, The contractile properties of human motor units during voluntary isometric contraction, *J. Physiol.*, **228,** 285–306 (1973).
4. G. L. Gottlieb and G. C. Agarwal, Dynamic relationship between isometric muscle tension and the electromyogram in man, *J. Appl. Physiol.*, **30,** 345–351 (1971).
4. D. A. Winter, Biomechanical model relating EMG to changing isometric tension, *Digest 11th Internat. Conf. on Med. Biol. Engng.*, Ottawa, August (1976).
6. P. A. Crosby, Use of surface electromyogram as a measure of dynamic force in human limb muscles, *Med. Biol. Eng. and Comput.*, **16,** 519–524 (1978).
7. S. J. Olney and D. A. Winter, Predictions of knee and ankle moments of force in walking from EMG and kinematic data, *J. Biomech.*, **18,** 9–20 (1985).
8. V. T. Inman, H. J. Ralston, J. B. Saunders, B. Feinstein and E. W. Wright, Relation of human electromyogram to muscular tension, *EEG and Clin. Neurophysiol.*, **4,** 187–194 (1952).
9. D. A. Winter, *The Biomechanics and Motor Control of Human Gait*, University of Waterloo Press, Waterloo, Canada (1987).
10. B. McFadyen, Anticipatory control for toe clearance during obstructed walking, Ph.D. thesis, University of Waterloo (1988).
11. H. J. Ralston and L. Lukin, Energy levels of human body segments during level walking, *Ergonomics*, **12,** 39–46 (1969).
12. D. A. Winter and D. G. E. Robertson, Joint torque and energy patterns in normal gait, *Biol. Cybernetics*, **29,** 137–142 (1978).
13. D. A. Winter, Overall principle of lower limb support during stance phase of gait, *J. Biomech.*, **13,** 923–927 (1980).
14. P. V. Komi, Relationship between muscle tension, EMG and velocity of contraction under concentric and eccentric work, in *New Developments in EMG and Clinical Neurophysiol.*, Karger, Basel (1973), Vol. 1, pp. 596–606.
15. R. Triolo, D. Robinson, E. Gardner and R. Betz, The eccentric strength of electrically stimulated paralyzed muscle, in *Proc. 9th Conf. of IEEE Engineering in Med. and Biol. Soc.*, Boston, November (1987).
16. D. A. Winter, Balance and posture in human walking, *Engineering in Med. and Biol.*, **6,** 8–11 (1987).
17. D. A. Winter, Kinematic and kinetic patterns in human gait: Variability and compensating effects, *Human Movement Science*, **3,** 51–76 (1984).
18. T. L. Wickiewicz, R. R. Roy, P. L. Powell, J. J. Perrine and V. G. Edgerton. Muscle architecture of the human lower limb, *Clin. Orthop. and Related Research*, **179,** 275–283 (1982).
19. P. D. Neilson, M. D. Neilson and N. J. O'Dwyer, Internal models and intermittency: A theoretical account of human tracking behavior, *Biol. Cybernetics*, **58,** 101–112 (1988).

Automedica, 1989, Vol. 11, pp. 175–199
Reprints available directly from the publisher only
Photocopying permitted by license only

RULE-BASED CONTROL OF A HYBRID FES ORTHOSIS FOR ASSISTING PARAPLEGIC LOCOMOTION

B. J. ANDREWS, R. W. BARNETT, G. F. PHILLIPS and C. A. KIRKWOOD
The Bioengineering Unit, University of Strathclyde, Glasgow, Scotland, UK

and

N. DONALDSON, D. N. RUSHTON and T. A. PERKINS
Medical Research Council, Neurological Prostheses Unit, London, UK

(Received September 1988)

The present stage of development of a laboratory prototype hybrid FES Orthosis is described. The system comprises a mechanically passive supracondylar knee ankle foot brace and a FES control system. The brace features protective constraint of the knee, ankle and foot joints. The combined or hybrid system was designed to enable the paraplegic patient to be able to take occasional "rest breaks" whilst standing between periods of upright activity to recover from the effects of physical exertion and FES induced muscle fatigue.

The hierarchical, finite state control system was developed using expert system techniques. Knowledge was represented by hand crafted production rules. We here indicate how artificial intelligence (AI) rule induction methods may be used in future versions of the control system.

Some preliminary results of our initial laboratory experience with the hybrid system, involving three spinal cord damaged volunteer subjects, are presented. Paraplegic subjects (A) and (B) were fitted bilaterally with surface electrode and implanted electrode versions respectively. Subject (C) who had sustained a C6/7 incomplete lesion was fitted unilaterally with a surface electrode version. In each case, quiet standing in parallel bars was repeatedly performed for periods in excess of one hour and was not found to be limited by FES induced muscle fatigue. Subjects (B) and (C), who had previously been trained to walk using purely FES systems, were able to use the hybrid system to extend their range of level ground walking without degrading the quality of their previously achieved FES gait patterns.

Keywords: Functional electrical stimulation, orthotics, spinal cord injury, expert system, feedback control.

1 BACKGROUND

Functional electrical stimulation (FES) has been demonstrated to be feasible in assisting paraplegic individuals to stand, transfer and undertake simple forms of walking [1–5]. For many paraplegics FES induced muscle fatigue and withdrawal reflex habituation are two of the factors that limit control and endurance of standing and ambulatory activities.

During FES walking paraplegics expend energy at a relatively high rate, typically 0.95 kcal/kg/min for cadence in the range 0.1–0.55 m/s [6]. This energy is expended by both the above lesion musculature of the trunk and upper limbs and the below lesion electrically activated musculature. For an able-bodied individual, such a rate of energy expenditure is approximately equivalent to running a mile in 14 minutes. Such activity can be sustained only for short periods without rest. To be independent, such rest breaks should be taken whilst standing, otherwise the patient would have to rely on a suitable chair being available, or else have to return to his wheelchair. Ideally, the rest break should enable the above lesion musculature to recover from its previous exertions and the below lesion, FES activated, musculature and withdrawal reflexes to recover from induced fatigue or habituation. Presently used clinical FES standing systems involve continued activation of the lower limb extensors resulting in rapid muscle fatigue. This may be further compounded in surface electrode applications to patients with sensory incomplete lesions who may, for comfort, prefer a higher rate of stimulation [7, 8].

1.1 *Hybrid FES Orthoses*

Tomovic, Vukobratovic and Vodovnik in 1972 [9] suggested how the three technologies FES, exoskeletal bracing and externally powered actuators may be usefully combined in assistive devices for the severely disabled. Applications have been reported, in which mechanical bracing components have been combined with FES to passively stabilise the lower limbs during body weight supporting postures, to circumvent the need for prolonged FES activation of the extensor musculature and thereby circumvent muscle fatigue. In these systems FES is applied in short bursts to provide for stabilization or assist with forward progression. A number of such combinations have been reported [10–14], some including externally powered components [15, 16]. We have adopted the term "Hybrid Orthoses" for such combinations following the original definition of Tomovic *et al.* [9].

The use of external mechanical components potentially offers a number of advantages in the design of practical FES based locomotion aids. For example, because of impaired protective pain mechanisms, some musculoskeletal structures become susceptible to injury or progressive deformity by excessive or prolonged force actions. Bracing may be used to reliably shunt such force actions away from delicate structures. Bracing may also incorporate mechanisms that limit degrees of freedom to simplify the control problem and regularise movements. The exoskeletal brace may include passive or externally powered locking mechanisms or actuators to supplement voluntary or FES induced movements [16] particularly where partial denervation is a limiting factor. The addition of powered brakes or dampers to the exoskeletal joints [15] may be used to improve motion control by regulating the underlying FES induced movements. Finally, the brace structures may be conveniently instrumented to serve as a sensor system as well as providing a convenient mounting structure for additional sensors, processing electronics and batteries without degrading donning and doffing times.

We report here the combined application of a knee stabilising brace and FES to spinal cord damaged individuals with extensive upper motor neurone paralysis of the lower limbs. The mechanical brace component of the hybrid orthosis combines the knee stabilising action of the floor reaction ankle foot brace [17] with the joint protection of the supracondylar knee brace [18].

1.2 *Floor Reaction Orthoses*

The stabilising effect on the knee, due to an anteriorly directed ground reaction force (GRF) when ankle dorsiflexion is restricted, has been used to clinical advantage in facilitating weightbearing with a minimum of quadriceps activity. Restricted dorsiflexion may be naturally due to an equinus contracture as a result of muscle shortening or spasticity in the ankle plantarflexors. Mechanically imposed restriction of ankle dorsiflexion has also been used clinically to stabilise the leg in cases of quadriceps weakness due to various neurological disorders such as poliomyelitis and incomplete spinal cord injury [18, 5, 17, 19]. In these applications, dorsiflexing movement was restricted either by surgical arthrodesis, ligament transfer or external bracing. The bracing applications were usually restricted to unilateral fittings. The patient must have sufficient voluntary strength and control of the affected leg to prevent knee buckling should the GRF shift behind the knee joint axis and to flex and extend the leg to make a forward step. Semi-rigid calf length boots or modern "high top" type sports shoes may also be used with similar but generally less effective stabilising effects. Indeed, it is our opinion, that restricted dorsiflexion due to natural causes or footwear/ankle braces may explain, in part, the wide variations in the maximum duration of FES assisted standing observed in paraplegic patients with comparable lesions and muscle properties [1].

1.3 *Rule-based Control*

The control system architecture adopted in the present system is based on the finite state approach to the control of prosthetic and orthotic systems [9, 20–25]. There are a number of possible modes of locomotion for paraplegic individuals using various combinations of mechanical walking aids such as crutches or rollators and leg braces. The preferred mode will depend upon a number of factors including: degree of functional disability; purpose; environmental situation.

The control of an assistive FES system is a problem in man–machine interaction or cybernetics. The patient has intact control of his body above the lesion; the machine effects some control to that part below the lesion. In complete spinal cord lesions, the patient can directly influence balance and posture by means of his preserved voluntary and reflex responses through his upper limbs and trunk musculature. In the case of incomplete lesions, the influence is extended through preserved lower limb motor control and sensory pathways to a degree dependent on the particular pathways preserved. The latter can be highly variable, almost individual, and enables the patient to learn to adapt, by means of compensatory movements, his preserved neuromusculoskeletal system to elevate his locomotor status. The entire system will therefore be controlled by both the patient's voluntary and reflexive upper body control and the robot-like sensor driven FES movements. The overall quality of control will depend on how well the former adapts to the latter.

Our approach to the synthesis of a controller was to decompose the multi-goal and multivariable control problem into subsystems organised in hierarchical levels [21] as illustrated in Figure 1. At the top level, the subject directly interacts with the control system through a command interface comprising manual switches and associated software. For example, in the present system the patient selects a particular locomotion mode using a multi-way switch. Each locomotion mode is controlled with reference to a finite state model of the process. This model serves to change the control strategy, as required, at different stages or phases of the locomotion cycle. Each stage in the gait cycle is represented in the model by a unique state. On entry into a state, an associated predetermined control strategy comprising a set of low level open- or closed-loop actuator controllers are enabled. This new state control strategy then supercedes the previously active state control strategy.

It is at the lowest level, the actuator control level, that directly interfaces with the multichannel FES stimulator and for some of the actuator control loops sensor feedback may be required. These actuator control loops have been based on

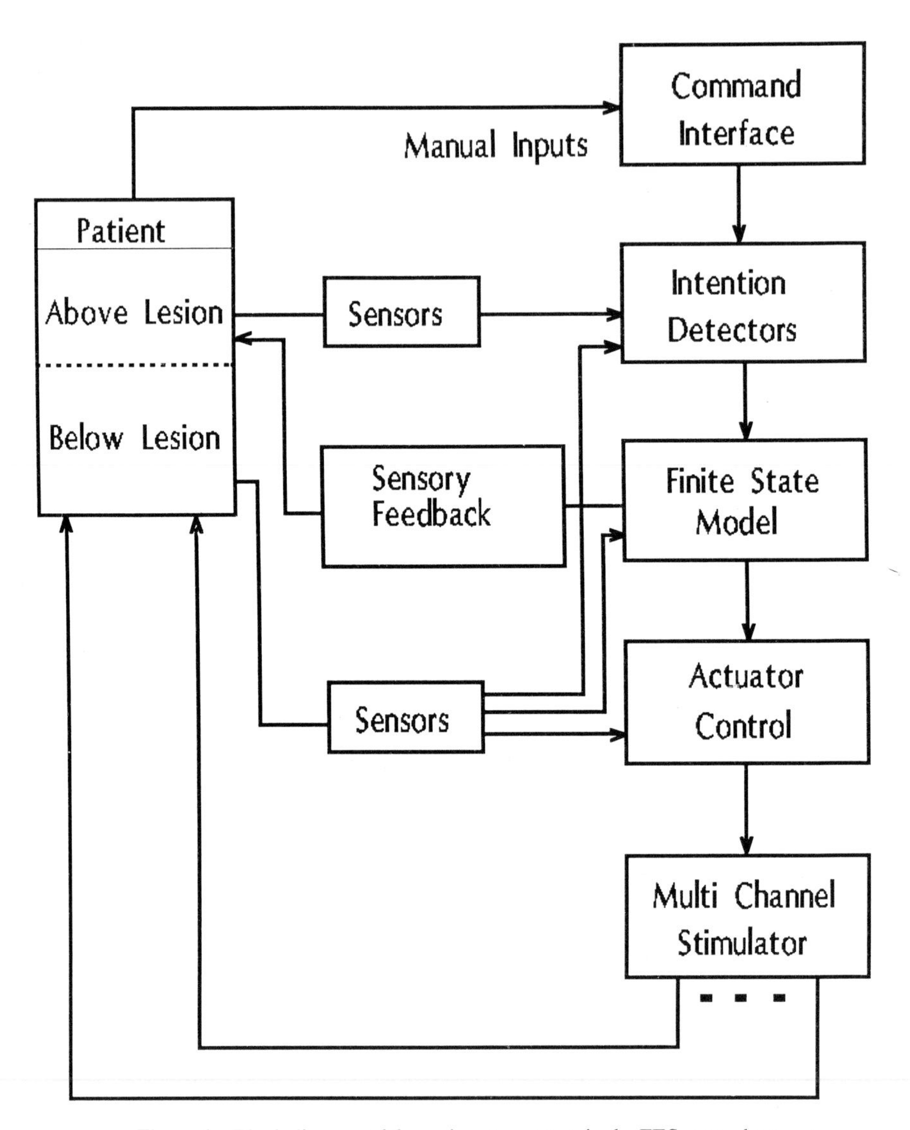

Figure 1 Block diagram of the main components in the FES control system.

artificial reflexes [22, 25–27] and more traditional control systems techniques [28]. Sensory feedback to the patient is provided by a suitable display to a sensate region above the level of the sensory lesion. We are presently exploring the value of a simple electrocutaneous feedback modality to assist the patient avoid standing postures that may cause the FES system to trip on in response to incipient knee buckling [11]. Sensors play a vital role in the control system. They are used in both the above and below lesion parts of the patient–machine system. These sensors are mainly used in two ways as indicated in Figure 1. Firstly, sensor signals are used for direct feedback at the actuator control level. Secondly, sensor data are input to rule-based pattern recognition algorithms that detect events in the gait cycle that are then used to update the finite state reference model of the locomotor task. Some of these events predict the patient's locomotor intentions by detecting voluntary preparatory movements that are uniquely associated with the intention.

The finite state model must change state, in real time, on the occurrence of key gait events that signal the need to change control strategy. The next state that the model can assume will be one of a permitted set. A transition will only occur when the first of a set of expected state transition events is detected.

Such state transition events are detected by rule-based pattern recognition algorithms operating, in real time, on the sensor signals. The above event and intention detectors embody detailed knowledge of the gait patterns leading up to these events. At present, such event detecting knowledge is represented by handcrafted If ... Then ... rules.

1.4 *Rule Induction from Examples*

An interesting, and so far unexplored, alternative to handcrafting event detection rules is to use rule induction methods, recently developed for machine learning in artificial intelligence, to automatically learn the rules from previously recorded sensor data. Research in the field of expert systems has indicated the superiority of computer induction versus handcrafting of rules [29].

To use induction algorithms such as ID3 [30], a collection of representative examples of the events to be detected or classified must be prepared. It is from this collection that the induction algorithm will learn-by-example and produce the required decision rules for the event detectors. In each example, the occurrence (instance) of a particular event (class) must be known *a priori*. For each training example of a known event a set of attributes that one assumes are related to that instance must be devised. Typically in the present application, these attributes are derived from some or all of the A-D converted output values of the sensor set sampled up to and including the time of occurrence of the event. Specifically, the steps in the process of inductive learning of decision rules is outlined by Quinlan [30, 31] as follows:

1) A collection of examples, termed the training set, is formed.

2) Each example in the training set is described in terms of a fixed number of attributes.

3) Each instance in the training set has an associated class value.

4) The algorithm seeks to characterize each class value in terms of its attribute values.

ID3 is an algorithm developed by Quinlan to perform this task. ID3 operates on the information theory principle of calculating the entropy of classification for each value of attribute.

A classification tree is then built which splits on the attribute with the most information (lowest entropy). The tree can then be used to produce a set of IF ... THEN ... rules which do not contain redundant attributes.

For our preliminary work in rule derivation from a collection of examples we have used the program ASSISTANT. This is an induction program developed at the Josef Stefan Institute, Ljubljana, Yugoslavia which allows the automatic learning of decision rules [32]. ASSISTANT is a member of the family of Top Down Induction of Decision Tree family and is based on the ID3 algorithm as follows:

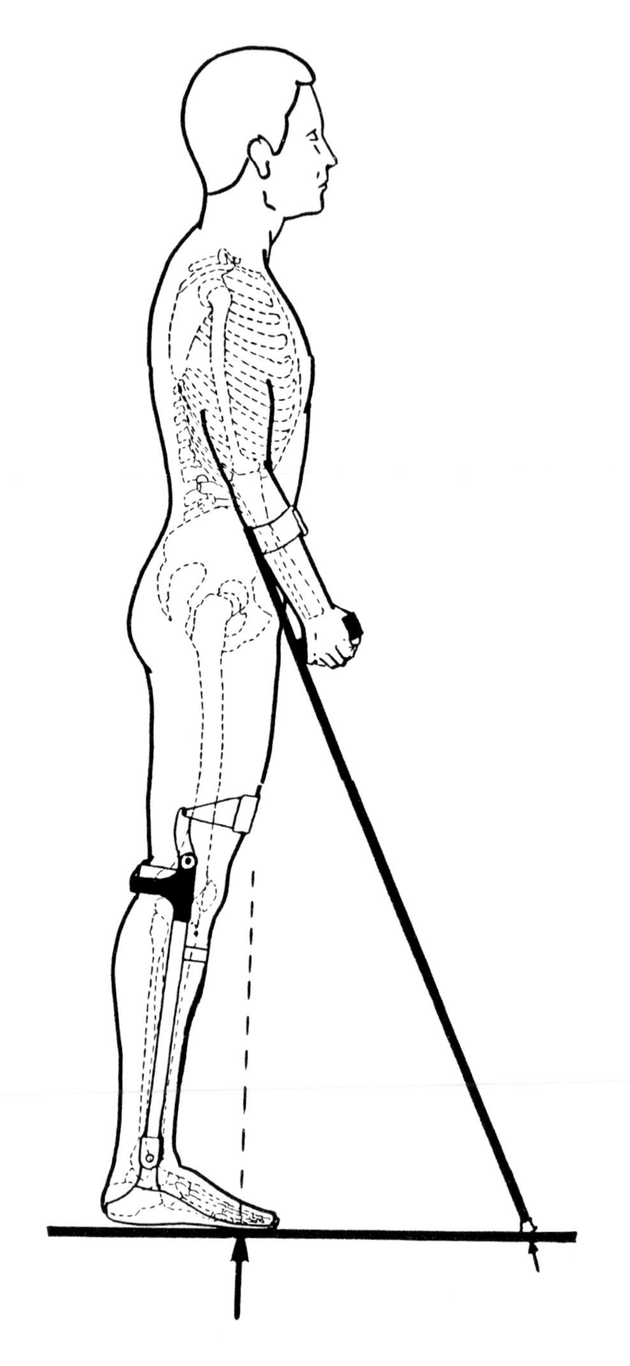

Figure 2 Schematic diagram of the "C" posture adopted during standing rest breaks. Also shown are the foot and crutch ground reaction forces.

IF (all the learning examples belong to a single class)

THEN (terminate with a leaf labelled with that class) ELSE

BEGIN
on the basis of the learning set choose the most informative attribute (using the entropy measure) for the root of the tree and partition the training set into subsets according to the value of the selected attribute;

FOR each value of the attribute DO recursively construct a subtree with the corresponding subset of examples.

END

ASSISTANT improves upon this basic algorithm by allowing for both continuous attributes (integer values) and providing for incomplete domains with Bayesian probabilities, with the inclusion of a NULL leaf in the decision tree.

We present here our preliminary results in applying rule induction to a simplified related experiment, that of classifying five different quasi-static postures of a normal subject from data collected using foot load sensing insoles.

2 THE HYBRID SKAFO ORTHOSIS

2.1 *The Brace Component*

The general design objectives for this orthosis were that: (i) It should enable rest breaks to be taken whilst standing quietly as indicated in Figure 2. This rest posture is often referred to as the "C" posture in which the patient leans slightly forward with the hips and knees fully extended. (ii) The mechanical brace should shunt potentially damaging force actions that may be applied to delicate joint structures and ligaments of the knee, ankle and foot. (iii) The mechanical brace component should be unobtrusive and should not be the limiting factor, when combined with FES, in achieving an acceptable level of static and dynamic cosmesis for level ground walking.

The lightweight but rigidly constructed prototype brace is shown in Figure 3. A single, laterally positioned, double action ankle joint was used having an adjustable dorsiflexion stop with the option of free or coil spring resisted plantarflexion. Typically, the ankle joint is stopped between 0 and 5 degrees of dorsiflexion. The rigid footplate inserts into standard footwear and extends to the forefoot region. The knee joints are free to flex but stopped in extension. The anterior supra-patellar and the posterior strap are adjusted so that the knee joint does not hyperextend. Prototype versions of the brace have been fabricated using carbon fibre composite plastic, thermoformed plastic or more conventional metallic components. The latter is illustrated in Figure 3b and c.

Whilst standing quietly, the paraplegic patient was trained to adopt the 'C' posture. Vertical stabilization was achieved using preserved muscle actions of the trunk and upper limbs. The hips were stabilised in full hyperextension against the ilio-femoral ligament. In this posture the GRF is applied in the metatarsal region and is directed in front of the knee joint and behind the hip joint. In this posture, the leg is mechanically stable without muscle activation. The dorsiflexion stopped ankle joint and rigid footplate also provide a measure of vertical stabilization. Any tendency to lean forward will shift the region of support ahead of the vertical projection of the centre of gravity, producing a moment tending to restore verticality.

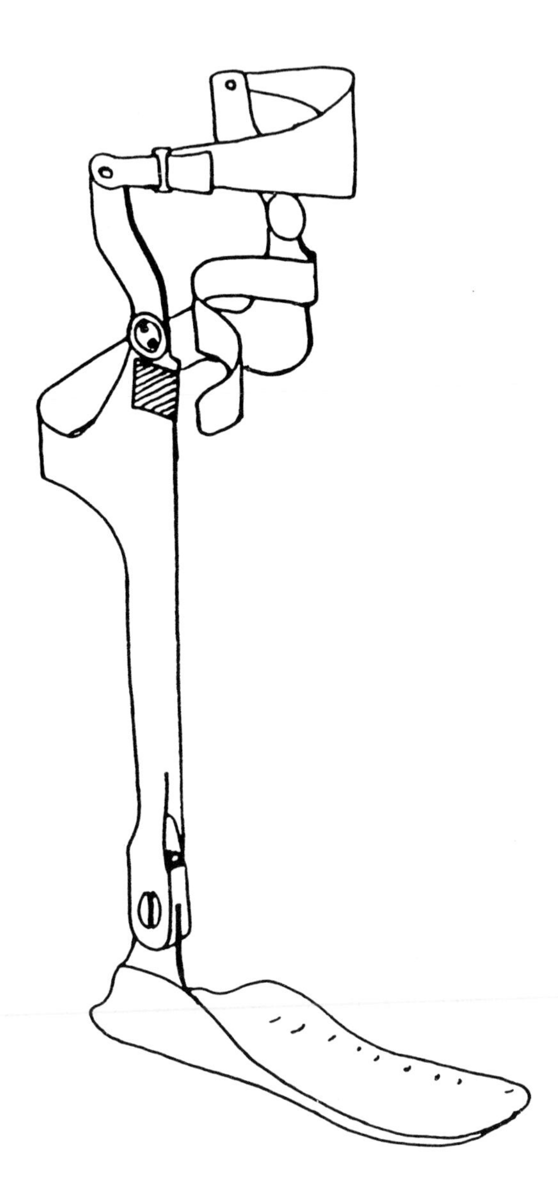

Figure 3(a) Schematic of the basic brace showing the limb contacting components.

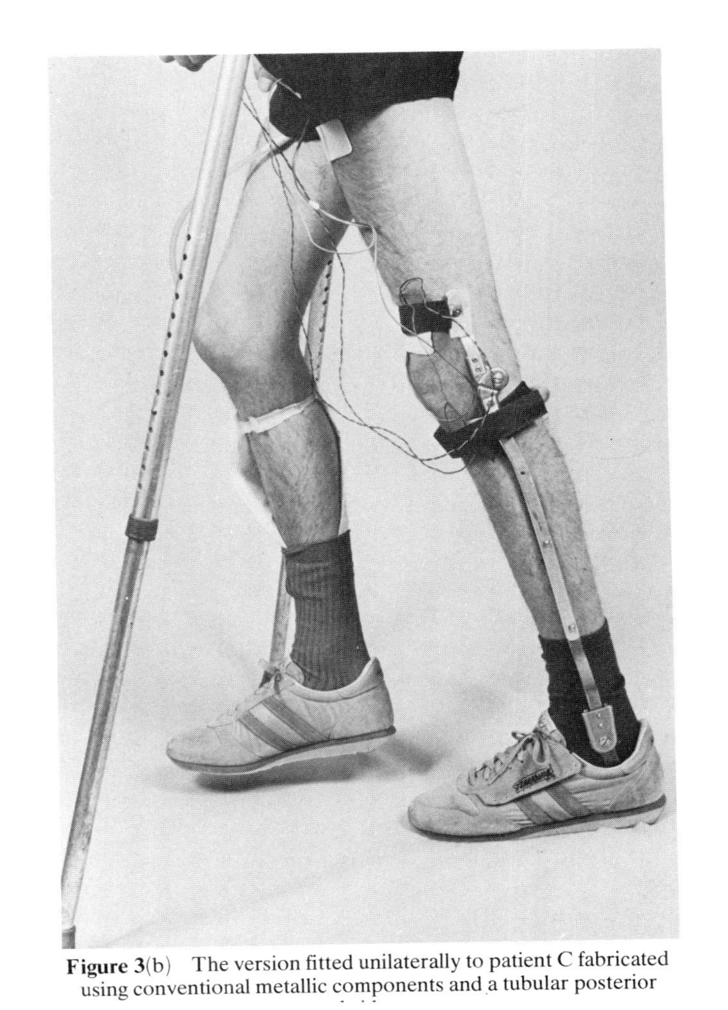

Figure 3(b) The version fitted unilaterally to patient C fabricated using conventional metallic components and a tubular posterior

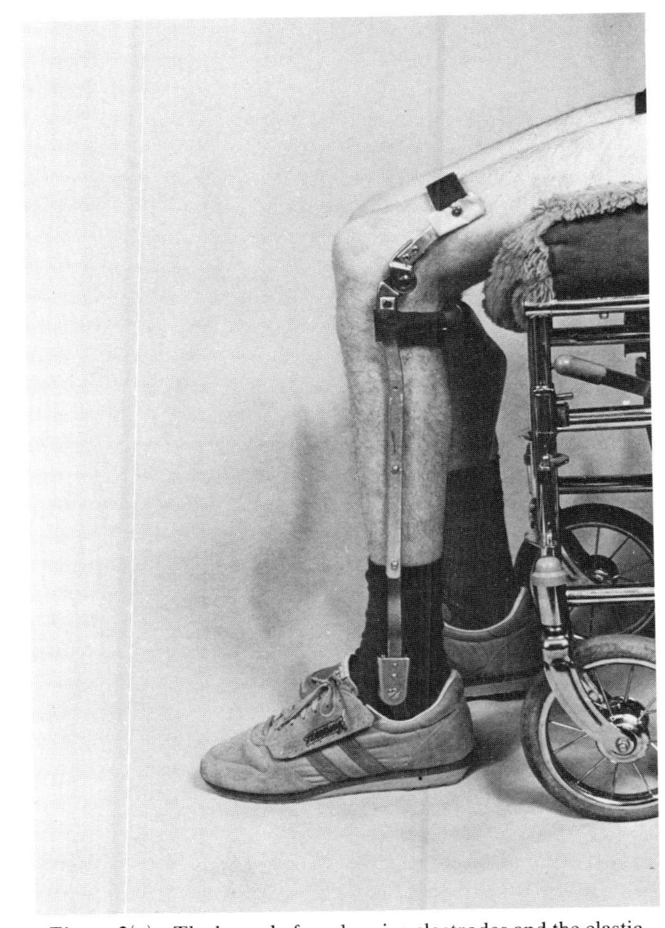

Figure 3(c) The brace before donning electrodes and the elastic patellar band.

If the GRF should shift behind the axis of rotation of the knee, then the knee would buckle. In response to such an event, the FES control system should immediately activate the knee extensors. In order to respond the control system must sense the buckling action. The supra-patellar and posterior straps are adjusted to prevent potentially harmful hyper-extension of the knee. Tension in the supra-patellar strap indicates the presence of resultant knee extending moment and if the knee flexes this tension will be zero. In the present experiments, the lateral upright of the brace, above the knee, was instrumented using strain gauges to sense anterior bending when the brace knee joint was fully extended and ankle joint stopped. This provided a voltage signal proportional to the tension in the supra-patellar strap. If the subject leans slightly forward, the centre of pressure will shift to the metatarsal region and the knee extending moment exerted by the GRF will be at a maximum. A zero signal was registered whenever the brace joint flexed indicating that the resultant knee moment was zero or had the sense of flexion.

2.2 *Electrical Stimulation*

The present hybrid system has been implemented using both surface and implanted electrodes. For the former electrical stimulation was delivered through self adhesive surface electrodes (Pals Plus, Axelgaard Manufacturing Co., Ltd., USA). Frequency and pulse width modulated monophasic and rectangular pulse trains were used (pulse width in the range 0.2–0.5 ms). Each stimulus pulse was current regulated in the range 50–120 mA. The 24 channel implanted stimulator used was developed by the MRC Neurological Prostheses Unit and has been described elsewhere [33]. Each output is a monopolar cathode, with a common anode on the body of the device. Although the pulse width is adjustable, 200 microseconds was used in this work while the amplitude was variable. In the control program, the amplitude is set on a scale of 0 to 255 and this modulates the transmitter voltage. The implanted receiver was designed to operate at critical coupling into a 500 Ohm load, via the multiplexer, and is therefore neither constant voltage nor constant current. For each electrode, the stimulus current will increase linearly with the set amplitude, unless the terminal voltage tries to exceed 12, where it is clamped. The electrodes are connected to the stimulator by Cooper cable [34]. The electrodes used on patient (B) were of two types: either 'nerve traps', in which the nerve is enclosed in a silicone rubber tube with the cathode wires; or "trocar" electrodes, which are made from Cooper cable ending with a few turns of Pt–Ir wire wound helically over the surface. After inserting these into the vicinity of the motor point, they are retained by barbs, also of Pt–Ir. Traps were used on the femoral nerve, for the quadriceps muscles, and on the lateral popliteal nerve to stimulate the ankle dorsiflexors, and to provide reflex hip flexion. Trocar electrodes were placed over the gluteus maximus, gluteus medius and the rectus abdominis muscles. Thus twelve electrodes were originally connected. After 2.5 years, before the present work on the hybrid system began, the barbs on the gluteal electrodes fractured, causing unattainably high thresholds for these muscles. Subsequently, surface electrodes were used, in conjunction with the implanted stimulator, for hip extension.

For our initial experiments the stimulus pattern delivered to quadriceps was predetermined and based on the following considerations. The stimulation is applied in response to potentially destabilizing events such as knee buckling or jackknifing at the hips. In both situations it is intended to quickly stop further joint flexion and then to fully extend and hold the joints without inducing rapid muscle fatigue until the patient restores the stable posture. We choose, empirically, a pattern that started with an initial high frequency burst (typically 80 Hz for 0.1 s) followed by a fixed low frequency pattern (typically 20 Hz) until the stimulation was again switched off. The computer implementation has been described in [11, 21, 33]. A typical physical setup is illustrated in Figure 4 showing patient C in state s2. The computers and stimulator hardware are mounted onto a trolley that is pushed behind the patient. The interconnecting cables are supported by an overhead gantry.

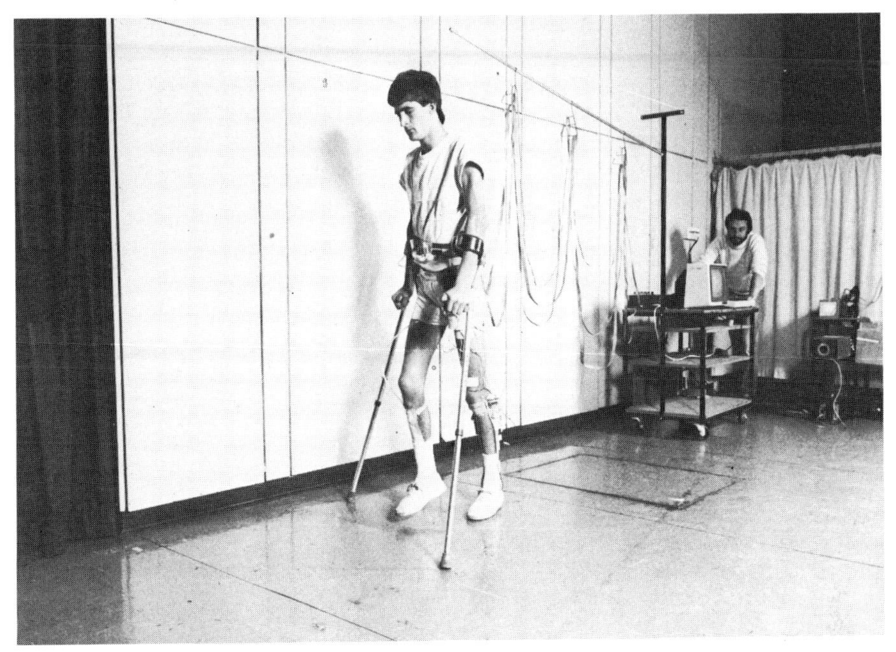

Figure 4 Typical physical set up for laboratory tests showing patient (C) in state s2 of a 4-point gait cycle.

3 PROLONGED STANDING

An appropriate knee stabilising scheme during stance would be to activate the quadriceps only in the absence of a significant resultant knee extending moment. In this way the quadriceps activity would be minimised during stance. The resultant knee extending moment comprises the sum of the moment due to the GRF and that due to the FES activation of quadriceps. In the experiments reported here a preset stimulus was applied to the quadriceps according to the following control rules. If knee stabilization is required then:

If (quads are switched off and $Vs < T_1$)
Then (switch on stimulation), rule (a)

If (quads are switched on and $Vs > Q + T_2$)
Then (switch off stimulation), rule (b)

where Vs is the strain gauge signal and T_1 and T_2 are adjustable threshold values. Q is the value of the maximum strain gauge signal produced by the quadriceps during an initial calibration procedure. Thus FES is switched on and off as the corresponding rule fires. The thresholds were introduced to switch on the quadriceps stimulation just before the knee buckles and to introduce hysteresis.

In our experiments, Q was determined for each leg during an initial calibration procedure, in which the subject was seated with the brace donned. The quadriceps were stimulated with the knee fully extended and the maximum strain gauge signal Vs recorded.

We refer to this on/off GRF activated control loop as an artificial knee extension reflex (KER) [26]. This does not imply that it mimics any natural reflex of the intact central nervous system, rather that it is a low level automatic pre-programmed response to a sensory input that can be either enabled or inhibited from within the control hierarchy.

A second artificial reflex was also used to assist the patient in maintaining the "C" posture with hyper-extended hips and lumbar spine by switching on the stimulus to the hip extensor and erector spinae musculature whenever the hip joint flexed. For this purpose flexible hip goniometers (type G180 supplied by Penny and Giles Ltd., UK) were used to sense hip flexion. The stimulation was applied to each leg independently when the hip joint flexed, and switched off when the hip was again extended beyond the set threshold according to the rules:

If (hip extension $> A$ degrees)
Then (switch off hip and spine extensors) rule (c)

If (hip extension $< B$ degrees)
Then (switch on hip and spine extensors) rule (d)

Where A and B are the preset upper and lower thresholds respectively. This reflex will be referred to below as the hip extending reflex (HER). The reflexes KER and HER were loosely coupled to improve stability by the condition that for each leg, if rule (d) fires then the rule (b) is not applied until after rule (c) fires.

4 FORWARD PROGRESSION

In the present implementation each locomotion task or mode, i.e., standing up and sitting down, double support standing, rollator gait, 4-point crutch gait, swing-through crutch gait is represented by a finite state model in the control system's repertoire. Swing through crutch gait is the preferred mode for some patients despite its high energy cost mainly for its speed of forward progression, simplicity of control and the degree of proprioceptive feedback. The 4-point gait, although slower, is often preferred for obstacle negotiation or movement within confined spaces. At present, in these modes, the legs are flexed indirectly using the preserved flexion withdrawal response. This reflex is known to be variable in response and subject to habituation as demonstrated in the next section.

4.1 *Habituation of the Withdrawal Reflex*

To study the endurance of the flexion withdrawal reflex, the patient sat in a saddle, with his trunk nearly upright and his test leg hanging freely. A goniometer on the test hip joint measured angle in the sagittal plane. In the first series of tests, the right lateral popliteal electrode was stimulated by bursts of pulses every four seconds for

1.5 min, simulating his gait cycle, to observe habituation, followed by a single burst every 30 s, during recovery. Eight tests were performed to include two amplitudes (200 and 255 on the scale of the control box), two pulse frequencies during the bursts (10 and 20 Hz) and two duty cycles (1 s and 1.5 s burst lengths).

Figure 5 shows two of the recordings, both at 10 Hz, with 1.5 s bursts; Figure 5a was made with the stimulation amplitude 255 while Figure 5b was at 200. Figure 5b illustrates the very irregular response which can occur when the reflex is weak. To give a measure of the habituation during the tests, we arbitrarily chose to take the greatest and the last flexion, after 1.5 min for inclusion in Table 1. The strongest flexion is seen at higher amplitude, with longer bursts but at the lower pulse frequency. Also, the stimulus parameters giving the strongest flexion response exhibited the least habituation at 1.5 min.

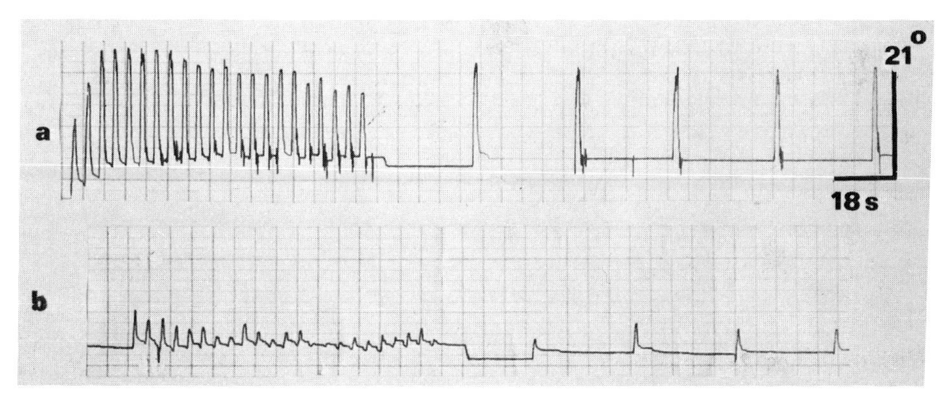

Figure 5 Habituation of hip flexion reflex; recordings of hip flexion against time.

Table 1

Burst length (sec)	10 Hz 200	255	20 Hz 200	255
1	5/0	22/7	8/2	17/6
1.5	10/4	26/14	12/1	16/6

Response amplitudes are tabulated, in degress of hip flexion. For each pair of figures, the first is the amplitude of the greatest hip flexion, while the second is the amplitude of the last hip flexion of the series.

On another day but with the same setup, the amplitude was 255, frequency 10 Hz and burst length 1 s, but the bursts occurred first at 6-s intervals, then at 4-s intervals. The responses are shown in Figure 6a and b respectively. After some facilitation, the reflex habituates and has fallen to about 60% after 90 s; however, there is a substantial recovery after 30 s rest but no significant improvement over 2 min thereafter. Whether the "gait" cycle was of 4 or 6 s made little difference. Similar rates of habituation were seen when the quadriceps were stimulated in antiphase to the flexion stimulus.

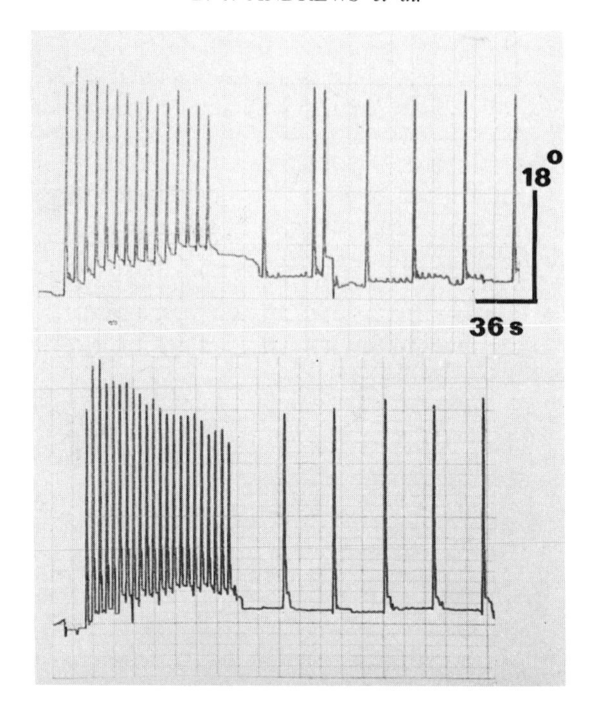

Figure 6 Habituation and recovery of hip flexion reflex; recorded hip angle.

4.2 *Control of 4-Point Gait*

The sequential phases of locomotion used in the finite state model for this gait pattern are illustrated in Figure 7 together with the corresponding finite states s1 to s5. The gait phase and state name are listed in Table 2 together with the control goals within for each state and the corresponding state control strategies presently implemented. The gait pattern proceeds as follows. The subject will normally commence from the double support state s1. Suppose that he wishes to make a right step, he first raises the left crutch clear of the ground, moves it forward and then replaces it on the ground. He then shifts his body weight over onto his left leg. Immediately following weight shift, the right leg can be flexed, using FES, thereby entering the initial swing state s2. To complete the swing phase the knee and hip were extended and the terminal swing state s3 entered in preparation for the double support state s1. On entry to the terminal swing state s3 the flexion stimulus was maintained, for a short period (typically 0.2 s), overlapping with the extension of the leg. This was found to be helpful in achieving the required stride length. In the present experiments flexion of the leg was achieved by eliciting the preserved flexion withdrawal reflex by stimulating the ipsilateral common peroneal nerve.

The patient has the option to control his gait either manually using handgrip switches or automatically. For the above right leg step, manual control proceeds as follows. In state s1 both hand switches are released. State s2 is entered by depressing the right handgrip switch and state s3 is entered upon releasing the switch. The double support state s1 is again re-entered immediately after the overlap period.

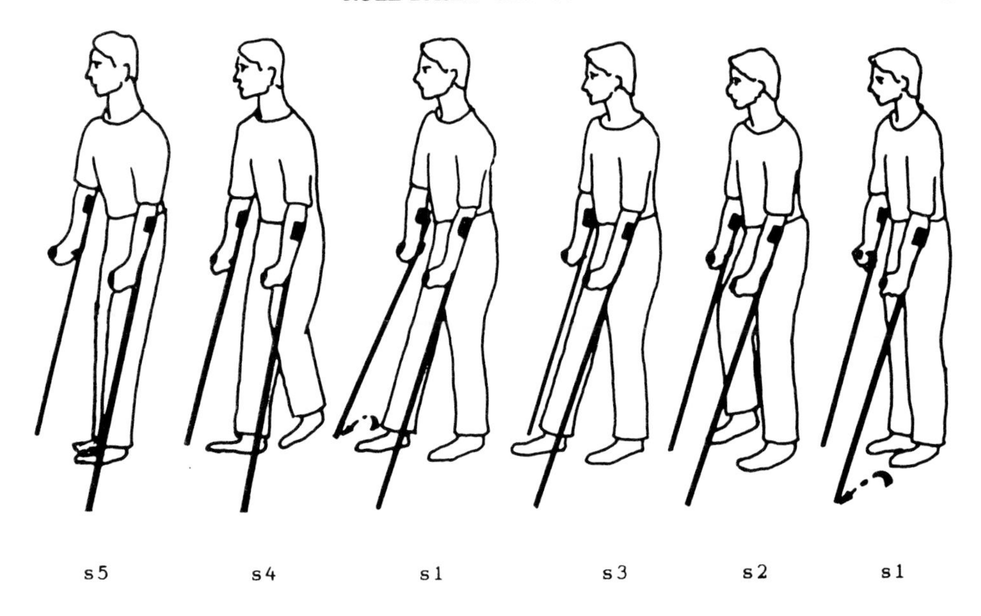

Figure 7 Schematic diagram indicating the typical sequence of stages (states) represented in the finite state model.

Table 2 States and state control strategies

State	Phase	FES Control Goals	Control Algorithms
s1	Double Support	Maintain L&R knee and hip joints extended.	Enable L&R KER and HER reflexes.
s2,[s4]	Initial Swing	Maintain L, [R] hip and knee joints in extension. Flex R, [L] leg.	Enable L, [R] KER and HER Apply R, [L] withdrawal stimulus
s3,[s5]	Terminal Swing	Maintain L, [R] hip and knee joints in extension. Extend R, [L] knee. Maintain hip flexion and ankle dorsiflexion.	Enable L, [R] KER and HER Apply stim to R, [L] Quads Maintain R, [L] withdrawal stimulus for short period (typically 0.2 s)

In addition to the above brace strain gauges monitoring knee extending force actions, and the flexible hip flexion–extension angle goniometers, two other types of sensor were used. A foot contact and load sensor comprising four force sensing resistive films (supplied by Interlink Electronics Inc., USA) were incorporated into a thin, flexible, low cost printed circuit insoles [20, 35]. The FSR's were positioned in the heel, medial and lateral metatarsal and big toe regions of the foot as shown in Figure 8a. To monitor limb load the output of each FSR was summed after A-D conversion. A simple crutch contact and load sensor based on the use of a single FSR was mounted onto the upper surface of the crutch or rollator handgrips as shown in Figure 8b. This sensor produced an analog signal that was related to the magnitude of the crutch load.

(a)

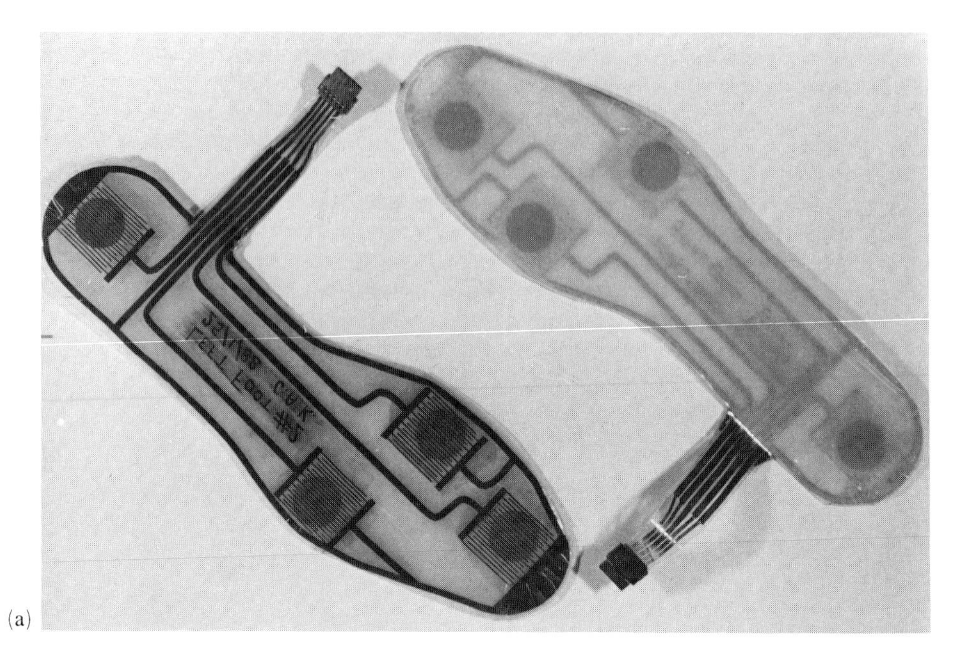

(b)

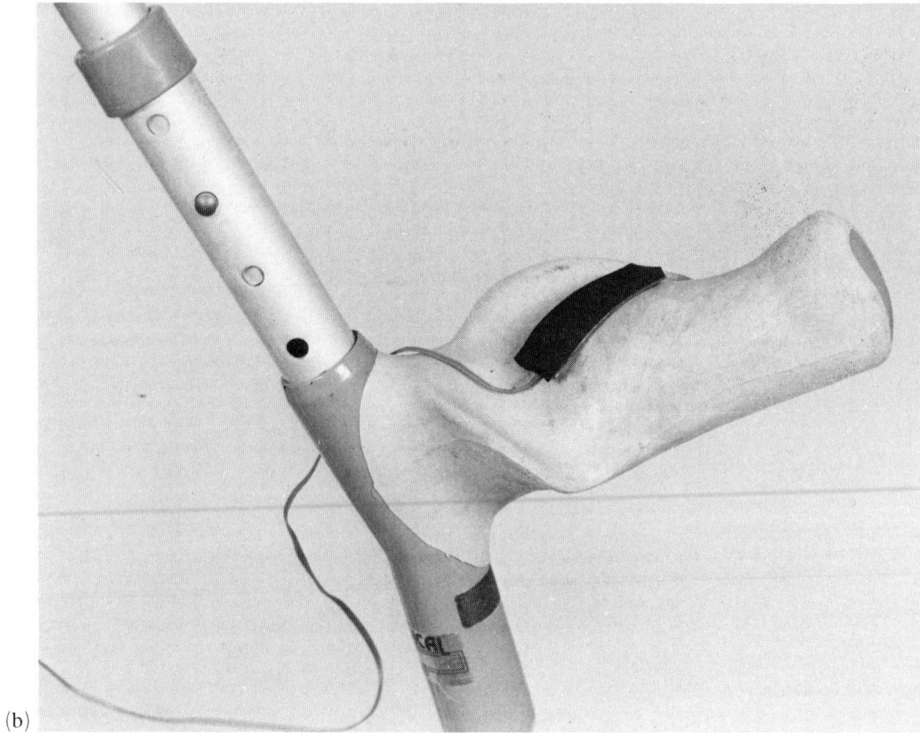

Figure 8 (a) Flexible printed circuit load sensing insoles. The four elements can be seen positioned at the heel, medial and lateral metatarsal and the big toe. (b) Crutch contact and load FSR positioned on the upper surface of the crutch handgrip.

The automatic control of stepping has been described elsewhere [11, 20, 21]. Briefly, this was implemented as indicated in the state transition diagram of Figure 9 in which locomotor task states are shown as circles and state transition events as squares. The cycle for making a step with the right leg proceeds as follows. The intention to step with the right leg, i.e, change from state s1 to s2 is detected by repeatedly asserting the following rule whilst in state s1 and concurrent with the state control strategy indicated in Table 2:

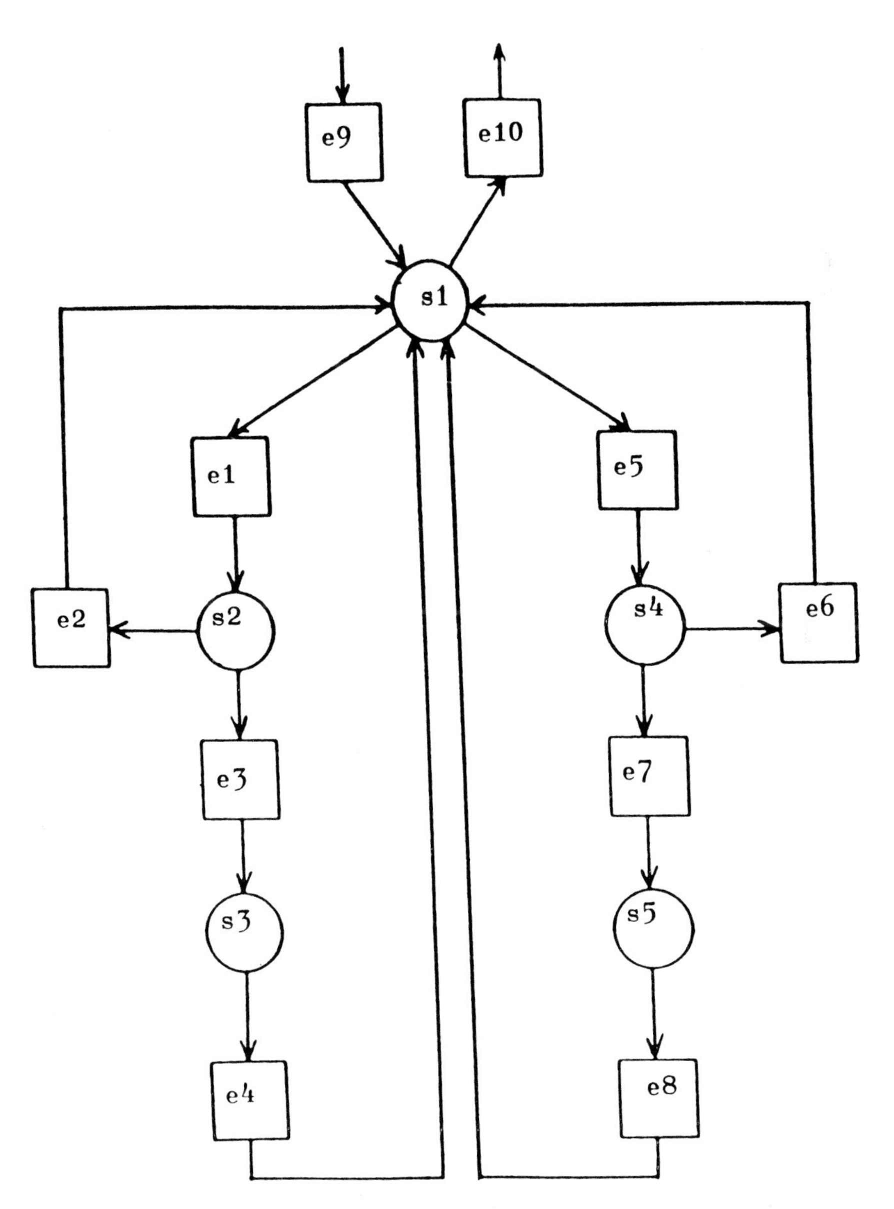

Figure 9 State transition diagram for 4-point gait. States are represented by circles and transition events by squares.

IF . . .
Condition (A)—the left crutch is raised for period greater than a preset interval t1 seconds.

AND IF . . .
Condition (B)—the left crutch contacts the ground after (A) is true.

AND IF . . .
Condition (C)—the left to right leg load ratio exceeds a preset threshold within a preset time window from either (C) or (B).

THEN the intention to make a right step has been detected, (e1) is true.

With each assertion the logical value of each condition is tested with the most recent analog to digital conversions of the involved sensors. For example, conditions (A) and (B) can be tested based on the transitions in the output signal of the left crutch load sensor; condition (C) was based on the ratio of the limb load insole signals.

A similar rule to detect the intention to make a left step is also repeatedly asserted whilst in state s1. The conditions of these intention detectors depend only on voluntary preparatory actions, characteristic of 4-point gait, that are normally undertaken subconsciously by the trained patient.

Potential conflicts were resolved by allowing state transition to occur on the detection of the first valid event. Thus the next transition to either s1 or s3 is made on whichever event e2 or e3 occurs first. The event e2 is a simple timeout and is true when a preset time has elapsed since entering s2 before e3 occurs. The event e3 is true when the right hip joint has reached a preset angle of flexion. State s3 is exited on the preset time period e4 set for the abovementioned flexion stimulus overlap.

5 PRELIMINARY PATIENT TESTS

We report here preliminary laboratory tests performed on the following three volunteer spinal cord damaged patients.

Patient A Male aged 22 yr, mass 57 kg, height 1.6 m, traumatic lesion T6/7 motoric and sensory complete, 1.5 yrs post injury.

Patient B Male aged 35 yr, mass 87 kg, height 1.6 m, traumatic lesion T6 almost-complete cord transection, 6 yrs post injury.

Patient C Male aged 24 yr, mass 70 kg, height 1.8 m, traumatic lesion C6 incomplete hemi-transection, 7 yrs post injury.

5.1 *Standing and Walking Tests*

Patients were all previously able to stand using purely FES systems. In the present tests, each patient was requested to remain standing in parallel bars. The test was terminated when either leg showed evidence of uncontrollable buckling, due to fatigue, or after one hour had elapsed. In the tests using only FES patients (A) and (C) below periodically increased the stimulus intensity up to a preset maximum

(Monophasic, rectangular, current regulated 120 mA, 20 Hz, pulse width 0.25 ms) in order to increase recruitment to accommodate for fatigue and prolong standing time. With the hybrid system the stimulus intensity was maintained at a preset level throughout the tests.

Patient A had used calipers regularly for standing. He stated that he liked the SKAFO braces and would use them regularly if they were made available to him. He was fitted bilaterally and with two channels of stimulation applied to quadriceps using surface electrodes. He had previously completed a program of FES muscle restrengthening exercises for his quadriceps. He was able to stand for periods up to 15 min using the 2-channel stimulator prior to being fitted with the SKAFO braces. When fitted with the braces he was able to remain standing for the one hour test period. Quadriceps activation was typically less than 5%. This activation was due to occasional gross posture changes (raising up on the bars using the upper limbs) performed to relieve discomfort/boredom. The degree of activation was considerably reduced (typically less than 1%) when the subject was asked to remain standing quietly, for periods up to 15 min, and if possible not to adjust his posture. At the present time this patient has not been trained to use an FES walking system.

Patient B had rejected callipers before his implantation. When he was first fitted bilaterally with braces he complained of discomfort and disliked them. However, after adjustment, when they fitted well, were stiffer and had the correct amount of dorsiflexion (approximately 3 degrees), he was sufficiently pleased that he said he preferred to walk with them than without them. For the standing tests, stimulation was delivered via implanted electrodes to the femoral nerves. He had previously used the FES system for standing and walking. Without the braces he was able to remain standing for periods up to 45 min. With the braces, the quadriceps were relieved as expected (assessed by visual inspection and palpation) and he was able to complete the test, however, we were not able to record quadriceps activation because of the particular experimental setup used. The patient had previously been trained to walk using FES and a rollator fitted with handgrip mounted finger switches. At present the patient has only used manual control in which he presses a handswitch to change over the stimulation from extension (KER and HER) to flexion and vice versa when the switch is released. There did not appear to be a significant change in either the quality of gait or cadence or step length or symmetry, with or without the braces, as judged from video recordings.

Patient C. This patient had a paralysed left leg having no functional control of the hip, knee and ankle. He had preserved some skin sensation and proprioception. In addition, his right leg was partially paralysed with reduced sensation. He also had partially paralysed trunk musculature and some weakening of the triceps. He had preserved sufficient strength and control to enable him to grip forearm crutches and once upright in them to remain so, supported by his right leg, for periods up to 20 minutes. However, he was not able to make a single step. He had previously used a conventional cosmetic KAFO, having a fixed ankle and lockable knee joint, for short range ambulation to supplement the use of his wheelchair. With the KAFO he could remain standing quietly for periods in excess of one hour. Since 1985 he has used a manually controlled 2-channel FES walking aid for his left leg using quadriceps stimulation for stance support and the flexion withdrawal for stepping [2]. The system was controlled by a handgrip mounted, finger operated, switch such that when the switch was pressed the stimulation changed over from quadriceps to flexion reflex and vice versa when the switch was released. This simple FES system provided good dynamic cosmesis and enhanced function, e.g., ability to reciprocally negotiate uneven ground and steps. On level ground his preferred 4-point walking speed was 0.41 m/s with good step length symmetry. Although he has used this system on a daily basis FES fatigue of quadriceps has limited its functional usefulness to him compared with his KAFO. He continues to use the system primarily for exercise purposes. In 1986 he was fitted unilaterally with the hybrid orthosis and was pleased with its cosmesis compared with his KAFO (Figures 3 and 4). He was able to stand for periods in excess of the test period with quadriceps activation typically less than 2% of the stance time. He has since used the system, in the laboratory, for level ground walking using both the manual and automatic stepping modes described above. The quality of his walking did not appear to be degraded, as judged by visual observation and his comments, his cadence was not significantly affected in manual mode but was slower (typically 0.3 m/s) when using the automatic mode. During the walking tests it was observed that he was confident with the familiar manual mode but less so with the automatic mode. He preferred the manual mode because of speed. In both modes, the KER was active during the stance phase of gait and the quadriceps stimulation was switched off shortly after foot flat. The quadriceps were activated for typically less than 15% of the stance phase. The patient also demonstrated an ability to walk slowly (typically 0.1 m/s), whilst taking short steps (typically 0.2 m) with the brace but without FES. However, in this case, he preferred to adopt a swing-to pattern using his good leg for speed and convenience.

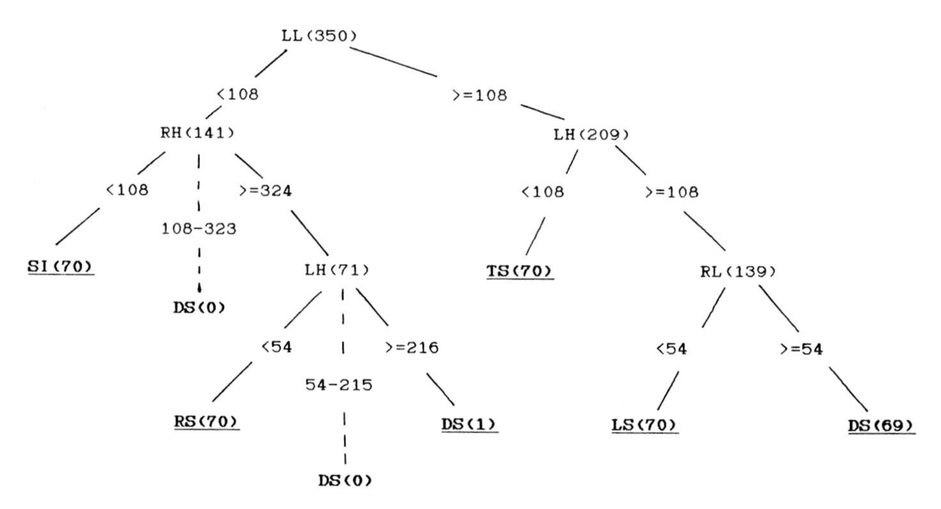

Figure 10 Decision tree for posture identification produced by the induction program ASSISTANT. States underlined, dashed lines represent a null leaf.

6 FURTHER DEVELOPMENT OF THE RULE BASE

Suppose that the control system has entered a state in which there are a number of exit events. We wished to explore the use of the induction program ASSISTANT in providing the required event detectors for a finite state model from a set of training examples of these events.

For the initial experiment we choose the following simplified situation based on tests involving one of the authors. Suppose that the range of possible outcomes are represented by the following five static postures.

1) Double leg standing (DS)
2) Left leg standing (LS)
3) Right leg standing (RS)
4) Tip-Toe standing (TS)
5) Sitting on chair (SI)

To characterise each of these classes a set of eight attributes was used namely the uncalibrated outputs from the left and right foot force sensing insoles [35] shown in Figure 7a:

1) Left heel (LH)
2) Left medial metatarsal (LM)
3) Left lateral metatarsal (LL)
4) Left big toe (LT)
5) Right heel (RH)
6) Right medial metatarsal (RM)
7) Right lateral metatarsal (RL)
8) Right big toe (RT)

The training set data was collected over 4 runs with each run collecting values for the five classes, i.e., 20 data files. Values were taken systematically from each run: 100 each from runs 1, 2, 3 and 50 from run 4 giving 350 examples (70 of each class). Four runs were used to give a good range of independent values with noise on the data being produced by sway, which further increased the range of values.

ASSISTANT produced the classification tree shown in Figure 10. This tree gave 100% classification accuracy whilst using only 4 of the 8 attributes. The classification tree can alternatively be represented by the following set of rules:

IF (LL < 108) AND (RH < 108) THEN SI (coverage 70)

IF (LL < 108) AND (RH > 324) AND (LH > = 216) THEN DS (coverage 1)

IF (LL < 108) AND (RH > 324) AND (LH < 54) THEN RS (coverage 70)

IF (LL > = 108) AND (LH < 108) THEN TS (coverage 70)

IF (LL > = 108) AND (LH > = 108) AND (RL < 54) THEN LS (coverage 70)

IF (LL > = 108) AND (LH > = 108) AND (RL > = 54) THEN DS (coverage 69)

There were also two NULL leafs, i.e., leafs covered by no examples and estimated using Bayesian probability:

IF (LL < 108) AND (108 < RH < 324) THEN DS (probability 1.00)

IF (LL < 108) AND (RH > 324) AND (54 < LH < 215) THEN DS (probability 1.00)

Further testing was provided by ASSISTANT splitting the data set into a training set (60%) and a testing set (40%). This random selection was repeated ten times with the following average results, calculated by ASSISTANT:

correct	weighted with inverse probability
absolute 99%	102%
relative 99%	102%

For this example there was therefore a very high predictive accuracy.

In principle, the same principle can be extended to multiple gait records to provide a training set with appropriate classes for the discrete events which are to be identified during the gait cycle. The rules thus obtained can then be used to form the basis of a finite state rule-based control system. As this method of rule derivation eliminates redundancy there is the possibility of designing a fault tolerant system where decision trees using different sets of attributes can be selected if a fault in a particular sensor is detected.

7 DISCUSSION

Finite state control The hierarchical finite state approach was found to provide a flexible structure for integrating the episodic changes in control strategies that were required during locomotion. The simple on/off FES switching control produced movements that were acceptably smoothed by the inertia of the limb segments. Presently finite state models are used mainly to select state control strategies, therefore, the minimum number of states in a locomotion cycle will correspond to the number of changes of state control strategies required. The complexity of the finite state model will be determined on the level of biomechanical knowledge of the gait pattern and how many stages can be reliably discriminated by event detectors and what state control strategies are presently possible or practical for a particular category of patient. In turn, this will depend on practical issues which include the availability of sensors and controllability of muscles.

Handcrafted rules The use of handcrafted rules was found to be a natural way for those expert in pathological gait to specify event detector requirements. The state transition graphs were found to be useful visual aids in showing which event detectors will be active during a particular system state and the next possible system states. The rules and facts (knowledge base) can be easily interpreted and modified without requiring extensive computer expertise. This is a promising basis for the development of software tools to interactively aid the manipulation of rules for developing and tuning the control system to the needs of individual patients.

Rule induction Machine learning potentially may affect a number of areas in the control of hybrid systems, for example, detection of events in ways that may not be intuitively apparent from the available sensor set. This may also help to optimise the choice of sensor set, particularly when the type of sensor and the permissable anatomic mounting sites are restricted for practical reasons. More advanced fault tolerant systems use built-in redundancy of components. For example, a limited number of sensor faults may be tolerated by implementing multiple schemes for the detection of the same control event using a different set of sensors. On detection of a sensor fault another event detector could be used or an alternative, perhaps simpler, finite state gait pattern switched in. Sensor faults could be detected by hardware means or by on-line detection of sensor checking events. In the latter, at the occurrence of a detected checking event, all non-involved sensors outputs would be checked against predetermined ranges. The sensor checking events need not correspond with control events and may be more numerous to fully utilise the redundancy in the sensor output information content.

Withdrawal reflex Our findings, using bursts of stimulating pulses at an implanted nerve trap electrode on the lateral popliteal, are not dissimilar to those of Dimitrijevic and Nathan [36, 37], who stimulated the plantar surface of the feet of paraplegics with repeated single stimulus pulses. Both show the general features of potentiation, habituation and then recovery of the reflex. We also found the reflex movements to be slow compared with the normal and the movements are delayed in the order ankle, knee and hip. These delays caused the gait of patient (C) to slow down when using the automatic stepping control with the present intention to step detector e5. When using manual control he clearly anticipated these delays and pressed the handswitch earlier.

Dimitrijevic and Nathan also observed that even with the stimulus well above threshold, there were still fluctuations in response from stimulus to stimulus. We did observe that maintaining the stimulation on, during state s2, until a desired hip flexion was achieved did have the effect of regularizing the step. Like us, they also observed a more rapid habituation at higher stimulus frequency. Our preliminary results also suggest that the reflex recovers quickly following a rest break. Further work is required to show if occasional rest breaks may also be beneficial in prolonging the effectiveness of the reflex during gait.

However, they also found that stimulation of the homologous site on the contralateral limb could sometimes cause enhancement of the reflex, but sometimes inhibition. Since this is what will happen during reciprocal gait, it would be interesting to know what effect alternating stimuli would have on endurance. We have not looked at that, nor did we investigate whether or not maintaining a steady burst frequency would influence the result. In another set of experiments, not yet completed, we alternated the site of the stimulation with the saphenous nerve using surface electrodes with the effect of delaying the onset of habituation. Much work remains to be done to optimise the use of stimulation to improve the endurance and dynamics of withdrawal reflex initiated stepping.

Laboratory test of hybrid orthosis The standing tests suggest that the brace may be used to circumvent quadriceps muscle fatigue. The KER was most often observed to occur when the patient assumed a vertical posture.

The level ground, laboratory gait patterns previously achieved by two of the patients were found not to be significantly affected by the presence of the brace. All the patients stated a preference for the cosmetic appearance of the brace compared with standard cosmetic KAFOs.

Our laboratory trials suggest that by using the brace the present limitations of muscle fatigue and reflex habituation may be avoided by facilitating sufficient rest breaks. However, further work is required to determine the minimum frequency and duration of such rest breaks for different forms of ambulation and the practicality of using the system outside the laboratory.

ACKNOWLEDGEMENTS

We should like to thank Mr P. A. Freeman FRCS and staff of the West of Scotland Spinal Injuries Unit at Phillipshill Hospital in Glasgow; also Dr R. G. Platts and staff of the Orthotics Department and Mr I. Bailey FRCS, and staff of the Spinal Injury Physiotherapy Unit of the National Orthopaedic Hospital Stanmore and Dr P. Mowforth of the Turing Institute at the University of Strathclyde. The financial support of the Scottish Home and Health Department is acknowledged. One of the authors, C. A. K., is a recipient of a Medical Research Council postgraduate studentship.

REFERENCES

1. T. Bajd, A. Kralj, J. Krajnik, R. Turk, H. Benko and J. Sega, Standing by FES in paraplegic patients, *Proc. 8th Internat. Symp. External Control of Human Extremities*, ETAN, Belgrade YU (1984), pp. 51–61.
2. T. Bajd, B. J. Andrews, A. Kralj and J. Katakis, Restoration of walking in patients with incomplete spinal cord injuries by use of surface electrical stimulation: Preliminary results, *Prosthetics and Orthotics Internat.*, **9**, 109–111 (1985).
3. G. S. Brindley, C. E. Polkey and D. N. Rushton, Splinting of the knee in paraplegia, *Paraplegia*, **16**, 428–435 (1978).
4. A. Kralj, Electrical stimulation of lower extremities in spinal cord injury, in *Spinal Cord Injury Medical Engineering* (Eds D. N. Gista and H. L. Frankel), C. C. Thomas, Ill USA (1987), pp. 439–509.
5. E. B. Marsolais and R. Kobetic, Functional electrical stimulation for walking in paraplegia, *J. Bone and Joint Surg.*, **69-A**, 728–733 (1987).
6. E. B. Marsolais and B. G. Edwards, Energy costs of walking and standing with functional neuromuscular stimulation and long leg braces, *Arch. of Physical Med. and Rehabil.*, **69**, 243–249 (1988).
7. L. L. Baker, D. R. McNeal, T. Wong and D. Smith, The effect of duty cycle on muscle fatigue using a stimulation frequency of 100 pulses per second, *Proc. 9th Ann. Conf. RESNA*, Minnesota USA (1986), pp. 276–278.
8. D. R. McNeal, L. L. Baker, S. McCaffrey and N. Lopez, Subject preference for pulse frequency with cutaneous stimulation of quadriceps, *Proc. 9th Ann. Conf. RESNA*, Minnesota USA (1986), pp. 273–275.
9. R. Tomovic, M. Vukobratovic and L. Vodovnik, Hybrid actuators for orthotic systems: Hybrid Assistive Systems, in *Proc. 4th Internat. Symp. on Control of Human Extremities*, Dubrovnik, ETAN Belgrade (1972), pp. 73–79.
10. B. J. Andrews and T. Bajd, Hybrid orthoses for paraplegics, *Proc.* (*Supple.*) *8th Internat. Symp. on Ext. Control of Human Extremities, Dubrovnik*, ETAN, Belgrade (1984), pp. 55–59.
11. B. J. Andrews, R. H. Baxendale, R. Barnett, G. F. Phillips, T. Yamazaki, J. P. Paul and P. A. Freeman, Hybrid FES orthosis incorporating closed loop control and sensory feedback, *J. Biomed. Eng.*, **10**, 2, 189–195 (1988).
12. R. J. Jaeger, G. M. Yarkony and E. J. Roth, Standing by a combined orthosis/electrical stimulation system in thoracic paraplegia, *Proc. ICAART* 88, Montreal, Canada (1988), pp. 336–337.
13. J. S. Petrofsky, C. A. Phillips, P. Larson and R. Douglas, Computer synthesised walking, *J. Neurol. & Orthop. Med. &Surg.*, **6**, 219–230 (1985).
14. L. Schwirlitch and D. Popovic, Hybrid orthoses for deficient locomotion, *Proc. 8th Internat. Symp. on Ext. Contr. of Human Extremities*, Dubrovnik, (1984), pp. 23–32.
15. J. M. Hausdorf and W. K. Durfee, Hybrid FES Gait Orthosis using controllable damping elements, *Proc. ICAART* 88, Montreal, Canada (1988), pp. 348–349.
16. D. Popovic and L. Schwirtlich, Gait restoration by active SMFO, in *Control Aspects in Biomedical Robotics*, Pergamon Press (1987), pp. 84–94.
17. J. Saltiel, A one piece laminated knee locking short leg brace, *Orthotics and Prosthetics*, June, 68 (1969).
18. H. R. Lehneis, New developments in lower limb orthotics through Bioengineering, *Arch. Phys. Med. &Rehabil.*, **53**, 303 (1972).
19. G. W. Yang, C. S. Chu, J. H. Ahn, H. R. Lehneis and R. M. Conceicaco, Floor reaction orthoses: clinical experience, *Orthotics and Prosthetics*, **40**, 33–37 (1986).
20. B. J. Andrews, R. H. Baxendale, R. Barnett, G. F. Phillips, J. P. Paul and P. A. Freeman, A hybrid orthosis for paraplegics incorporating feedback control, *Proc. 9th Advances in External Control of Human Extremities*, Dubrovnik, Yugoslavia (1987).
21. B. J. Andrews, Rule based control of hybrid FES orthoses, *Proc. 1st IFAC Symp. on Modelling and Control in Biomedical Systems*, Venice, Italy, April (1988).
22. G. A. Bekey and R. Tomovic, Robot control by reflex actions, in *Proc. of the 1986 IEEE Int. Conference on Robotics and Automation*, Vol. 1, IEEE Computer Society Press (1986), pp. 240–247.
23. R. Tomovic and R. McGee, A finite state approach to the synthesis of bioengineering control systems, *IEEE Trans. on Human Factors in Electronics*, **HFE 7**, 122–128 (1986).
24. R. Tomovic and D. Popovic, Bioengineering actuator with non-numerical control, in *Proc. IFAC Conf. on Orthotics and Prosthetics*, Columbus, Ohio, Pergamon Press (1982), pp. 145–151.
25. R. Tomovic, Control of assistive systems by external reflex arcs, *Advances in Control of Human Extremities*, Published by Yugoslav Committee for ETAN, Belgrade, Yugoslavia (1984), pp. 7–21.

26. B. J. Andrews and R. H. Baxendale, A hybrid orthosis incorporating artificial reflexes for spinal cord damaged patients, *J. Physiol.* **38,** 19 (1986).
27. R. Tomovic, D. Popovic, and D. Tepavac, Adaptive reflex control of assistive systems, *Advance in ECHE* 9, Published by Yugoslav Committee for ETAN, Belgrade, Yugoslavia (1987), pp. 207–213.
28. H. Chizeck, R. Kobetic, E. B. Marsolais, J. J. Abbas, I. Donner and E. Simon, Control of functional neuromuscular stimulation systems for standing and locomotion in paraplegics, *IEEE Trans. on Biomed. Eng.* (In press).
29. R. S. Michalski and R. L. Chilausky, Knowledge acquisition by encoding expert rules versus computer induction from examples, *Int. J. for Man-machine Studies,* **12,** 63–87 (1980).
30. J. R. Quinlan, Discovering rules by induction from large collections of examples, *Expert Systems in the Micro-electronic Age* (Ed. D. Michie), Edinburgh University Press, Edinburgh (1979).
31. J. R. Quinlan, Induction of decision trees, *Machine Learning,* **1,** 81–106 (1986).
32. B. Cestnik, I. Kononenko and I. Bratko, Assistant 86: A knowledge-elicitation tool for sophisticated users, in *Progress in Machine Learning* (Proc. of EWSL 87) (1987).
33. N. de N. Donaldson, A 24-output implantable stimulator for FES, *Proc. 2nd Vienna International Workshop on Functional Electrostimulation* (1986), pp. 197–200.
34. P. E. K. Donaldson, The Cooper Cable: an implantable multiconductor cable for neurological prostheses, *Med. &Biol. Eng. &Comput.,* **21,** 371–374 (1983).
35. C. A. Kirkwood and B. J. Andrews, A Flexible printed circuit board for monitoring patterns of foot loading, in *Proc. ICAART* 88, Montreal, Canada (1988), pp. 488–489.
36. M. R. Dimitrijevic and P. W. Nathan, Studies of spasticity in man: 4. Changes in flexion reflex with repetitive cutaneous stimulation in spinal man, *Brain,* **93,** 743–768 (1970).
37. M. R. Dimitrijevic and P. W. Nathan, Studies of spasticity in man: 5. Dishabituation of the flexion reflex in spinal man, *Brain,* **94,** 77–90 (1971).

Automedica, 1989, Vol. 11, pp. 201–207
Reprints available directly from the publisher only
Photocopying permitted by license only

FOOT-CONTROLLED ARTIFICIAL ARM WITH COMPLEMENTARY ACTIVATION BY BODY POWER

SUJOY KUMAR GUHA, SNEH ANAND and JITENDRA SHARMA

Centre for Biomedical Engineering, Indian Institute of Technology and All India Institute of Medical Sciences, New Delhi 110 016, India

(Received September 1988)

An innovative concept of foot-controlled electronic artificial arm is presented in the paper. Toe flexion initiates hand grip and foot inversion controls supination. The correspondence of movements between the artificial arm and the foot eliminates the "robotic feel" encountered in earlier models. Power requirement is met by two means, body power stored in a spring and rechargable batteries. Fast motion when the fingers are approaching the object but not in contact is derived from the electrical power and thereafter on contact body power store automatically takes over. A differential gear arrangement allows connecting both energy sources for independent as well as combined action. A laboratory model confirms the feasibility of obtaining prehension, pronation and supination with foot controls in a manner which feels natural.

Keywords: Arm prosthesis, transducer, foot control, body power, differential gear, prehension, supination.

1 INTRODUCTION

Design of controlled and powered artificial arm for amputees have been improving and various models are currently available. Yet there remains a wide gap between the features desirable from the users point of view and the facilities provided. The lacunae are manifested both in respect of the control as well as regarding the power source. Electromyographic (EMG) systems have been remarkably refined over the past years [1–5] and form the basis for many upper extremity prostheses. Nevertheless, the problem of obtaining several different independent EMG control signals with a number of distinguishable and reproducible levels remain. In hot and humid environments prevalent in India as well as large parts of the world, the unpredictability in the performance introduced by sweating brings in a problem [6] which has not been completely solved. There is therefore a clear need to investigate other control modalities.

A completely different approach was proposed earlier by Alderson in 1954 [7]. In this methodology the foot was utilized to obtain control signals for a pneumatically powered prosthesis. Pressure switches were placed at three different points in the sole of the shoe. Activation of these switches by foot pressure controlled the artificial arm. As such the principle was workable but did not find user acceptance because there was no natural correspondence between the foot movement and the hand motion. An undesirable "robotic feel" arose and led to failure of the concept in practice. However, the technique has great potential because the foot can provide dextrous movements enabling fine control of the artificial arm. In the present research this approach was examined and modified so as to obtain a "natural feel" and eliminate the delinked robotic character of the original implementation of the concept.

Any artificial limb energized by nonbiological sources of power such as electrical batteries and pressurized gas cylinders require frequent recharging of the power source. In the past some biological sources of energy such as that derived from the repetitive pressing and release of a hydraulic cylinder fitted to the heel of the shoe during walking has been proposed. Although conceptually attractive they have not been practically used. In these schemes the power that can be obtained is limited and the energy storage capacity is also small. Therefore, if all movements of the artificial arm are powered from the energy stored by body power the reservoir of energy gets rapidly depleted. The user has to frequently put in an effort to charge the storage reservoir. With the hydraulic cylinder placed in the heel this step presents difficulties when not walking and the other types too demand socially unacceptable conspicuously awkward movements. It is more appropriate to have complementary artificial energy source together with a body power source. This approach was followed in the work being reported. Northrop Aircraft Inc., California in their final report [7] on prosthetic development up to 1947, also suggested a scheme for combining body power with electrical power. A limitation was that the subjects had to consciously select one or the other source of power for every movement of the artificial arm. When body power was selected each arm movement was effected by motion of some other part of the body. Considerable attention and psychological strain was introduced and the method has failed in respect of user acceptance. Here, complementary body power is proposed but with a difference. The objective is to do away with amputee attention to the switch over from one power mode to another and also to dissociate body energy storage motions with each arm movement.

2 MATERIALS AND METHODS

2.1 *Subject Selection*

In general, the greater the extent of the disability, i.e., the higher the level of amputation, the greater the demand for versatility in the control. Foot control can be developed to provide control over a wide variety of movements. At the initial stage of formulation of the new approach attention has been directed to two basic actions, that of prehension and pronation, since these are valuable operation modes for all levels of amputation except for low forearm amputees in whom the pronation facility in the artificial arm is not required.

2.2 *Control Scheme*

Flexion of the digits of the foot at all the interphalangeal joints and the metatarsophalangeal joints is very similar to hand action. In other words, curling the toes and the forefoot is akin to flexing the fingers to obtain a grip on some object. The control scheme formulated utilizes this correspondence to obtain a natural feel in the control. A set of strain gauge elements are placed on the sole surface of a stocking extending from the region where the toes fit to about halfway along the sole towards the heel. As the subject flexes the toes the transducer chain also bends so as to have a concavity on the underside of the transducers whereby the resistance of the transducing element increases. This information is transmitted via cables to the electronics placed in a package carried on a hip belt. In turn, signals from the electronics activate the actuators in the artificial arm to give corrresponding movements of the artificial fingers for gripping. This system has the advantage that there is a distinct correlation between the action of the foot and that of the artificial arm and hence in the design the so-called "robotic feel" is minimal. During walking, standing or even sitting down the toes may be extended. The transducer resistance changes in an opposite manner as compared to the variation on flexion. An electronic threshold logic can readily distinguish between these changes and preclude false gripping motion of the hand on toe extension or toe extension can also be made to open up the fingers beyond a certain state occupied when the foot is in the relaxed condition. A transducer design shown in Figure 1

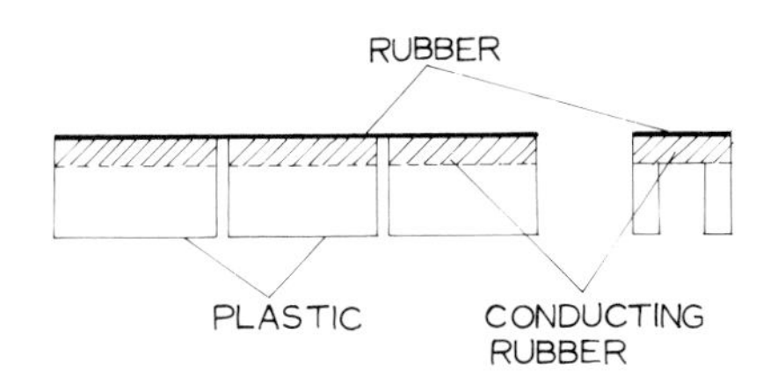

Figure 1 Stretch transducer using conductive rubber.

further reduces artifacts. A very inexpensive transducing element, conductive foam rubber has been used. Resistance of this rubber increases on extension. Strips of rubber have been incorporated in a chain of hard plastic shells and placed close to the top of these shells. A layer of ordinary nonconductive rubber is affixed to the upper surface of the plastic shells to hold the chain in place. When the sole is forced to bend so that there is a concavity below, lower margins of the hard plastic shells strike each other and the upper margins are forced to separate so the rubber located near the upper margins extend and the resistance goes up. On the other hand if the sole bends with the concavity above then upper margins of the hard shells strike each other and the rubber does not extend significantly, although the lower margins of the shells separate. Thus a selectivity to flexion and extension is built into the transducer design. Furthermore, inversion and eversion of the foot has no effect on the transducer.

A control for pronation and supination is also required. Manipulations for this purpose need to be relatively independent of the motions from which finger control is obtained. That is, the subject should be able to do any one of the control activities alone or together. The foot inversion movement has been used for the purpose. A transducer similar to the one placed in the sole is placed within layers of the stockings such that the transducer chain is on the medial side of the leg passing over the medial malleolus. In Figure 2 the position of the transducer chain is marked with white strips on the stocking. The conductive rubber in the transducer is set near to the skin surface so that in a foot inversion movement the rubber is stretched and the transducer resistance increases. This resistance rise is the trigger for a supination turn of the artificial hand. Since inversion tends to bring the sole of the foot facing up it bears similarity to hand supination and so there is a correspondence between the foot and artificial arm motion giving a natural feel. Toe flexion for gripping and inversion of the foot are fairly independent acts which may be done together too.

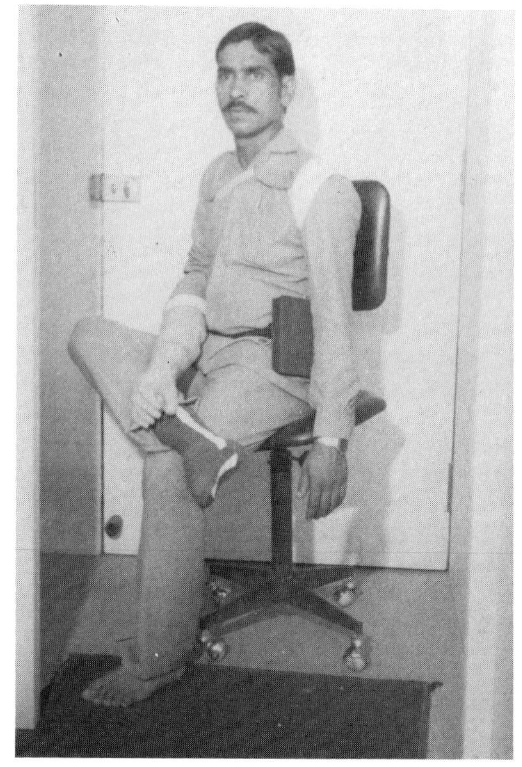

Figure 2 Subject with artificial arm and stocking with transducer. Transducer chain position marked with white strips.

2.3 *Actuators*

Adequate coordination between foot and hand motions can only be realized if the speed of movement of the output device, that is the artificial fingers and the wrist, is as fast as the change in the control signal generated by the foot. This demand is usually not met in conventional systems. The high force requirement at the output terminal is derived by a gear reduction system which slows down the speed. A new system has been designed to overcome the difficulty. High speed during the period that the fingers have not come in contact with the object to be held is obtained by an electrical drive with low gear ratio coupling. During this period the motor is not working against a load and so the energy consumption is low. When the finger comes in contact with the object and thereafter the soft padding on the fingers begin to get compressed higher force is needed. The speed can at this time be slow. Even so the energy consumption is high. A switchover to body energy storage is appropriate at this stage.

In order that the dual drive concept works there has to be some mechanism by which the electrical motor drive and the body power storage drive can independently move the output and have additive effect. Yet they should not interfere with each other. Some clutch devices have been proposed in the past for similar applications but they are very cumbersome and maintainability is poor.

Presently a well known mechanism, the differential gear, which is common in the rear transmission of automobiles has been adapted. If one of the rear wheels of a car is held fixed and the other jacked up, with the rotation of the propeller shaft the other wheel will rotate. If the propeller shaft is fixed, rotation of one rear wheel will produce rotation in the other wheel in the opposite direction. In the artificial arm the equivalent of the propellor shaft is the drive from the electrical motor. Body power goes to the equivalent of one rear wheel and the equivalent of the other rear wheel is the output to the hand. Hence, the two drives can work independently one at a time or together.

Body power is stored in a spiral spring which is wound through a nonreversible ratchet mechanism similar to that of a clock. A pin roller arrangement (Figure 3) converts a repetitive cord tension into a rotation to wind the spring. The cord tension is given by action of the opposite shoulder or other motions can also be selected. The spring is wound at times convenient to the user and the time of winding has no relation to the activation of the hand. The output shaft of the spring is coupled to the differential gear by a bevel gear. The spring output shaft motion through a chain of gears which includes a non-self-reversible worm and worm gear is converted into a low torque control. The spring output shaft will not exert a driving torque unless this low torque shaft is rotated at high speed. No energy is consumed in holding the spring system at equilibrium. A small electrical motor with a worm gear controls the low torque control. This motor is termed the "control motor" (Figure 3).

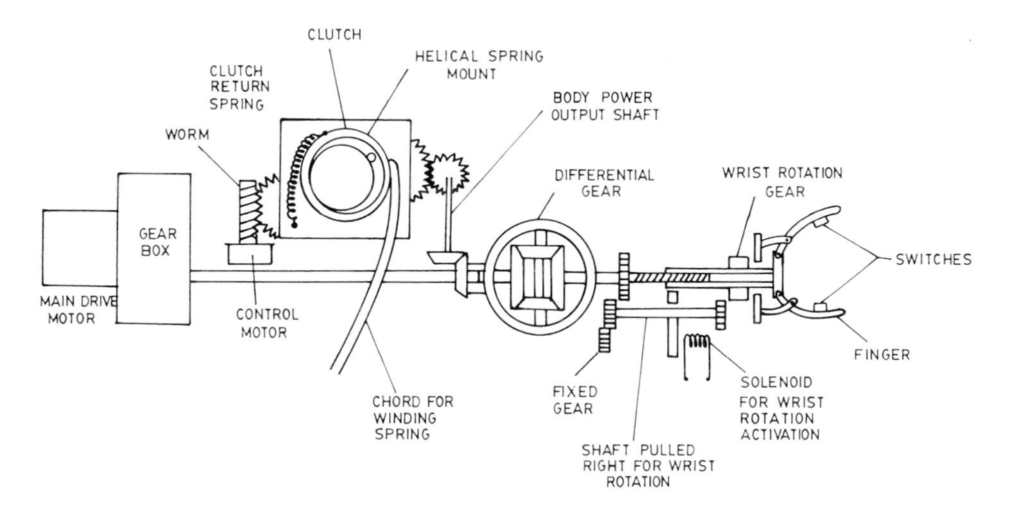

Figure 3 Schematic of the artificial arm. Solenoid for wrist activation is shown below the main drive axis for clarity. In the actual arm the solenoid is positioned above the drive axis.

In operation the main drive motor begins to operate through the differential gear to bring together the fingers. When the finger contact with the object is established load begins to increase and the electrical current to the motor begins to rise. A current level logic senses the rise and switches off the main motor. At the same time the control motor is switched on and rotation of this motor allows the spring torque to be transmitted to the output via the differential gear. This rotation does not get transmitted back to the main motor drive by virtue of the differential mechanism.

Thus, the high initial speed as well as switchover to final high torque from body power is obtained without user attention. The main motor energy consumption is low since it is not working against a load and the control motor too draws low power. Battery energy is conserved and, most importantly, the switchover from one power source to the other does not require user attention and the user need not apply body power for every hand motion.

Wrist rotation is not essential to this scheme but has been provided. The fingers are closed by means of a pantograph-like arrangement. For wrist rotation the entire linkage is rotated. The rotation does lead to a small closing or opening of the fingers but the ratios are so set that this closing or opening is negligibly small for 180 degree rotation of the wrist. The wrist rotation is normally kept checked. When required, a solenoid in intermittent pulses disengages the lock and engages the main drive to the wrist rotation. As soon as the foot from the inverted state is brought back to the normal position the solenoid is deactivated totally and a spring locks the wrist rotation. Spring torque can come into play during pronation. Directions of rotations are such that in the present design spring torque cannot come into action during supination. Although the subject may be trained to perform tasks in such a manner that supination is not against a load it is desirable to improve the design so that spring torque can assist both pronation and supination.

2.4 *Other Controls*

Currently the operation is in the open-loop mode. The subject has no feedback of the force exerted in the grip. Only two switches are provided in the fingers to switch off both the main drive motor and the control motor when the force exceeds a certain level. Thereafter, the foot positions have to change toward the normal state before the drives begin to reverse. The subject must learn to move his foot control surfaces at a speed which approximately matches the speed of the drives.

3 RESULTS AND DISCUSSION

A prototype of the artificial arm is shown in Figure 4, on a subject in Figure 2. The mechanisms are still not sufficiently compact to give a user-acceptable arm. The

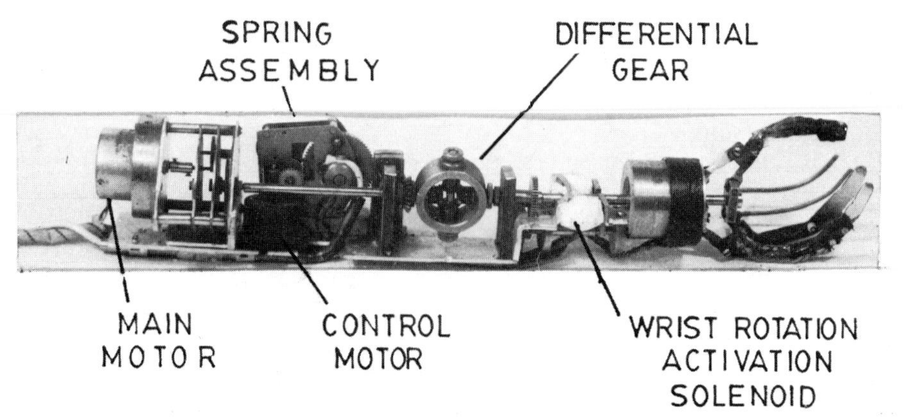

Figure 4 A photograph of the artificial arm mechanism.

basic concept has been demonstrated to be workable. Besides reduction in size the main inprovement required is in the quality of the energy storing spring. One of the better clockwork springs has been tested but for practical design more energy storage would be desirable. Such springs are within the scope of current technology. With recent developments in prosthetic manufacturing technology [8] it should also be possible to obtain better linkage with the opposite shoulder or other part to give traction for winding the spring.

Position feedback can also be incorporated so as to have complete similarity between foot and hand motions but the additional complexity involved does not seem to be really warranted. Experience with artificial arms shows that subjects can work quite effectively with visual feedback. However, variable speed drive matching the speed of change in the control signal levels will no doubt be an advantage. As compared to myoelectric control the distinguishable levels available are not limited and so adequate trigger signals for the variable speed can be obtained.

The problem of sweating encountered in myoelectric control is not applicable here. But just as close fit of the EMG electrode holder to the body surface is important in the new design, the stocking incorporating the transducer must also fit well with the contours of the subject's feet and ankle. Deviations produce motions somewhat different than what a subject would expect. Of course, this artifact may be taken care of by closed loop control.

REFERENCES

1. E. Graupe, Control of upper limb prostheses with several degrees of freedom, *Bull. Prosthet. Res.*, **BPR 16/17,** 226–235 (1975).
2. S. C. Jacobson, D. F. Knutti, R. Johnson and H. H. Sears, Development of Utah artificial arm, *IEEE Trans. BME,* **29,** 249–267 (1982).
3. N. Hogan, Myoelectric prosthesis control: optimal estimations applied to EMG and cybernetic consideration for its use in man machine interface, Ph.D. thesis, M.I.T., Cambridge, Mass (1976).
4. D. F. Lovely, B. S. Hudging and R. N. Scott, Implantable myoelectric system with sensory feedback, *Med. Biol. Eng. Comput.,* **23,** 87–89 (1984).
5. L. Phillipson and R. Sorbye, Control accuracy and response time in multiple state myoelectric control of upper limb prosthesis initial results in non disabled volunteers, *Med. & Biol. Eng. & Comput.,* **25,** 289–293 (1987).
6. G. C. Ray and S. K. Guha, Equivalent electrical representation of the sweat layer and gain compensation of the EMG amplifier, *IEEE Trans. BME.,* **30,** 130–132 (1983).
7. S. W. Alderson, The electric arm, in *Human Limbs and Their Substitutes* (Eds. P. E. Klopsteg and P. D. Wilson), McGraw-Hill Book Company Inc., New York (1954).
8. D. Jones, Impact of advanced manufacturing technology on prosthetic and orthotic practice, *J. Biomedical Engineering,* **10,** 179–183 (1988).

Automedica, 1989, Vol. 11, pp. 209–220
Reprints available directly from the publisher only
Photocopying permitted by license only

A MASTER-SLAVE TYPE MULTI-CHANNEL FUNCTIONAL ELECTRICAL STIMULATION (FES) SYSTEM FOR THE CONTROL OF THE PARALYZED UPPER EXTREMITIES

NOZOMU HOSHIMIYA

Faculty of Engineering, Tohoku University, Sendai 980, Japan

and

YASUNOBU HANDA

Department of Anatomy, Tohoku University School of Medicine, Sendai 980, Japan

(Received September 1988)

A microcomputer-based Multichannel Functional Electrical Stimulation (FES) system has been developed which can restore motor function to the paralyzed upper extremities. In order to realize fine control of the finger, wrist and elbow, increase of the number of stimulation channels is inevitable. Recently, we have proposed a new strategy for the generation of the multichannel stimulation patterns. This method was fundamentally effective in the pulse amplitude modulation (PAM) method. On the other hand, a pulse width modulation (PWM) type FES system may be suitable for multichannel system if it has the same controlling capabilities as a PAM system by this method, since with the PWM system it is easy to reduce total size and weight. In this article, the following items are described.

1) Development of a master–slave type multichannel FES system which is composed of two 16-bit microcomputers in order to generate 48-channel stimulation patterns in both the PAM and PWM method.

2) Clinical evaluation of the restored hand movements realized by the multichannel stimulation patterns based on EMG activities of the normal subjects.

3) Interactive capabilities with which the patient can actively participate in realizing fine control of the paralyzed upper extremities.

Keywords: Functional Electrical Stimulation (FES), multi-channel stimulation, paralyzed upper extremity, neural prostheses, master–slave type, iteractive system.

1 INTRODUCTION

Multi-channel Functional Electrical Stimulation (FES) is very efficient for the restoration of the paralyzed upper extremities. We have been developing several microcomputer-based FES systems and have applied them to clinical usage, i.e., restoration of motor function of the paralyzed upper extremities [1–5].

In these systems, pulse-amplitude modulation (PAM) has been adopted for the regulation of the recruitment characteristics. The system required many D/A converters (or demodulators) and massive isolators with good linearities so that it was not easy to reduce total size and price when the number of stimulation electrodes increased. On the other hand, a pulse-width modulation (PWM) type

Correspondence to: Dr. Nozomu Hoshimiya, Dept. of Electrical Communications, Faculty of Engineering, Tohoku University, Sendai 980, Japan.

FES system may be suitable for a multi-channel system if it has the same controlling capabilities as a PAM system, since the PWM system does not need D/A converters and can be isolated by small optical isolators. Crago *et al.* reported that similar contraction characteristics were observed between PAM and PWM methods in animal experiments [6]. But we have no clear evidence on human subjects (patients) with multi-channel stimulation to the nerve trunk. Then it is important to investigate the control capabilities of PAM and PWM in the paralyzed upper extremities for the purpose of developing a more advanced multi-channel FES system.

Recent progress of the FES study has enabled us to control neuromuscular system very finely by introducing EMG-referenced stimulation data, e.g., a patient could extend the elbow, open the hand, grasp a cup, flex the elbow and then drink the canned tea [7–9]. But the patient sometimes wishes to modify or improve the FES-induced movements for activities of daily living (ADL). Therefore, it is desirable to include flexible capabilities (here we call as "self-learning capabilities") in a FES system to modify stimulation pattern in accordance with the patient's desire. By this self-learning system, patients can actively participate in realizing a practical and feasible FES system.

In this paper, the following three items are described:

1) Design and main performance of a master–slave type multi-channel FES system generating 48 channel stimulation patterns in both PAM and PWM methods.

2) Experimental results on the FES-control of the fingers/wrist/elbow of a quadriplegic patient with both PAM and PWM methods.

3) Interactive learning capabilities which enabled a patient to modify the stimulation pattern by herself.

2 PRINCIPLE OF FES SYSTEM

2.1 *Principle*

Figure 1 shows a schematic diagram of the overall system for restoration of the hand function of a quadriplegic patient. The functional electrical stimulation (FES) system generates multichannel stimulation pulse trains in accordance with a certain kind of residual volitional signal of the patient, e.g., voice, electromyogram (EMG), flexion angle of the elbow. Electrical stimulation is given to each neuromuscular system through the percutaneously indwelling electrode. Although the upper extremity is a multijoint system with multi-degrees of freedom, objective sophisticated movements can be realized by spatio-temporarily well organized multi-channel stimulation. In other words, the FES system is one kind of artificial central nervous system for motor control.

The stimulation electrode is the Caldwel–Reswick type (a helically-wound coil) made of multistrand stainless steel braid with Teflon coating. Previously, we often used 50 micron × 7 strand SUS 316 braid (AM systems Co. Ltd), but recently developed 25 micron × 19 strand SUS 316L braid (Nippon Sei-sen Co. Ltd) has now been mostly used [10]. The tip of the electrode was deinsulated, and the electrode was injected through the skin to fix near the peripheral nerve fibers innervating the objective muscle (motor point).

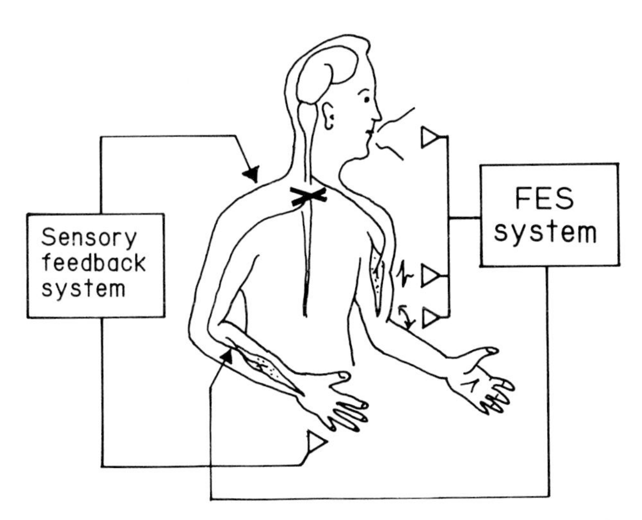

Figure 1 Schematic diagram of the overall system for restoration of the hand function of a quadriplegic patient.

The sensory feedback system shown in Figure 1 is an apparatus which gives substitutional sensation of grasping power/displacement to the senseless patient by means of electrical stimulation to the position with remaining sensation (e.g., shoulder surface). It has been developed because of its importance for the quadriplegic patient but cannot yet be integrated as a whole system [11].

2.2 *Stimulation Data*

According to recent studies, it was proved that EMG-based stimulation data were quite useful in a PAM (pulse amplitude modulation) type multi-channel FES system for the restoration of the hand function [9]. In this method, multi-channel stimulation data were created by trapezoidal approximations of the averaged EMG waveforms recorded from normal subjects and then some modification was done by inputting the threshold and maximum values for each neuromuscular system. It was interesting that this method of open-loop FES could reproduce almost the same movements of normal EMG volunteers in the paralyzed upper extremities of C4 and C5 quadriplegic patients. It is noteworthy that coactivation of antagonists, synergists, stabilizers, in addition to prime movers of which data were obtained from the EMG analysis, realized not only well-coordinated and graded movements but also joint stabilities of the upper extremities [9]. An example of the stimulation data is schematically shown with corresponding finger movements in Figure 2. Ordinate and abscissa correspond to stimulation data and memory address, respectively. Stimulation data mean amplitudes of the stimulation pulse trains in a PAM system. Maximum stimulation data 250 corresponded to a pulse amplitude of -10 volts in PAM experiments. In our system, a memory address could be assigned by an analog voltage of a sensor placed on the patient (e.g. displacement sensor, angle sensor, etc), i.e. volitional control could be realized by this new allocation method [2]. In Figure 2, we can see EMG-based stimulation data of FDS (flexor digitorum superficialis) during hand movement illustrated at the bottom of the figure. It is easy to understand that stimulation data increase on the right hand

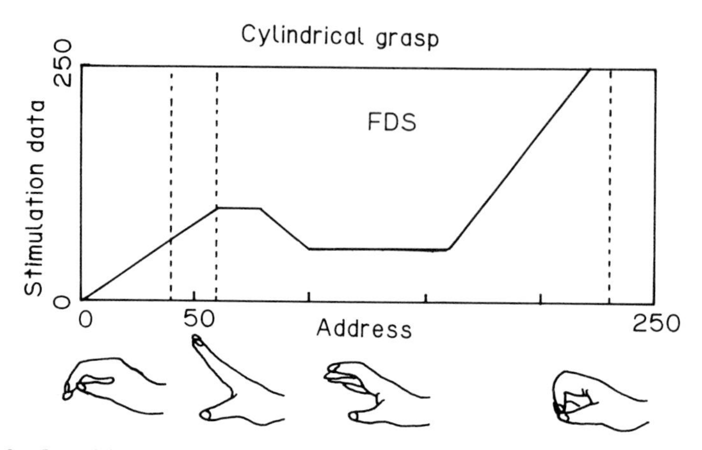

Figure 2 One of the multichannel stimulation patterns and corresponding hand movements.

side (beyond 160 of address) in accordance with the finger flexion, but it is quite interesting that increase of FDS stimulation is necessary during hand opening which is primarily produced by ED (extensor digitorum). Coactivation of many muscles is nonintentionally involved in the stimulation data.

3 SYSTEM CONFIGURATION

In order to realize fine control of a multijoint system of the upper extremities, a FES system should include more than 30 channel outputs, so that it is actually impossible to perform high grade interactive functions during multi-channel pulse generation by a single microcomputer.

Figure 3 shows a block diagram of a master–slave type multi-channel functional electrical stimulation (FES) system which we developed. It was composed of two microcomputers, i.e., a commercially available 16-bit microcomputer NEC PC-9801VM (CPU : μPd70116) as a master computer, and a 16-bit microcomputer board INP 186 (CPU : 80186) as a slave computer. The main functions of PC-9801 were (1) data processing of input signals from an A/D converter, (2) selection of the stimulation data, (3) interactive communication through a CRT display, and

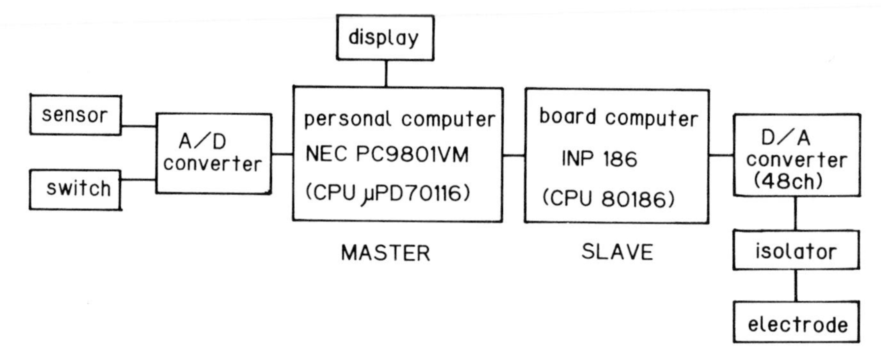

Figure 3 Block diagram of the master–slave type FES system.

(4) data transfer to the slave computer. The slave computer (INP 186) was devoted to generate multi-channel (48 channels) stimulation pulse trains with reference to the data given by the master. Both PAM type and PWM type stimulation pulse trains could be generated. The main and slave microcomputers were interrupted every 50 ms, and the data were transfered through a parallel 1/0 line. Pulse repetition period (50 ms) was determined so as to produce enough muscle force with less fatigue [1,9]. The system was based on the RMX-86 operating system, and all programs were written by PL/M-86 language.

Figure 4 is a timing diagram of the data transfer and stimulation pulse generation in a PWM method. Since a pulse width of 1 ms was required in the maximum case in the PWM method, each stimulation pulse was generated with 1 ms offset in time at 50 ms interval. Data for each channel (channel no. and stimulation data) were divided into two 8 bits, and they were transfered at an interval between any two stimulation pulses only when data had to be changed.

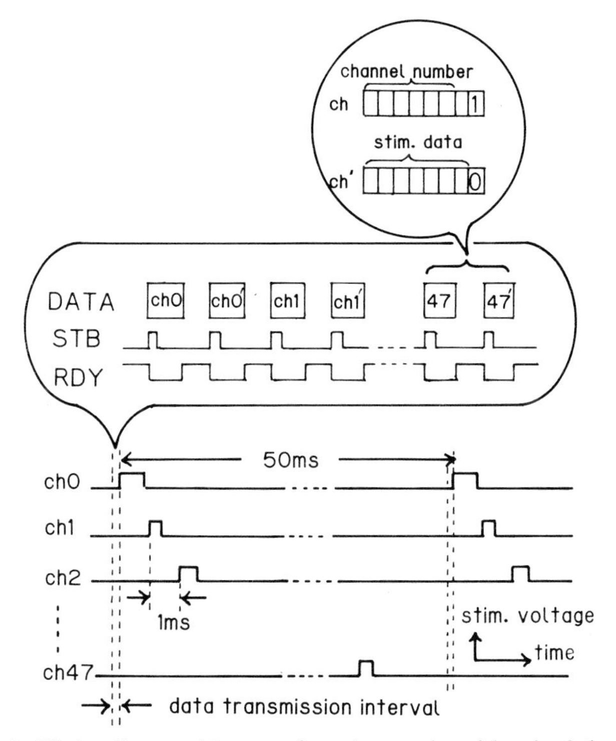

Figure 4 Timing diagram of data transfer and generation of the stimulation pulses.

In Figure 5 an outline of the job flow in the system is shown. We could select the block to be controlled at "block selection", and could select grasping pattern at "menu selection" after "hand" block was assigned. After these selections we could select one of four modes: "training mode", "ADL mode", "Test mode" and "max-min mode". In "training mode", this FES system worked as a therapeutic electrical stimulation (TES) system. In "ADL" mode, the patient could do several kinds of movements for ADL, e.g., a patient could extend her elbow, grasp a can, flex the forearm and then could drink a can of tea. In "max-min mode", threshold level and

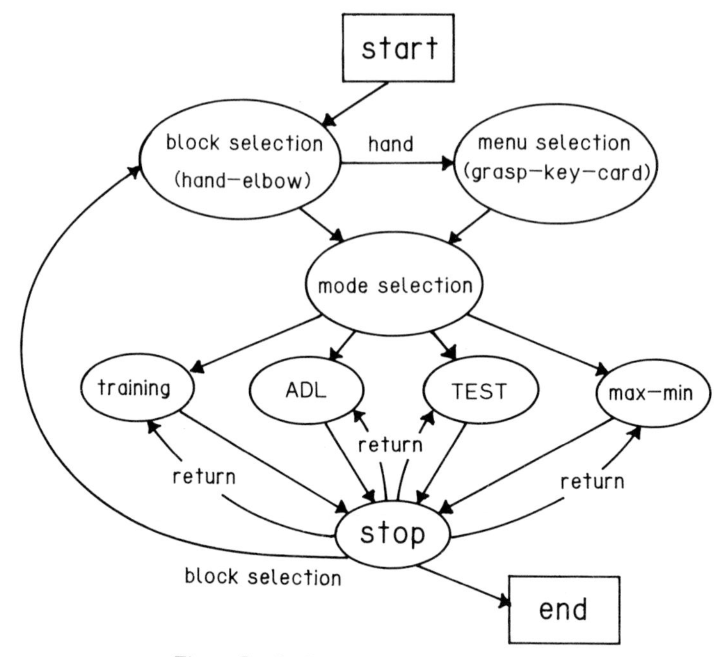

Figure 5 Outline of the job flow in the FES system.

maximum intensity of the stimulation pulses were determined in order to refine the EMG referenced standard stimulation data to fit the individual neuromuscular system.

4 COMPARISON BETWEEN PAM AND PWM MULTI-CHANNEL STIMULATION IN A QUADRIPLEGIC PATIENT

In order to evaluate the possibilities of clinical application of the PWM method based on EMG referenced creation method of the multi-channel stimulation data, an experimental comparison was performed in both PAM and PWM methods in a C4 quadriplegic patient by using the master–slave type FES system. The stimulated neuromuscular systems were as follows (28 muscles including finger, wrist and elbow):

AdP: Adductor pollicis,
AbPB: Abductor pollicis brevis,
AbPL: Abductor pollicis longus,
BiLg: Long head of the biceps brachii,
BiS: Short head of the biceps brachii (2 channels),
Br: Brachialis,
BrR: Brachioradialis,
ECRL: Extensor carpi radialis longus,
ECRB: Extensor carpi radialis brevis,
ECU: Extensor carpi ulnaris,
ED: Extensor digitorum (2 channels),
EPB: Extensor pollicis brevis,

FCR: Flexor carpi radialis,
FCU: Flexor carpi ulnaris,
FDS: Flexor digitorum superficialis (2 channels),
FDP: Flexor digitorum profundus (2 channels),
FPL: Flexor pollicis longus,
FPB: Flexor pollicis brevis,
OpP: Opponens pollicis,
PL: Palmaris longus,
TrLg: Long head of the triceps brachii,
TrLt: Lateral head of the triceps brachii,
TrMd: Medial head of the triceps brachii (2 channels).

Controls for fingers/wrist and for elbow were independently performed [9, 12]. Realized movements are shown in Figure 6. There was only a little discrepancy in

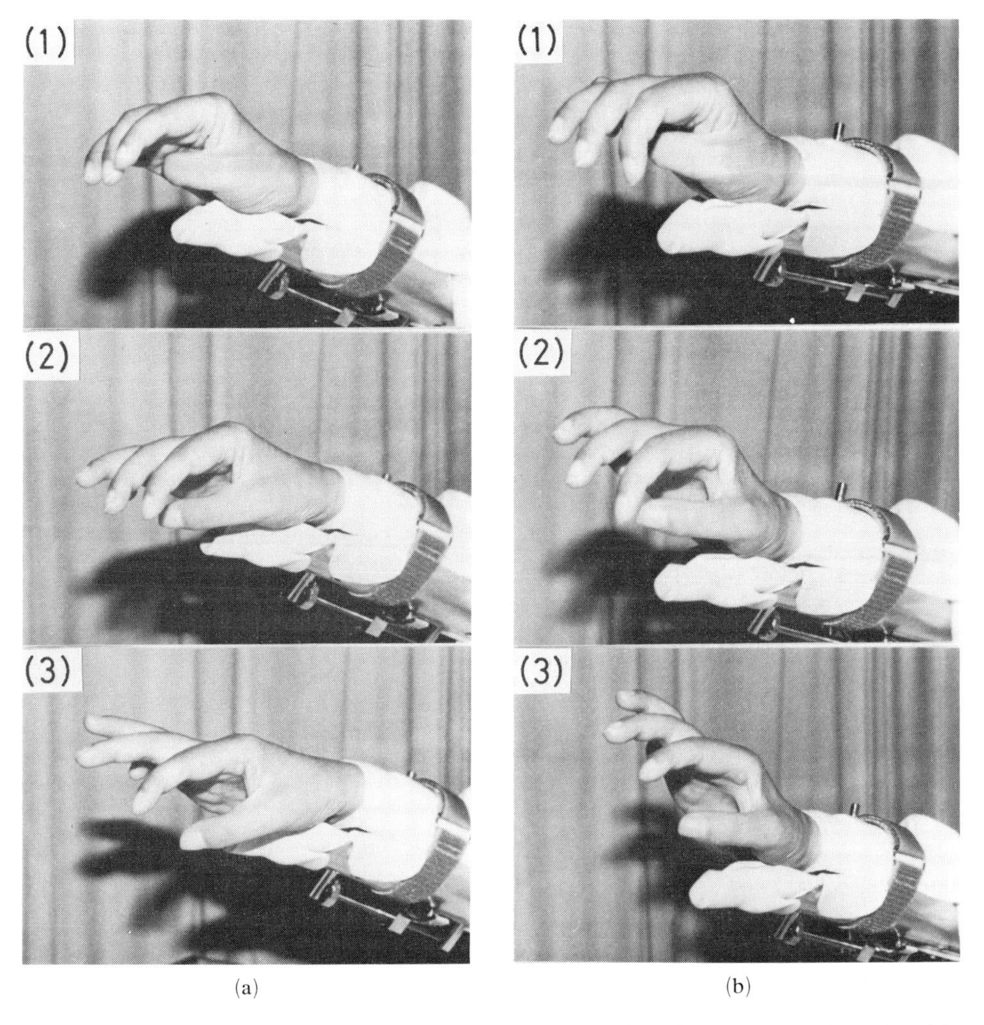

Figure 6 Comparison of the realized movements of a quadriplegic patient between PAM method (a) and PWM method (b).

the movements between the PAM method and the PWM method, but we have concluded that almost the same movement can be realized by both the PAM and PWM stimulation methods, both of which are based on the trapezoidal approximation of the averaged EMG signals. In Figure 7, three typical movements of the paralyzed finger/wrist system without stimulation (a), at opening position (b) and at grasping position (c) are shown.

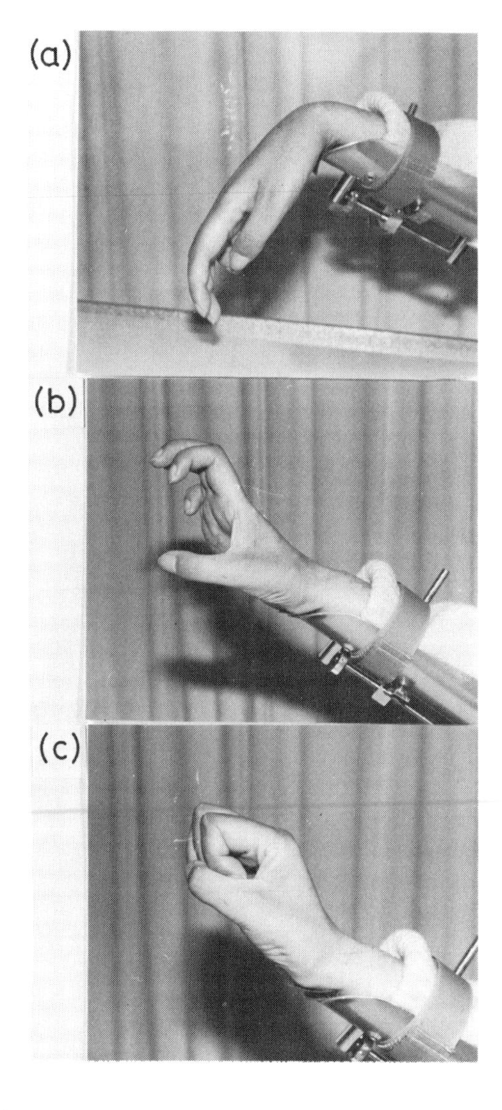

Figure 7 Three typical shapes of the fingers/wrist system without stimulation (a), at opening position (b) and at grasping position (c) induced by FES.

5 INTERACTIVE MODIFICATION CAPABILITIES OF THE STIMULATION PARAMETERS

After a certain period of FES application, the patient wishes to modify or improve the FES-induced movements for ADL by himself/herself for the following reasons:

1) to realize more sophisticated or finer movement to satisfy requirements in ADL,

2) to change parameters according to changes of the muscle characteristics, e.g., improvement in spasticity or in muscle force after continuous FES application.

In the first clinical trial to permit the patient to access the FES system, a quadriplegic patient could modify stimulation parameters which were stored in the master computer to fit her desired movements.

First the objective block in the upper extremity was selected (Figure 8a). An arrow moved repeatedly from bottom to top and the patient gave a one bit signal when the arrow came to the position which she wanted to select. For example, when the muscle group controlling thumb movement was selected, the shape of the thumb and fingers was displayed on the screen, then the direction to be modified was selected (Figure 8b). She should select one among flexion, extension, adduction, abduction and opposition. For example, she selected flexion. Then she

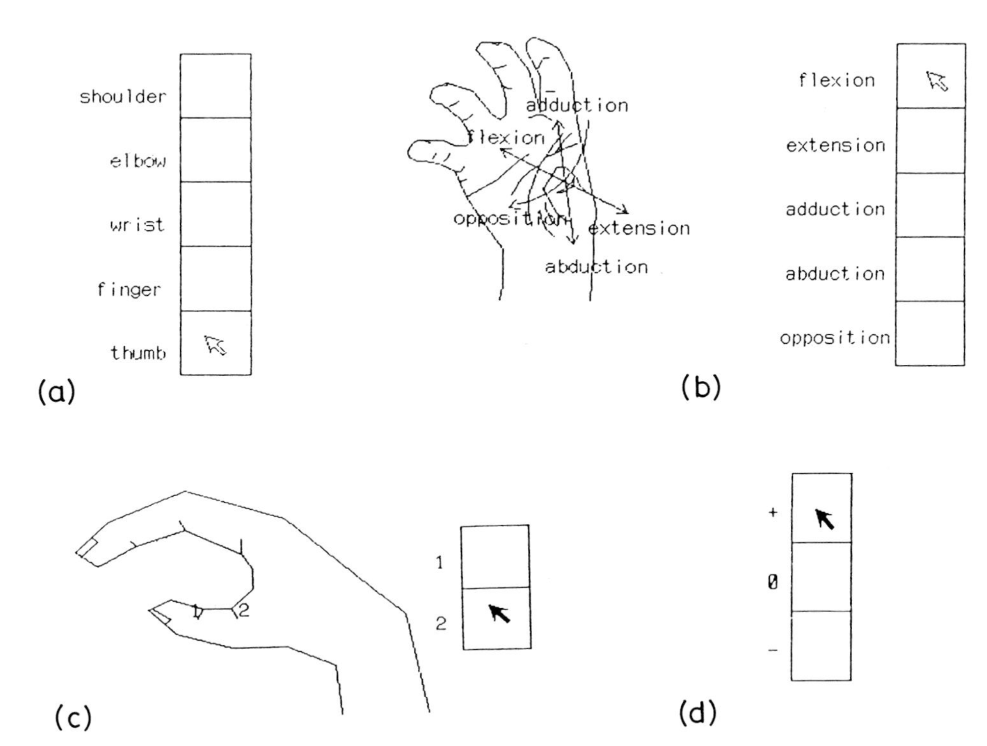

Figure 8 Examples of the displays in the interactively modifiable FES system. (a) For selection of the objective block, (b) illustration of the thumb movements and a display for selection of movement, (c) for selection of joint, (d) for selection of increase (+), decrease (−) or reset (0).

should select one from the IP (1) and MP (2) joints (Figure 8c). For example she selected the MP joint (2). In this case, the system picked up the data file "FPB" (primary flexor of the MP joint of the thumb), "FPL" (supplementary flexor) and "EPB" (antagonistic extensor) from the data files. Then she could change the stimulation level within an allowable range which had been designated by a doctor and a biomedical engineer. This was done by selecting "+" or "−" in Figure 8d. Modification was performed on the maximum stimulation value (pulse amplitude for PAM method, pulse width for PWM method). Modification was repeatedly carried out by changing the stimulation level by a prescribed step (3–5%) and by testing the effect using "Auto" mode program. If misoperation occurred, she could reset the system to the initial stimulation condition by selecting "0" in Figure 8d. Figure 9 summarizes the modification procedure in this interactive capabilities of this FES system.

Each modified process can be recorded so that the doctor can periodically evaluate the information. This capability may be useful for realization of high grade restoration of the function of the upper extremities, and for the design of the training schedule.

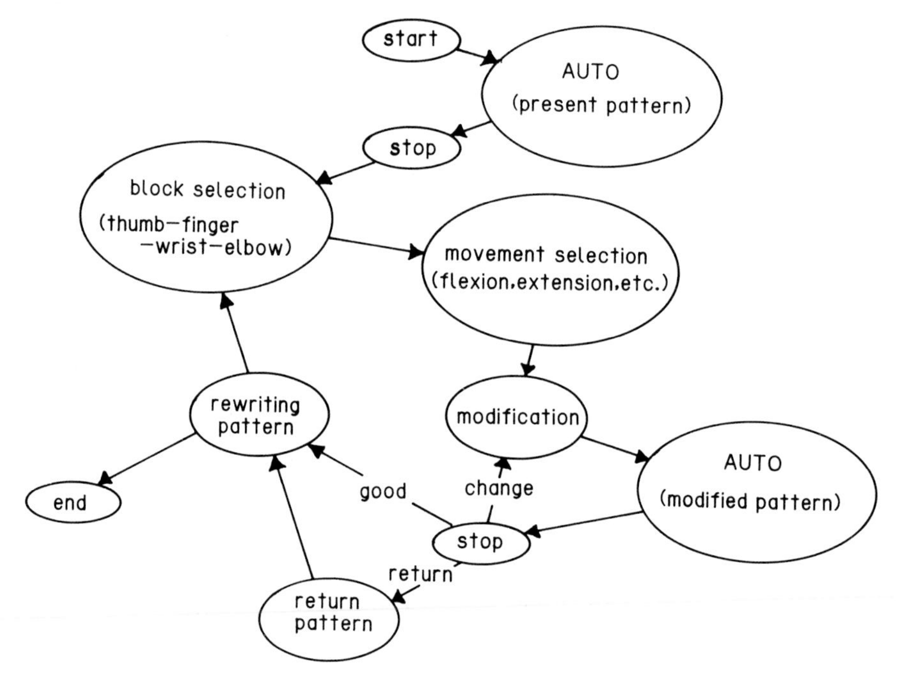

Figure 9 Outline of the flow in the interactive processing.

6 CONCLUDING REMARKS

A master–slave type multi-channel FES system enabled us to carry out PAM type and PWM type stimulation in up to 48 channels. In clinical experiments using this system, it was confirmed that the EMG-referenced creation method of the multi-channel stimulation data was quite efficient and practical in both PAM and PWM type methods. But there was no clear relation between the pulse amplitude threshold and pulse width threshold, and between the pulse amplitude maximum and pulse width maximum, in two modulation methods. In clinical usage, simple manual adjustment for each neuromuscular system may be inevitable, since distance between an electrode and objective nerve fibers is different, and also each nerve trunk is different in diameter. We can conclude that control capabilities between PAM and PWM are nearly the same so that the EMG-referenced strategy can also be adopted in a PWM type system. Although we emphasized the high performance FES system in this paper, development of a small portable type [13] or an implantable type [14] will also be important in ADL usage.

Interactive modification capabilities were first introduced in this system. We do not have enough experience, but a patient was willing to adapt and use it actively. We believe that this approach is quite important in the next generation of FES systems.

In future, a more elaborate FES system, for example a multi-layered control system, will be practically available. This master–slave type system is a first step of this approach.

ACKNOWLEDGEMENTS

The authors wish to thank members of our FES research group for their sincere cooperation, especially to Miss M. Yasojima and Dr R. Futami for their technical support and valuable discussion. This work is greatly indebted to Miss M. Suejima. This work was supported by the Ministry of Education, Science & Culture of Japan under a Grant-in-Aid for Developmental Scientific Research No. 60850073 (1985–1986) and No. 61890007 (1986–1987), under a Grant-in-Aid for Scientific Research No. 62460130 (1987–1989), and Japan Research & Development Cooperation (1985).

REFERENCES

1. Y. Handa, Y. Shimada, S. Komatsu, A. Naito, M. Ichie, Y. Nakatsuchi, R. Yagi, Y. Sugimoto, K. Iijima, R. Futami and N. Hoshimiya, Electrically induced hand movements and their application for daily living, in *Proc. 8th Int. Symp. ECHE (Dubrovnik)* (1984), pp. 169–180.
2. Y. Handa, T. Handa, Y. Nakatsuchi, R. Yagi and N. Hoshimiya, A voice-controlled functional electrical stimulation system for the paralyzed hand, *Jap. J. Med. Electr. Biol. Eng.*, **23**, 292–298 (1985).
3. Y. Handa, M. Ichie, T. Handa, R. Yagi and N. Hoshimiya, Control of the paralyzed hand by a computer-controlled FES system, in *Proc. of IEEE/7th Ann. Conf. EMBS (Chicago)* (1985), pp. 322–326.
4. N. Hoshimiya, K. Iijima, R. Futami, Y. Handa and M. Ichie, A new FES system for the paralyzed upper extremities, in *Proc. of IEEE/7th Ann. Conf. EMBS (Chicago)* (1985), pp. 327–330.
5. Y. Handa, A. Naito, M. Ichie, T. Handa, N. Matsushita and N. Hoshimiya, EMG-based stimulation patterns of FES for the paralyzed upper extremities, in *Proc. 9th Int. Symp. ECHE (Dubrovnik)* (1987), pp. 329–337.
6. P. E. Crago, P. H. Peckham and G. B. Thrope, Modulation of muscle force by recruitment during intramuscular stimulation *IEEE Trans. Biomed. Eng.*, **27**, 679–684 (1980).
7. M. Ichie, Y. Handa, A. Naito, T. Handa, N. Matsushita and N. Hoshimiya, EMG analysis of the thumb and its application to FNS, in *Proc. of IEEE/8th Ann. Conf. EMBS (FortWorth)* (1986), pp. 538–540.
8. N. Matsushita, Y. Handa, M. Ichie, A. Naito, T. Handa and N. Hoshimiya, Analysis of wrist movements and its application to FNS, in *Proc. of IEEE/8th Ann. Conf. EMBS (FortWorth)* (1986), pp. 618–619.
9. Y. Handa and N. Hoshimiya, Functional electrical stimulation for the control of the upper extremities, *Med. Prog. Technol.*, **12**, 51–63 (1987).
10. Y. Handa, N. Hoshimiya, Y. Iguchi and T. Oda, Development of percutaneous intramuscular electrode for multi-channel FES system, *IEEE Trans. Biomed. Eng.* (in press).

11. N. Hoshimiya, T. Izumi, M. Tsuruma, R. Futami, T. Ifukube and Y. Handa, Electro-cutaneous phantom sensation as a sensory feedback method, in *Proc. 9th Int. Symp. ECHE* (*Dubrovnik*) (1987), pp. 341–351.
12. N. Hoshimiya, A. Naito, M. Yajima and Y. Handa, A multi-channel FES system for the restoration of motor functions in high spinal cord injury patients: a respiration-controlled system for multi-joint upper extremity, *IEEE Trans. Biomed. Eng.* (in press).
13. Y. Handa, K. Ohkubo and N. Hoshimiya, A portable multi-channel FES system for restoration of motor function of the paralyzed extremities, *Automedica*, this issue.
14. B. Smith, P. H. Peckham, M. W. Keith and D. D. Roscoe, An externally powered multichannel implantable stimulator for versatile control of paralyzed muscle, *IEEE Trans. Biomed. Eng.*, **34**, 499–508 (1987).

Automedica, 1989, Vol. 11, pp. 221–231
Reprints available directly from the publisher only
Photocopying permitted by license only

A PORTABLE MULTI-CHANNEL FES SYSTEM FOR RESTORATION OF MOTOR FUNCTION OF THE PARALYZED EXTREMITIES

YASUNOBU HANDA

Department of Anatomy, Tohoku University School of Medicine, Sendai 980, Japan

KIYOSHI OHKUBO

NEC San-ei Co. Ltd., Tokyo 183, Japan

and

NOZOMU HOSHIMIYA

Faculty of Engineering, Tohoku University, Sendai 980, Japan

(Received September 1988)

A portable multi-channel FES system was developed in order to restore the motor function of the paralyzed extremities in stroke and spinal cord injury patients. The system consists of an 8-bit microcomputer and 3 channels of A/D and 16 channels of D/A converters. In addition, a stimulation data creating system (SDC system) consisting of a personal computer (NEC PC-98LT) was also developed. Standard stimulation data for controlling the paralyzed upper or lower extremities were created from integrated EMG data during motion for activities of daily living in normal subjects and were stored in the memory of SDC system. Stimulation data for each patient were automatically created by inputting the maximum and threshold stimulating voltages of individual muscles to the standard data through a keyboard of the SDC system, and were sent to the memory of the portable FES system. This system realized versatile usage of the portable FES system for restoration of motor function in various types of paralysis of the extremities. Furthermore, a very complicated multi-joint system with many degrees of freedom such as the paralyzed upper extremities in C4 quadriplegics could be easily controlled with high reliability by this system. This paper describes system design, function and clinical applications of the portable FES system.

Keywords: Functional electrical stimulation, portable FES system, multi-channel stimulation, paralyzed extremity, neural prostheses.

1 INTRODUCTION

Programmed electrical stimulation can restore motor function of paralyzed upper extremities [1–4] as well as lower extremities [5, 6] induced by upper motor neuron disorders. Recent advances of computer technology have enabled us to create very complicated multi-channel stimulation patterns for controlling the paralyzed upper extremities with multi-degrees of freedom [2–4]. For precise and well-coordinated control of movement of the upper extremities in three-dimensional space, a spatio-temporal sequence of stimulation is required. Recently, our research revealed that

Mailing address: Dr. Yasunobu Handa, Dept. of Anatomy, Tohoku University School of Medicine, Sendai 980, Japan.

multi-channel stimulation patterns based on muscle activity (electromyography, EMG) during motions of the upper extremities in normal subjects were practically useful for the restoration of functional movements of the upper extremities [2–4]. These multi-channel FES systems were developed by using commercially available personal computers, and these were clinically applicable to a certain extent. For the widespread use in activities of daily living (ADL) by the patient, miniaturization of the system might be essential.

In this paper, system design, specifications and some examples of clinical application are described on a practical portable FES system which we recently developed.

2 SYSTEM CONFIGURATION

Figure 1 is a photograph showing the overall system. The right-hand side is a portable FES system (stimulator), and the left-hand side is a stimulation data creating system (SDC system) which supports the portable FES system. The SDC system consisted of a portable type personal computer (NEC PC-98LT) with a printer in an attaché case, and the portable system was originally developed by

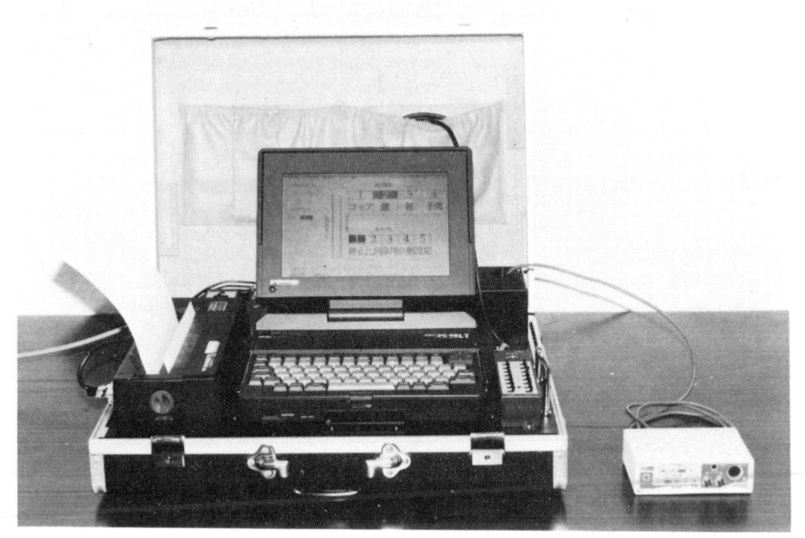

Figure 1 Portable multi-channel FES system (right) and stimulation data creating system (SDC system) (left).

modifying our previous system [7]. In Figure 2, a block diagram of the portable FES system is shown together with a SDC system. The portable FES system was composed of an 8-bit microcomputer (Z80 compatible CPU) and 3 channels of A/D and 16 channels of D/A converters. Standard stimulation data for controlling the paralyzed upper or lower extremities were created by a trapezoidal approximation of integrated EMG data during motions for ADL in normal subjects, and were stored in the memory of the SDC system. Here, stimulation data meant amplitudes of pulse trains (pulse width, 0.2 ms; pulse interval, 50 ms), i.e., pulse amplitude

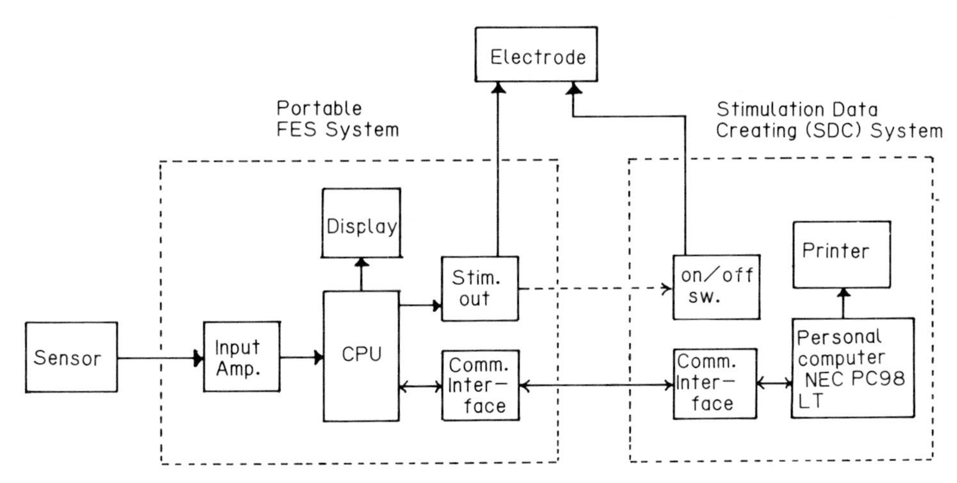

Figure 2 Block diagram of the portable FES and SDC systems.

modulation (PAM) was used for control of the recruitment characteristics of the neuromuscular system. Stimulation data for each patient were automatically created by inputting the maximum and threshold of stimulating pulse amplitudes for the individual neuromuscular system through a keyboard of the SDC system, and were sent to the memory of the portable FES system through serial communication channel (1 channel, 9600 baud).

After the individual stimulation data were established, the portable FES system could be separated from the SDC system for ADL usage. Stimulation was

Table 1 Specifications of a portable FES system

(Input for volitional control)	
Number of analog inputs	2 channels
Number of logical inputs	1 channel
(Stimulation output)	
Number of outputs	16 channels
pulse width	0.2 ms (constant)
pulse interval	50 ms (constant)
pulse amplitude	0-- 15 volts
(Display)	
Modes (class of movements)	1,2,3,4
Operation status	Stop, Proportional, Hold, Auto, Setting
Battery capacity warning	
(Auditory monitoring by earphone)	
Sound generation	at logical input at change of mode at change of status
(Size and Weight)	
Size	9 × 14.5 × 4.5 cm
Weight	*ca.* 500 g
(Battery)	
U-3 battery	4

performed by the portable FES system through percutaneous intramuscular electrodes. Basic function of the portable FES system was the same as that of our recent system [4, 7]. Specifications of the portable FES system are summarized in Table 1. The portable FES system was 9 × 14.5 × 4.5 cm in size and about 500 g in weight. It had 16 output channels for stimulation and 3 input channels for control signals from the patient. Some new facilities were included in the SDC system. There were six kinds of graphic displays on a liquid crystal display of the SDC system, i.e., Entry display, Test display, Control display, Auto display, Monitor display and Graph display. In Table 2, specifications of these 6 displays are summarized. The SDC system has a small printer to provide hard copies of these displays.

Table 2 Contents of six presentation on a SDC display

1. ENTRY display
   - Graphic modes (class of movements): 4 modes.
     - (1) Cylindrical grasp
     - (2) Key grip
     - (3) Parallel extension grip (grasp for newspaper)
     - (4) Auxiliary

2. TEST display
   - Stimulation amplitude.
     - Maximum: 15, 7.5, 3.75, 1.88, 0.94 (volts)
     - Threshold: adjustable within a range of maximum.
   - Stimulation channels: 16

3. CONTROL display
   - Stimulation channels: 16
   - Repetition rate of stimulation:
     - 5, 25, 50, 75, 100 s/repetition or
     - 1-256 s (selectable)

4. AUTO display
   - Stimulation channels: 16
   - Repetition rate of stimulation:
     - 25, 50, 75, 100 s/repetition
   - Rest time:
     - 25, 50, 75, 100 s or
     - 1-256 s (selectable)
   - Training time:
     - 15, 30, 45, 60 min or
     - 1-99 min (selectable)

5. MONITOR display
   - Graphic mode monitor: 4 modes
   - Operational status monitor:
     - Stop, Proportional, Hold, Auto, Setting
   - Control input monitor:
     - ON/OFF input, Proportional input
   - Hold position monitor:
     - Position at which mode was previously switched from "Proportional" to "Hold" mode.

6. GRAPH display
   - Simultaneous display numbers of stimulation data: 8
   - Number of presentation: 2
   - (Total number of stimulation data: 16)

In Figure 3, five typical hard copies of the displays of the SDC system are shown. Figure 3(a) is a copy of the "ENTRY" display. Date, Patient's ID No., Patient's name, Modes (class of motions) No.1–4 with data creation date were displayed, and then the system asked the patient to select the mode. In this case, mode No. 3 was selected as shown in the display.

Figure 3(b) is a copy of "TEST" display. This is one example (FCR) of 16 stimulation patterns. One set of stimulation data for one neuromuscular system was stored in a memory (RAM) with 250 address areas as shown. The dotted line is an example of a standard stimulation data, and the solid line is an example of actual stimulation data after modification by inputting the threshold and the maximum amplitude for the individual neuromuscular system. Maximum dynamic range (−15 volts), threshold voltage (1.08 volts minimum) and maximum voltage (7.02 volts maximum) are displayed in the figure. Figure 3(c) is a copy of the "CONTROL" display. The dynamic range of actual stimulation amplitudes of 16 channels were displayed, and a bold vertical line was also displayed on the same display during proportional control mode in order to show the assigned memory address by a volitional signal. "AUTO" display is not shown here, but is quite similar to the "CONTROL" display. The difference is that automatic training conditions are shown. Figure 3(d) is a copy of "MONITOR" display. Presently working "Mode" (class of movement) and "Status" were simultaneously shown with black color. A bold vertical solid line on the left, and a bold vertical line on the right indicated volitional analog input for proportional control, and analog input level at which "hold" status had been specified, respectively. Figure 3(e) is a copy of "GRAPH" display. Eight stimulation data out of 16 data files were simultaneously shown. Dotted lines were standard patterns obtained from trapezoidal approximation of EMGs, and solid lines were actual stimulation data for individual neuromuscular system. (note: these patterns are different from clinically used patterns. For clinical data, see the references [4, 8].

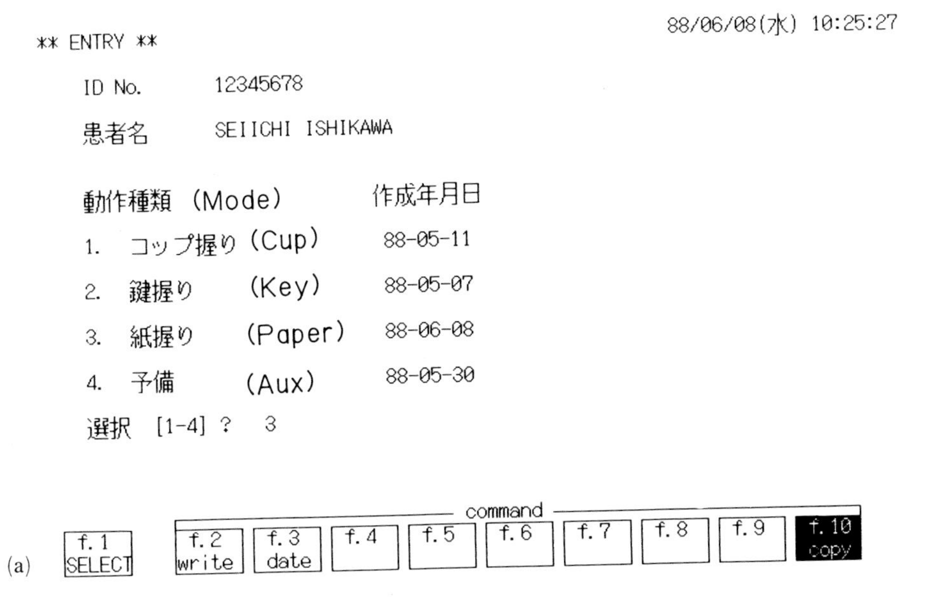

Figure 3 Display information of the SDC system. (a) ENTRY.

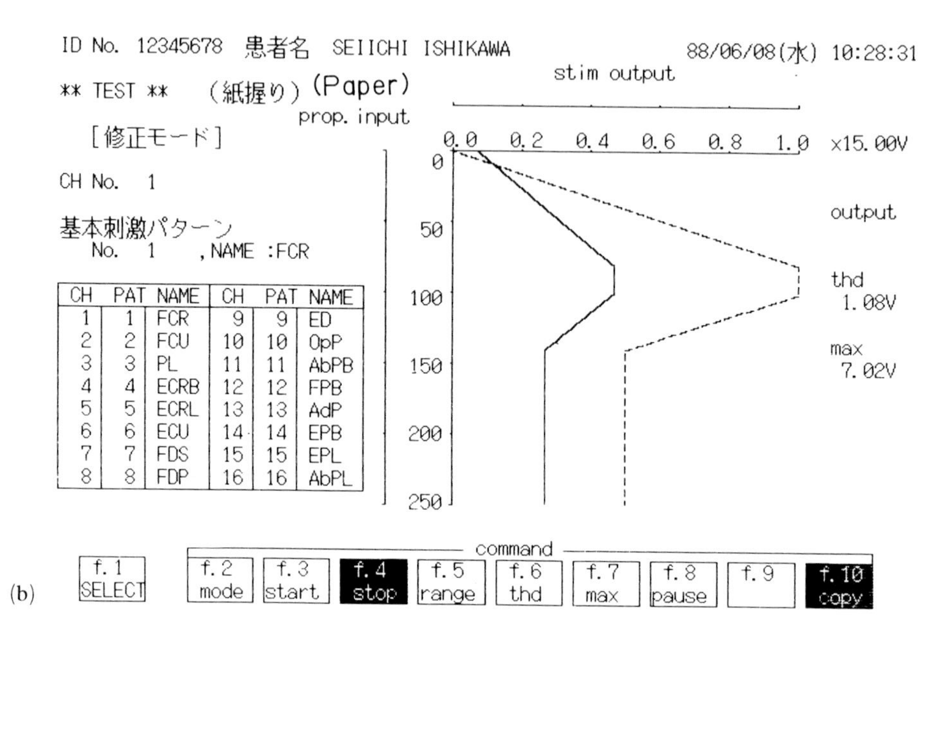

(b)

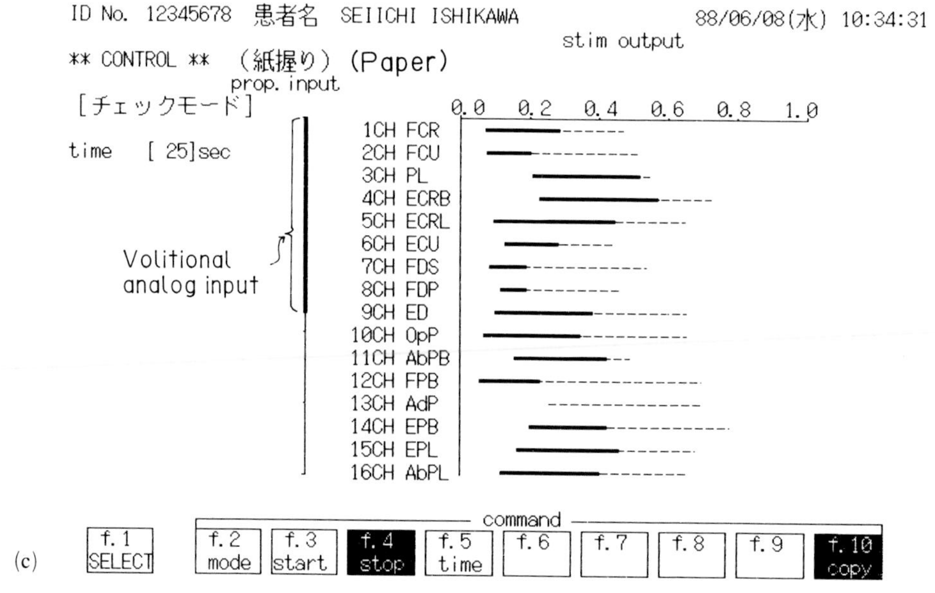

(c)

Figure 3 (continued) (b) TEST, (c) CONTROL.

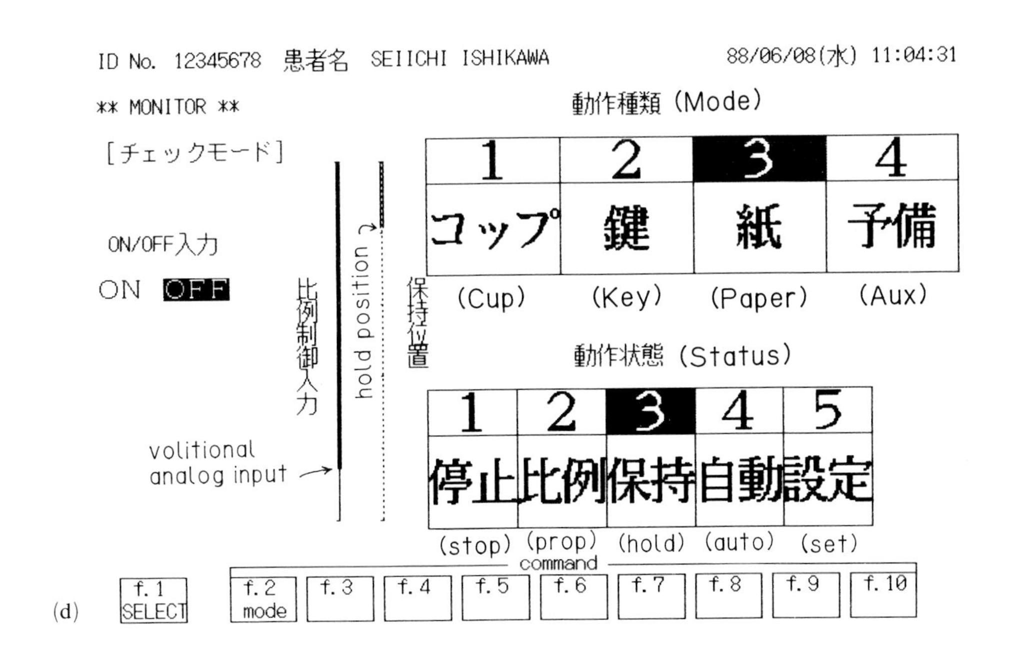

(d)

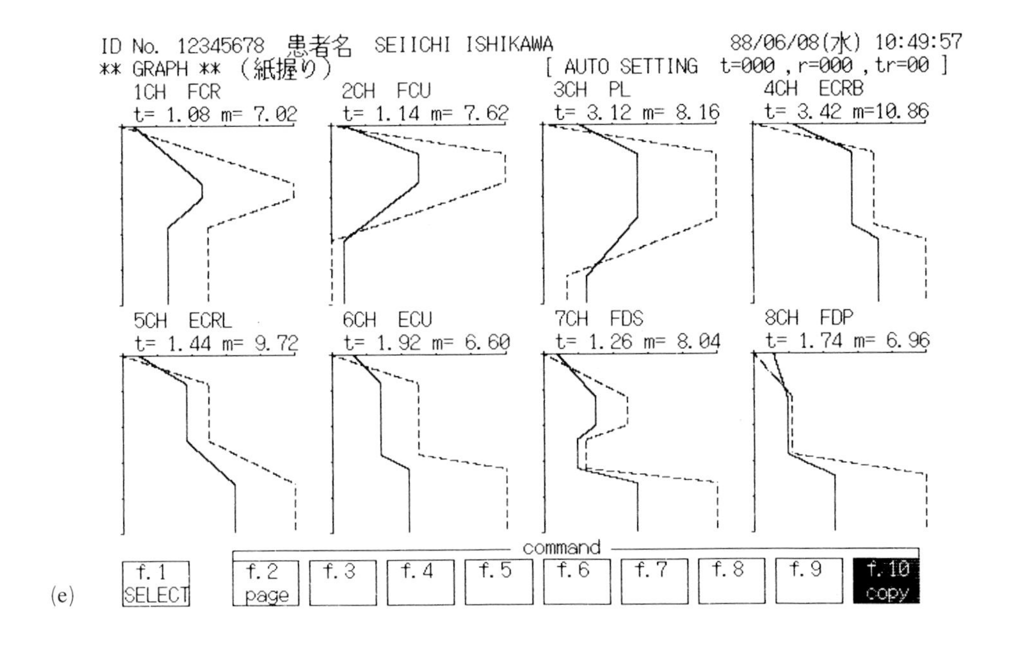

(e)

Figure 3 (continued) (d) MONITOR, (e) GRAPH.

3 CLINICAL APPLICATION

This portable FES system was applied to stroke and spinal cord injury (SCI) patients. Some of the stroke patients with hemiplegia could not utilize their paretic upper limbs for ADL because of difficulty in hand opening. Flexion of the thumb and fingers and flexion/extension of the elbow were volitionally performed by themselves. In these cases, application of programmed stimulation data for hand opening through percutaneous intramuscular electrodes could restore upper limb function for ADL such as drinking, writing and so on [Figure 4(a)]. Figure 4(b) shows an example of application to the paralyzed upper limb of a C4 quadriplegic patient. As shown in this figure, a respiration sensor which consisted of two mouthpieces (A) and a box (B) containing two pressure sensors (Dinamation Co. Ltd.: Figure 5) was used for detecting control commands by expiration and inspiration of the patients. Opening and grasping of the hand and flexion and extension of the wrist were controlled by the combination of the signals from the respiratory sensor [8]. She could apply lipstick which could be grasped by the portable multi-channel FES system. The portable system can be seen on her waist in Figure 4(b).

Drop foot during the swing phase of walking in stroke and SCI patients was also corrected by the portable FES system. Furthermore, synchronous operation of four sets of the portable FES systems realized standing up from a chair in a paraplegic patient is shown in Figure 4(c).

Thus it has been proved that essential minimum ADL movements could be realized by this portable multi-channel FES system.

Figure 4 Examples of clinical applications of the portable FES system. (a) Drinking by FES-control in hemiplegic patient, (b) Applying lipstick grasped by FES-induced grasp in a quadriplegic patient, and (c) FES-induced standing in a paraplegic patient.

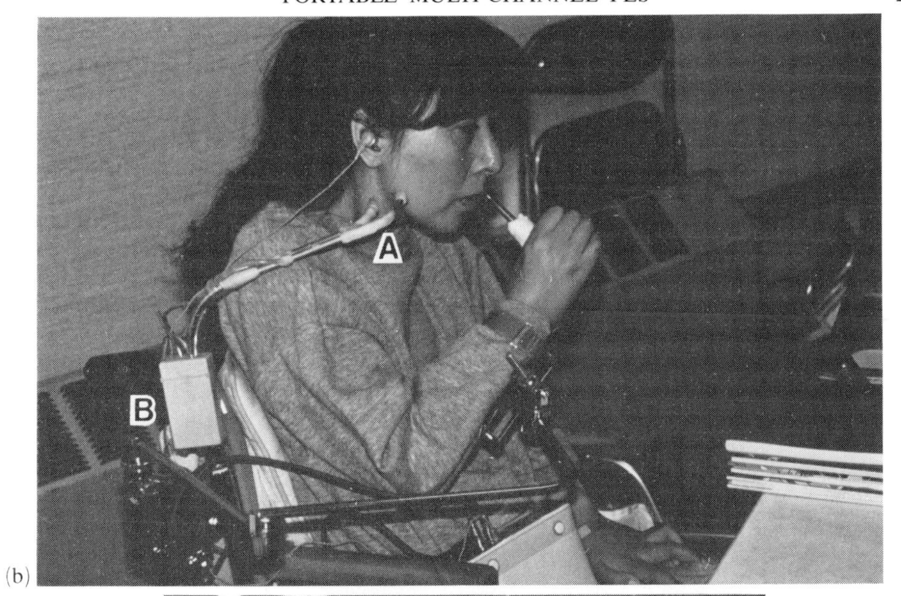

(b)

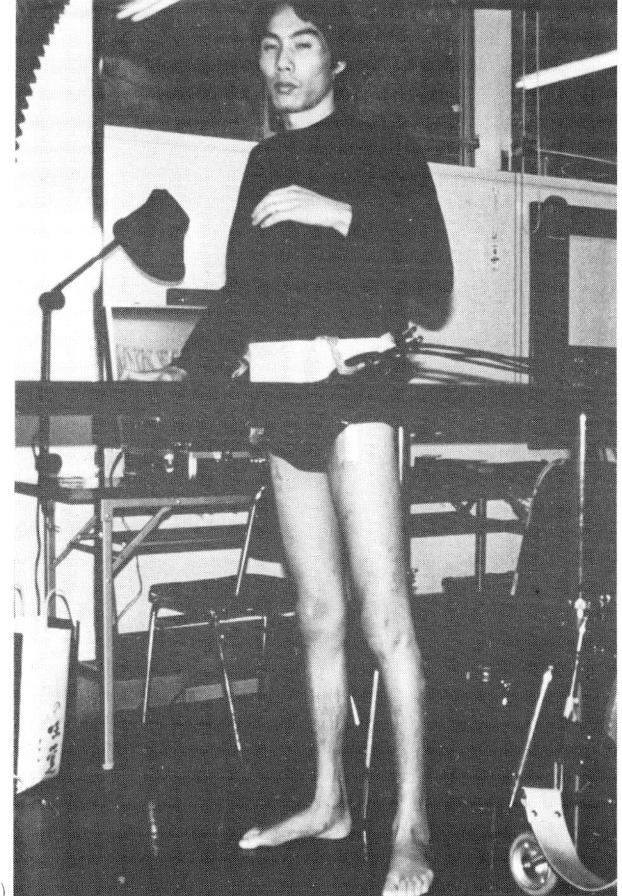
(c)

Figure 4 (continued)

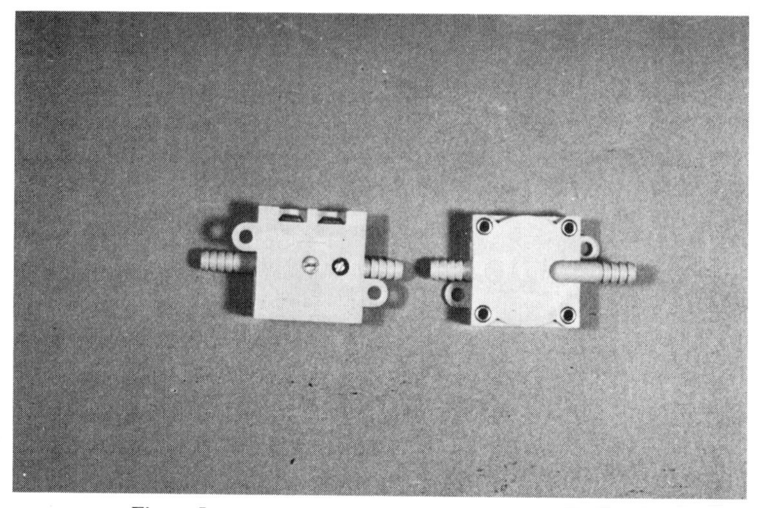

Figure 5 Pressure sensors used in a respiratory sensor.

4 CONCLUSION

It was found that combination of the portable FES system and stimulation data creating system (SDC system) had fundamentally the same function as the personal computer based FES system for laboratory use which we reported elsewhere [2–4]. The function of the SDC system is as follows:

1) Measurement of threshold and maximum stimulation voltages of individual muscles to be controlled.

2) Stimulation data creation for each patient.

3) Test stimulation for checking and improving FES-induced movements.

4) Transfer of the stimulation data to a RAM of the portable FES system.

For daily usage of the FES system in ADL, miniaturization has been required. The newly developed portable FES system is small and light enough to install on the waist of the patient. Power supply is maintained by two sets of rechargeable Ni–Cd batteries. Clinically, the paralyzed limbs of the stroke and SCI patients were mostly controlled by this FES system with 16 channel outputs. In the case of more severe patients, however, more than sixteen output channels are necessary. For example, simultaneous or sequential control of the hand–wrist–elbow system in the C4 quadriplegic was provided by the 30 or 48 channel FES system for laboratory use [8, 9]. In this case, synchronous usage of two sets of the portable FES system is required. Standing in the paraplegic patients was achieved by 4 sets of the portable FES systems with 64 output channels. Synchronous operation of these devices was performed by simultaneous input of control commands to individual devices. For practical usage of such a multi-device system, establishment of a communicating system among these portable FES systems (for instance, a multi-layered control system) is essential. Along with this, further miniaturization is required for installing the multi-device system on the patient.

ACKNOWLEDGEMENTS

The authors wish to thank Mr. Seiichi Ishikawa and collaborators of NEC San-ei Co. Ltd. for their technical support during development of the portable FES system. Thanks are also given to members of our FES research group for their cooperation. This work is greatly indebted to Miss M. Suejima. This work was supported by the Ministry of Education, Science & Culture of Japan under a Grant-in-Aid for Developmental Scientific Research No. 60850073 (1985–1986) and No. 61890007 (1986–1987), and under a Grant-in-Aid for Scientific Research No. 62460130 (1987–1989), and also Japan Research & Development Cooperation (1985).

REFERENCES

1. R. H. Nathan, The development of a computerized upper limb stimulation system, *Orthop.*, **7**, 1170–1180 (1984).
2. Y. Handa, T. Handa, Y. Nakatsuchi, R. Yagi and N. Hoshimiya, A voice-controlled functional electrical stimulation system for the paralyzed hand, *Jap. J. Med. Electr. Biol. Eng.*, **23**, 292–298 (1985).
3. Y. Handa, M. Ichie, T. Handa, R. Yagi and N. Hoshimiya, Control of the paralyzed hand by a computer-controlled FES system, in *Proc. IEEE/7th Ann. Conf. EMBS (Chicago)* (1985), pp. 322–326.
4. Y. Handa and N. Hoshimiya, Functional electrical stimulation for the control of the upper extremities, *Medical Prog. thr. Tech.*, **12**, 51–63 (1987).
5. E. B. Marsolais and R. Kobetic, Functional walking in paralyzed patients by means of electrical stimulation, *Clinical Orthopaedics*, **175**, 30–36 (1983).
6. J. S. Petrofsky and C. A. Phillips, Closed-loop control of movement of skeletal muscle, *CRC Critical Reviews in Biomedical Engineering*, **13**, 35–75 (1985).
7. Y. Handa, T. Handa and N. Hoshimiya, A portable FNS system for the paralyzed upper extremities, in *Proc. IEEE/8th Ann. Conf. EMBS (Fort Worth)* (1986), 65–67.
8. N. Hoshimiya and Y. Handa, A multi-channel FES system for the restoration of motor functions in high spinal cord injury patients: a respiration-controlled system for multi-joint upper extremity, *IEEE Trans. Biomed. Eng.* (in press).
9. N. Hoshimiya and Y. Handa, A master–slave type multi-channel functional electrical stimulation (FES) system for the control of the paralyzed upper extremities, *Automedica*, this issue.

Automedica, 1989, Vol. 11, pp. 233–245
Reprints available directly from the publisher only
Photocopying permitted by license only

A PRACTICAL APPROACH TO MULTIFUNCTIONAL ABOVE KNEE PROSTHESIS

SUJOY K. GUHA

Centre for Biomedical Engineering, Indian Institute of Technology and All-India Institute of Medical Sciences, New Delhi 110016, India

and

KRISHAN K. CHAUDHRY

Dept. of Applied Mechanics, Indian Institute of Technology, New Delhi 110016, India

(Received September 1988)

An above knee prosthesis has been designed to provide the functions of walking, squatting, kneeling and sitting cross-legged. Although multi-functional, the construction is simple and can be made with components which are readily available. Assembly as well as maintenance can be carried out in small workshops with simple machines and no special molding setup. Parameter sensitivity in the function is not high and so tolerance specifications can readily be met. The prosthesis is particularly suitable for use in rural areas.

Keywords: Above knee prosthesis, squatting, modular, sitting cross-legged, simple design, rural fabrication and use

1 INTRODUCTION

Social structure in developing countries is so evolving that all facilities for the handicapped have to be provided. Both urban and rural population have to be covered and yet there are limitations on the access to advanced technology emerging from specialized manufacturing centers. The only solution lies in developing designs which are based on commonly available components and can be implemented in small workshops which have only simple machines and tools. A survey carried out in India showed that all over the country there are workshops which cater to the needs of repair of agricultural machines and tools, trucks and tractors as well as the ubiquitous bicycle. Machines available are simple types of centre lathe, drilling machine, grinding machine and either electrical arc or oxyacetylene gas type welding equipment. A design should be so simple that an artificial leg can be fabricated, maintained and repaired by the rural craftsmen with materials which are commonly used for the maintenance of equipment routinely handled by such workshops, that is, the agricultural machinery, automobiles and rural vehicles like bullock carts.

The above requirements and constraints immediately rule out the use of specially molded and cast metal and plastic components. There is a problem in fabrication by welding and using simple workshop facilities that the dimensional accuracies and tolerances achieved are seldom very good. If the design is such that the performance is unduly sensitive to the size and accuracies, then obviously it is not possible to adopt the general workshop type of fabrication procedures. Therefore,

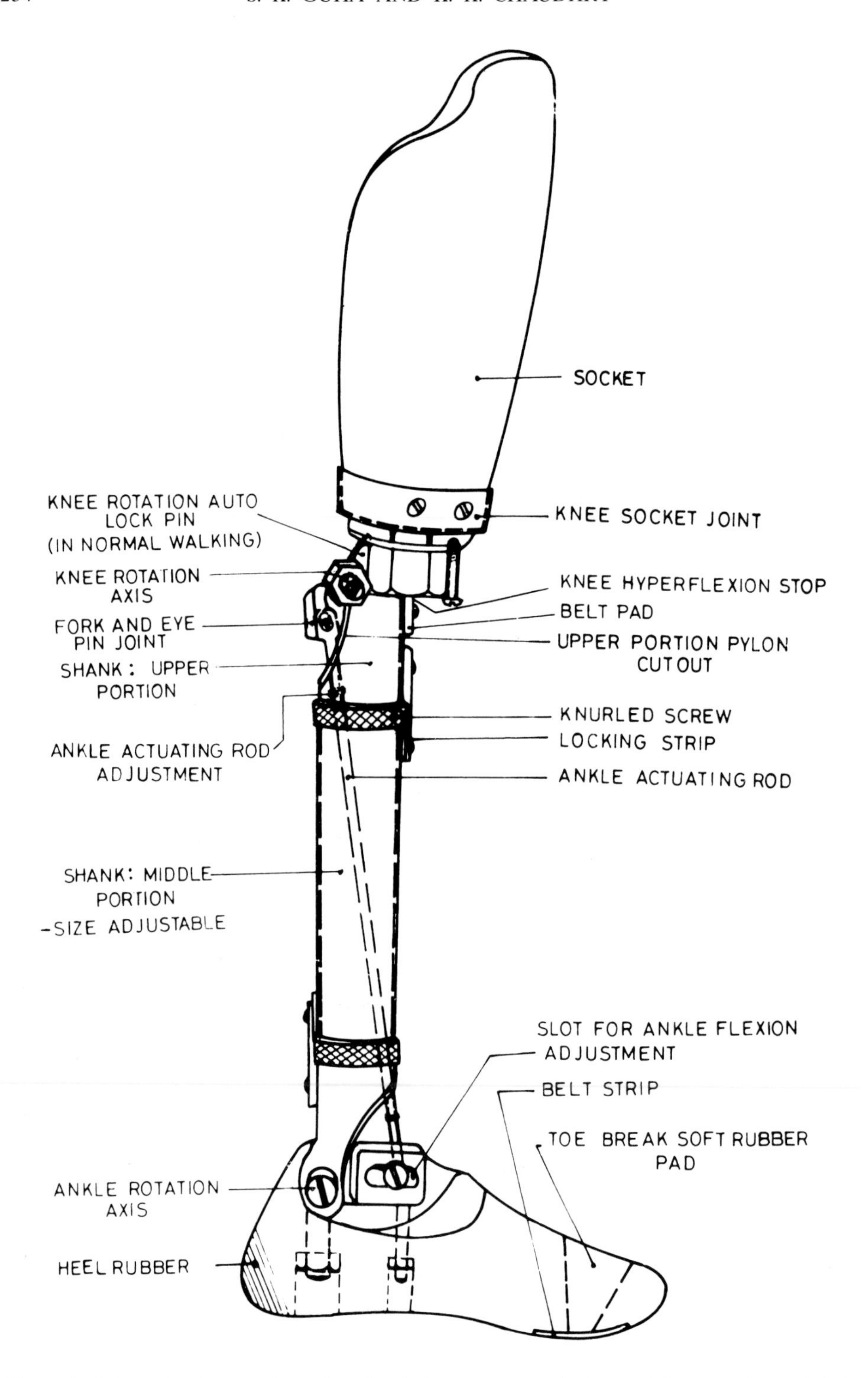

Figure 1 Schematic diagram of multifunctional above knee prosthesis permitting squatting and sitting cross-legged.

one essential requirement of the design is that the function as well as durability should not be unduly affected by marginal deviations from the absolutely ideal manufacturing specifications. In spite of all these limiting factors the functional capability to be incorporated in the design has to be more than usually available in conventional limbs. Not only walking and sitting on a chair should be possible but living conditions require the squatting, sitting cross-legged and kneeling postures to be made available. The present design and technology has been formulated keeping in mind all these requirements.

2 MATERIALS AND METHODS

A drawing of the prosthesis made is shown in Figure 1. The basic construction is around the pylon made of thin walled steel piping of wall thickness 1 mm and 4 cm outer diameter of the type used for making automobile shock absorbers. The shank middle portion is made in four different sizes to fit persons of varying heights. In addition, on the upper and as well as the lower end of this piece, there is a screw thread over which the connector to the knee and the ankle are screwed. Fine adjustments of length can be made with these two screw arrangements. Following adjustment, accidental rotation at these screws is checked by means of knurled nuts with screw threads inside as well as locking strips which have a sliding arrangement to accomodate leg length variations. Linkage between the knee and the ankle is established by means of an ankle actuating rod which runs from a fork and eye pin joint in the knee piece to a metal fixture on the ankle which has a slot for ankle flexion adjustment. Thus, a four bar linkage mechanism is established.

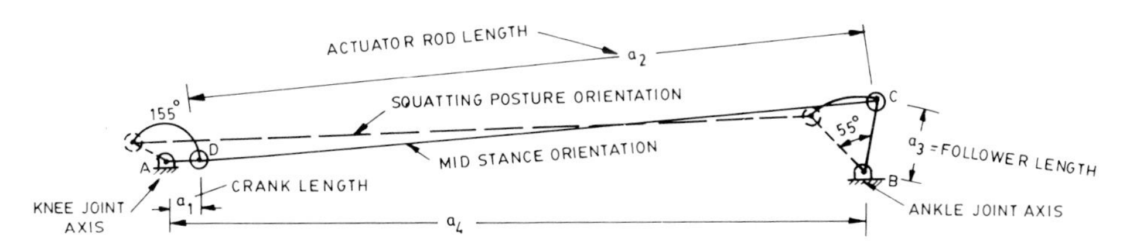

Figure 2 Schematic diagram of the four bar linkage mechanism.

2.1 *Analysis of Four Bar Mechanism*

In the planer four bar mechanism provided in the prosthesis, the knee and the ankle rotation axes are fixed base points. The crank portion is in the knee joint and the follower link is on the ankle joint end. The adjustable actuating rod connects both the crank and the follower (Figure 2). The crank follower mechanism can be analyzed using the Freudenstein equation [Eq. (1)] which is as follows [1]:

$$K_1 \text{Cos } \phi - K_2 \text{Cos } \psi + K_3 = \text{Cos}(\phi - \psi), \tag{1}$$

where

$$K_1 = \frac{a_4}{a_3}; \quad K_2 = \frac{a_4}{a_1}; \quad \text{and} \quad K_3 = \frac{(a_1^2 - a_2^2 + a_3^2 + a_4^2)}{2a_1 a_3}.$$

The lengths of the various links of the four bar mechanism can be easily designed by simple hand calculations using three accuracy points which are the three positions of the crank angles ϕ and the corresponding three follower angles ψ. The solution of the Freudenstein equation using the three crank and follower angle pairs gives the proper values of a_1, a_2, a_3 and a_4, which are the lengths of the four bar linkage mechanism.

As the three accuracy points in the four bar mechanism of the present prosthesis are the knee as well as ankle angles, these should be chosen in such a manner so as to match the prosthetic gait to the normal human locomotion. Lamoreux [2] has reported that the knee flexion angles during the stance phase are: 0° at heel strike, 15° at foot flat, 10° at mid stance and 35° at heel off. The corresponding angles at ankle joint are: 0° at heel strike, 10° planter flexion at foot flat, 4° dorsiflexion at mid stance, 15° dorsiflexion at heel off. In addition, the knee flexion is 155° in the squatting position and the corresponding ankle dorsiflexion is 55° at this position. In the present design the three accuracy points are: angular positions of pairs of knee angles and the corresponding ankle angle for normal standing/mid stance phase, the beginning of the swing phase and the squatting phase, the matching being with respect to these parameters.

Comparison of prosthetic gait with normal gait The four bar mechanism mimics the near normal human locomotion at mid stance, the beginning of the swing phase and the squatting posture by virtue of the use of three above mentioned accuracy points. At heel strike, the angular setting of the foot is at 10° planter flexion. The soft rubber heel provides 5–7° of dorsiflexion during heel strike phase. Thus the ankle angle is 3–5° planter flexion during heel strike which is close to the normal ankle angular configuration. Furthermore, at foot flat the ankle takes up the normal angular setting of 10° planter flexion. At mid stance the rubber pads at the toe break get compressed with a resultant 4° dorsiflexion of the foot. Lastly, at the heel off posture the rubber pads get compressed further with the resultant 15° dorsiflexion. Hence, the three accuracy points and the three adjustments derived from the soft rubber cushion heel and the toe break rubber pads gives six positions where the prosthetic gait is matched with the normal gait.

2.2 *Knee Joint*

A uniaxial knee joint is preferred on account of simplicity as well as adequacy. In order to check lateral movements the joint is made of screw and nut with a simple lubrication port. The screw part of the knee joint is welded onto a large size nut. Another short screw of diameter 1.75 cm passes through this large nut and serves as a connector to the thigh socket. By virtue of this arrangement, the socket can rotate to provide the sitting cross-legged facility. The advantage of this arrangement is that the entire knee joint assembly is common to all sizes of the prostheses. Also, the components are standard items widely used in machinery.

In the normal standing position as well as in the early phase of bending of the knee as shown in Figure 3, the rotation of the thigh socket must be restricted. This requirement is met by providing a rotation locking strip as shown in Figure 4 (lower diagram). The shape of the locking strip is so designed that after a certain knee flexion of about 35°, the rotation of thigh is enabled but does not take place abruptly. Thigh rotation can be increased gradually as knee flexion increases. Such a mode is essential to ensure that when going into sitting cross-legged posture, the thigh rotation does not occur all at once with the possible consequence of the subject falling or becoming unstable. When rising from the sitting cross-legged position again, the thigh rotation is reversed automatically by a spring shown in Figure 3 (upper diagram). Once again, the contour of the locking strip ensures that the return is gradual and well correlated with knee flexion.

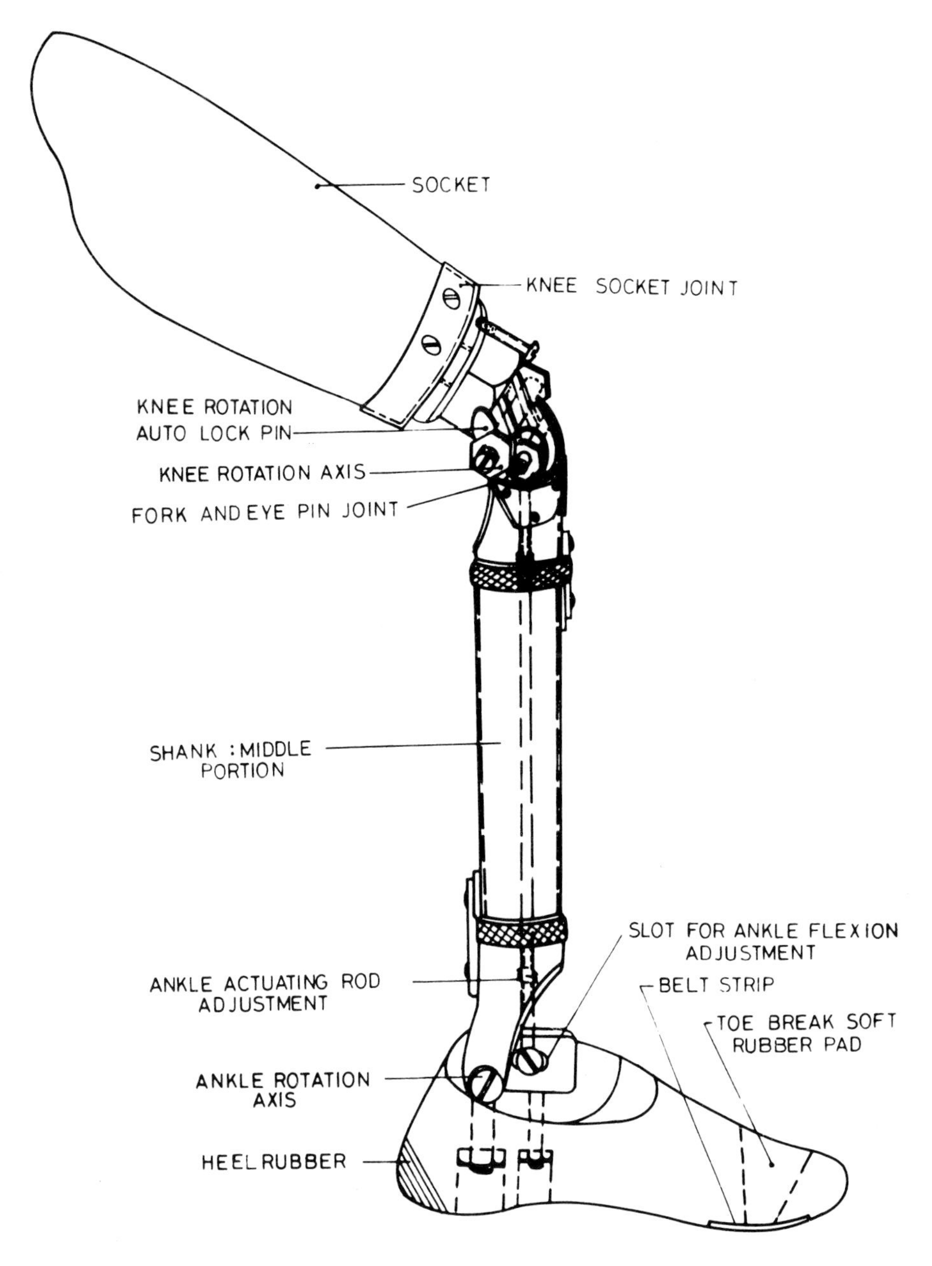

Figure 3 Schematic diagram of multifunctional above knee prosthesis in the early phases of bending the knee to achieve squatting and sitting cross-legged.

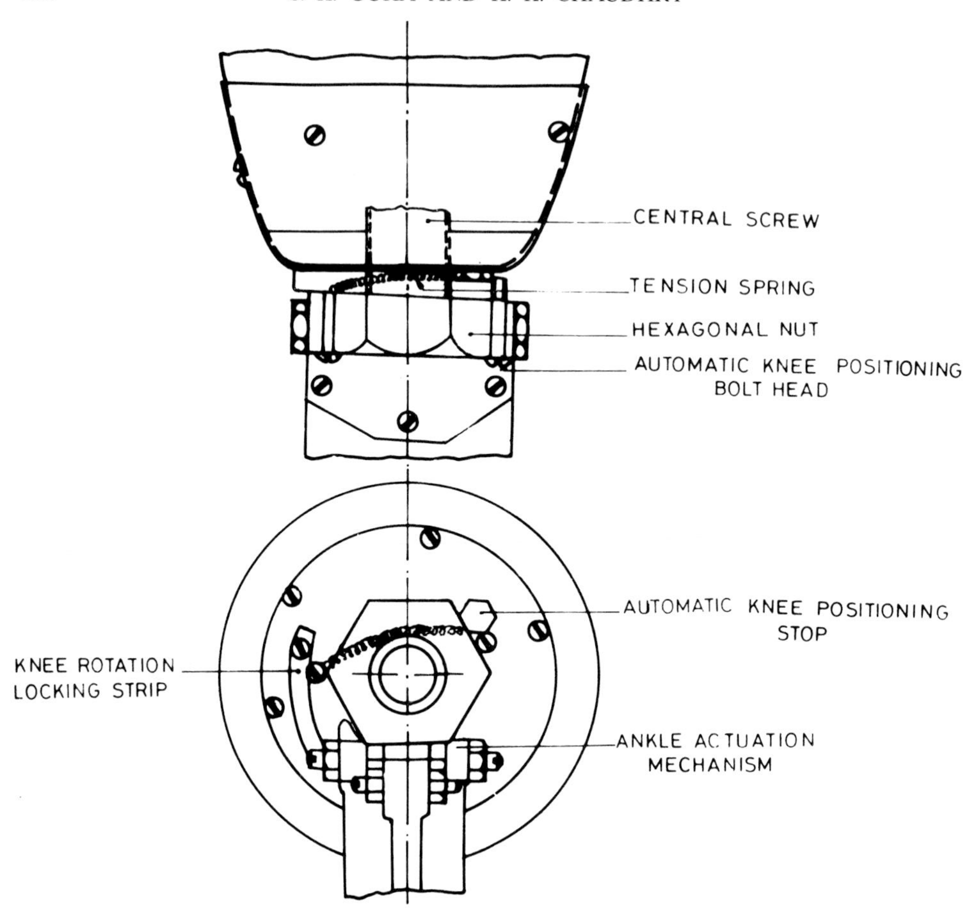

Figure 4 Details of the knee joint of the multifunctional above knee prosthesis.

2.3 *Foot–Ankle Assembly*

Ankle flexion can occur around the ankle rotation axis (Figure 1). The foot is pulled up by the ankle actuating rod. If this rod is advanced in the slot for ankle adjustment the ankle flexion will occur more rapidly as the knee flexes. Positioning the rod more posteriorly reduces the ankle flexion. The foot is of conventional design with a toe-break soft rubber pad and heel rubber (Figures 1 and 3).

3 LIMB FITTING AND USE

Both suction socket as well as the more conventional sockets without partial vacuum holding facility are suitable for the design. The alignment and length adjustments are carried out in the same manner as for the conventional prostheses. The special adjustments are in respect of the length of ankle actuating rod and the ankle flexion adjustment slot. There is adjustable padding on the top of the shank piece which can be suitably positioned to check hyperextension to the extent desired. This padding also helps to check noise.

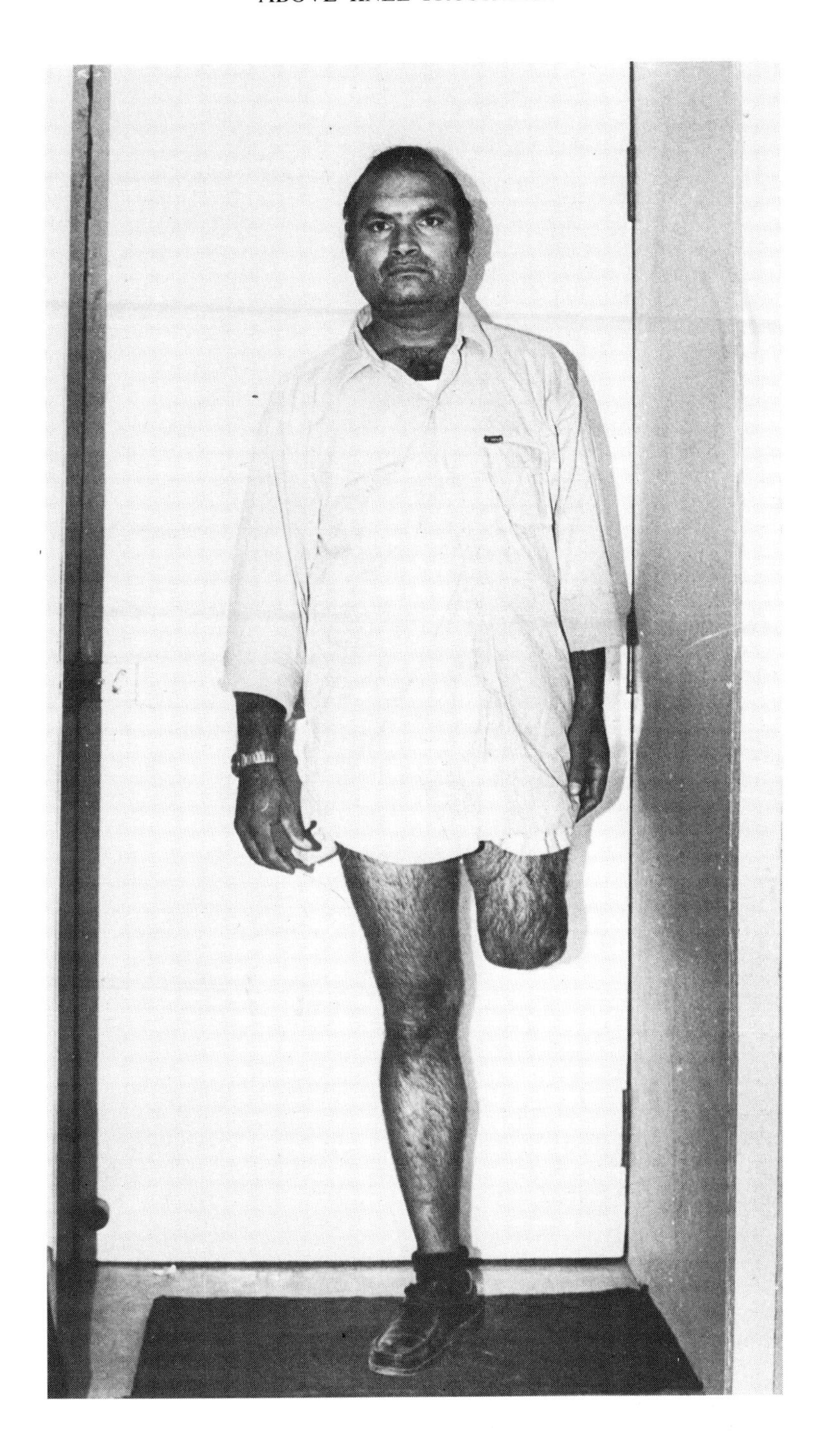

Figure 5 An above knee amputee.

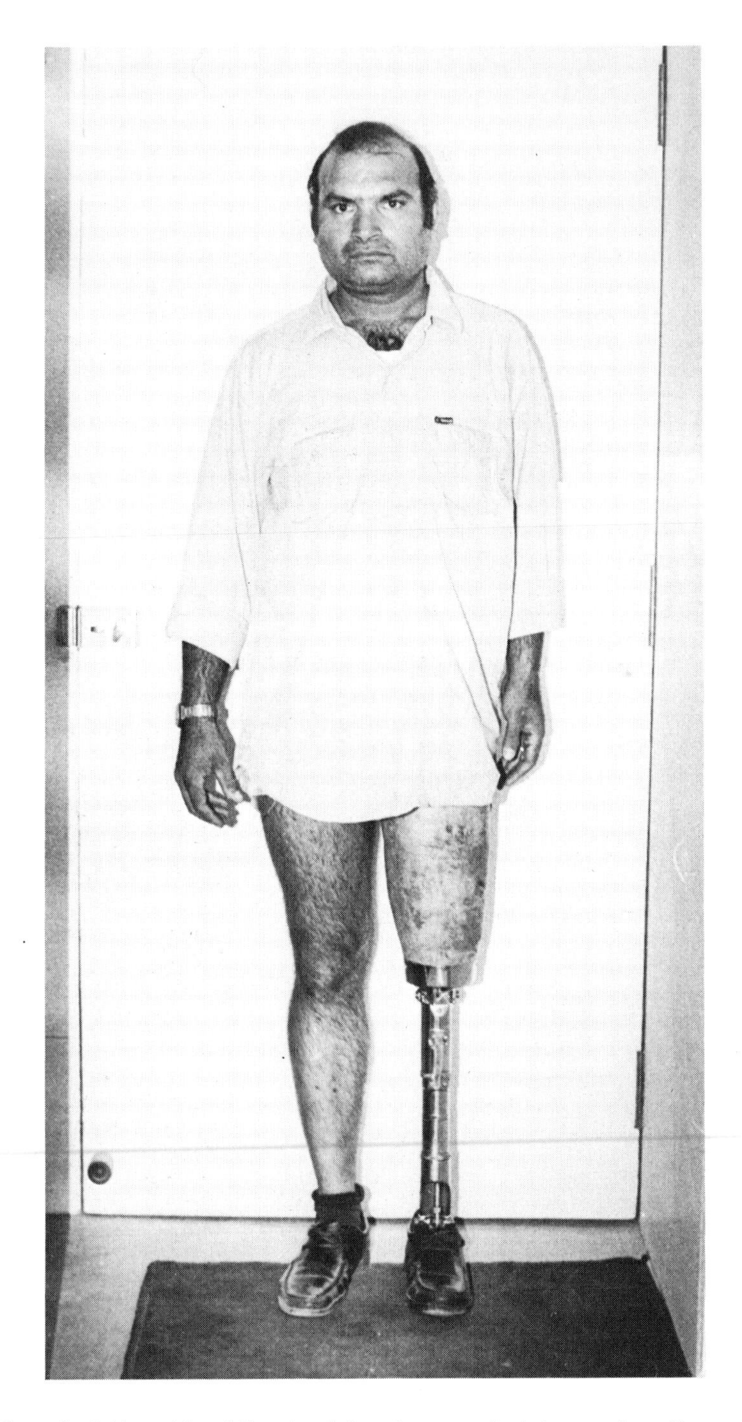

Figure 6 Subject with multifunctional above knee prosthesis in normal standing posture.

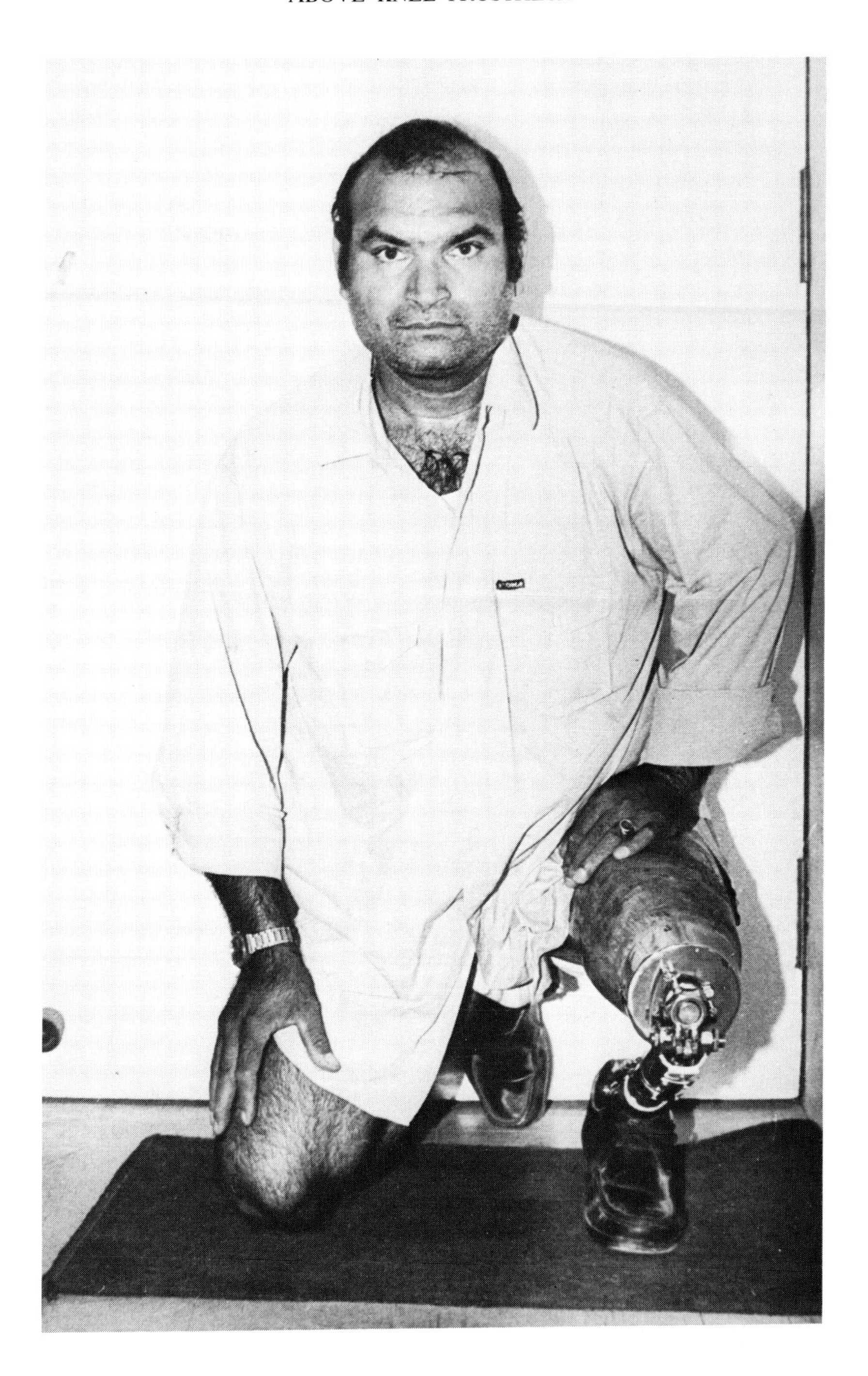

Figure 7 Subject with multifunctional above knee prosthesis in unsymmetrical squatting posture.

Limbs of this design can be fitted to subjects of widely varying stump length. Figure 5 shows a subject with a moderately long stump and of heavy build. Normal standing (Figure 6), unsymmetrical squatting (Figure 7), symmetrical squatting (Figure 8), sitting cross-legged (Figure 9) and kneeling (Figure 10) are all possible. In actual use the knee joint is protected by a leather cover which serves to check damage to the trousers at the knee joint. The cover has been removed to indicate the components clearly. No manipulation by hand is required in any of the movements and the motions are near normal. When the person sits cross-legged, there is no recoil effect arising from the spring tension and so the person can sit for a long time quite comfortably.

4 CONCLUSIONS

A simple modular design with adjustability to fit persons of different heights has been achieved. Seventy percent of the components are common to all sizes of the limb. The materials used are those that are commonly available and the assembly

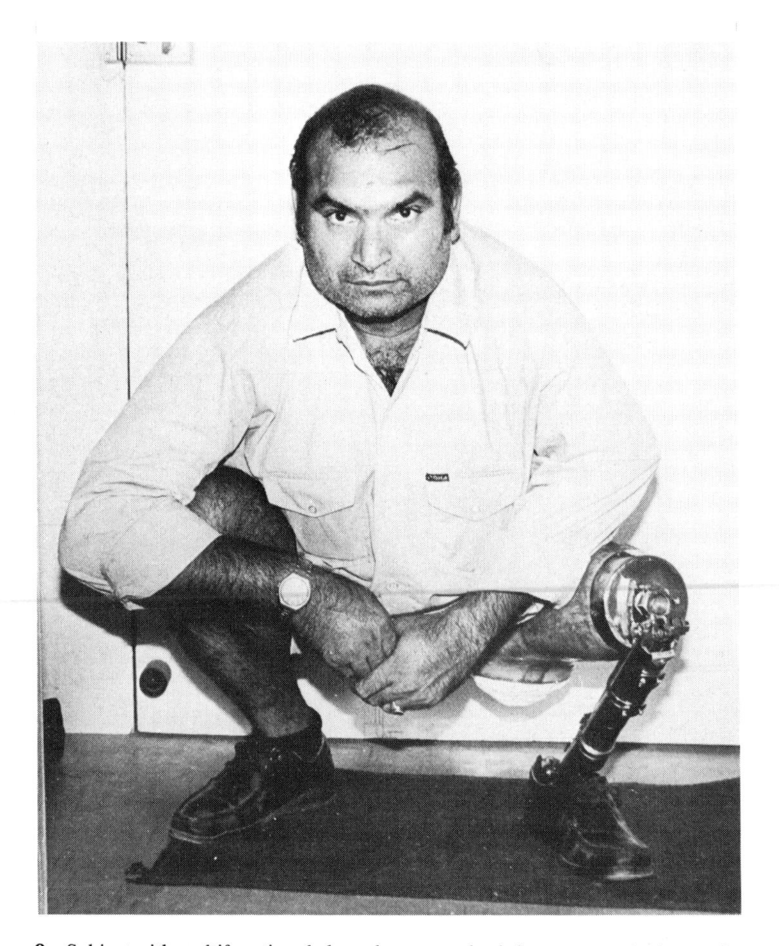

Figure 8 Subject with multifunctional above knee prosthesis in a symmetrical squatting posture.

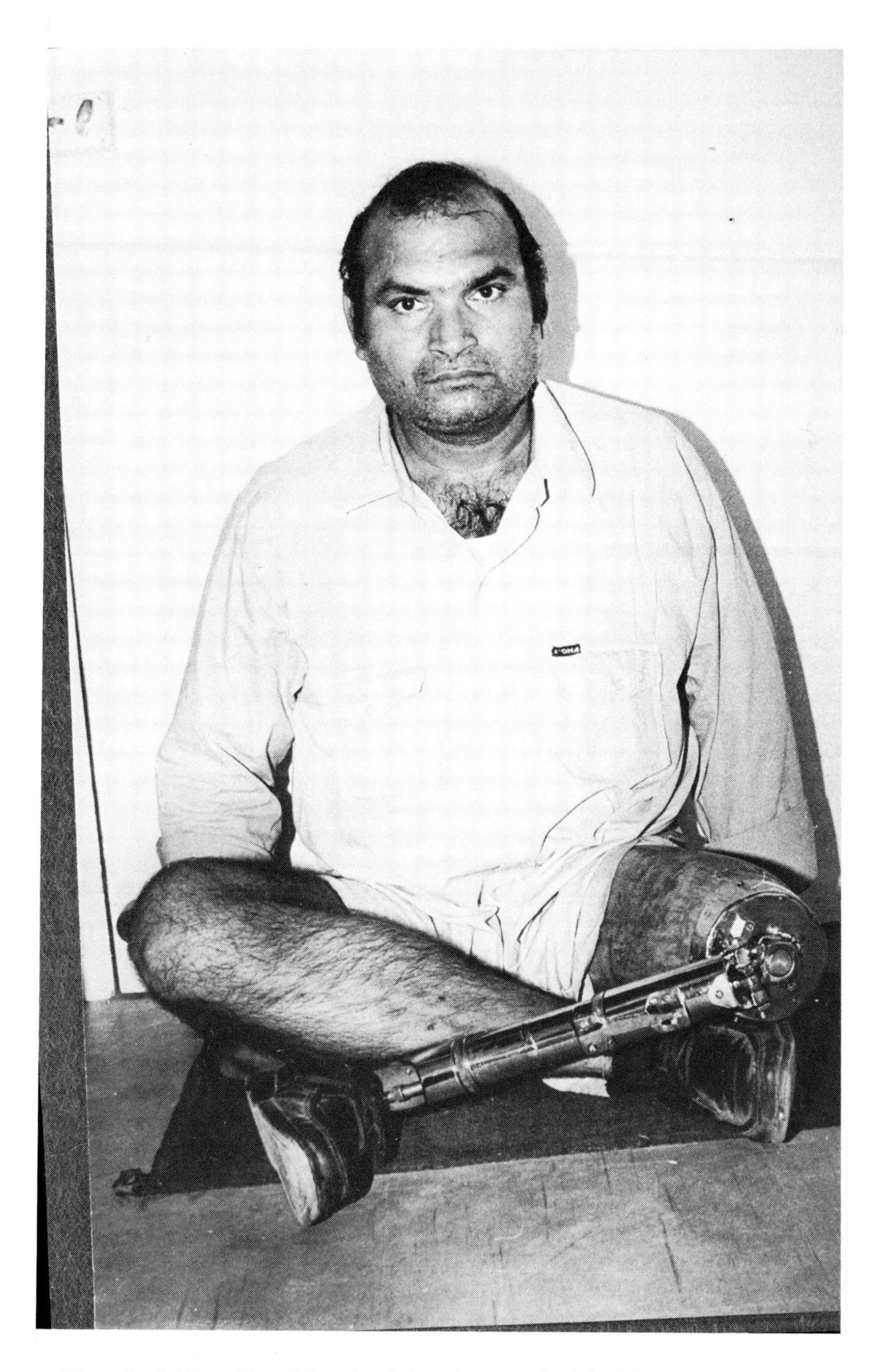

Figure 9 Subject with multifunctional above knee prosthesis in sitting cross-legged posture.

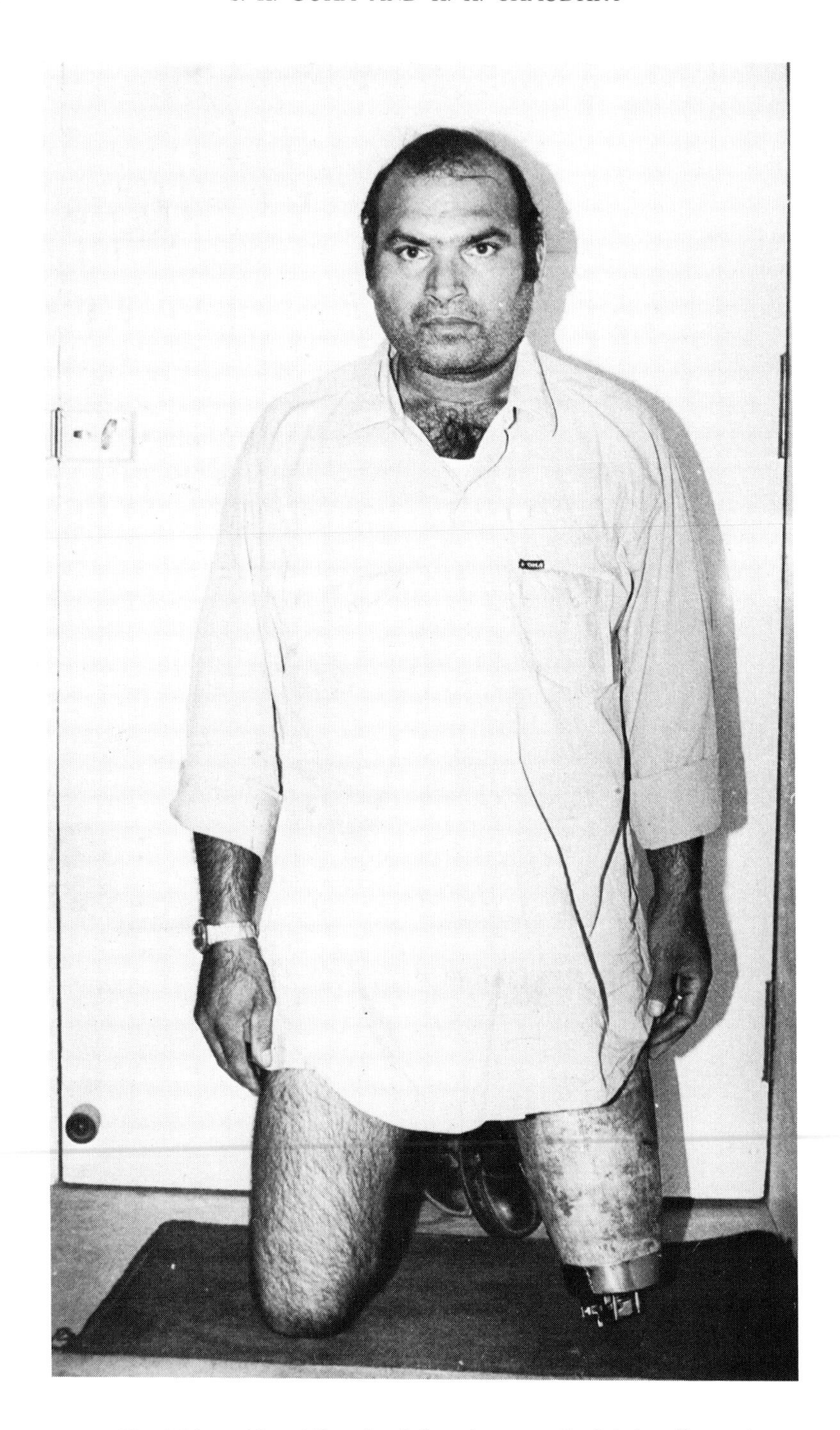

Figure 10 Subject with multifunctional above knee prosthesis in kneeling posture.

requires only a drilling machine and a welding machine. No specially molded parts are required. These features constitute an improvement over earlier squatting prosthesis models [3, 4]. The kinematics is such that walking, squatting and sitting cross-legged movements all appear near normal. Maintenance involves only periodic lubrication of the screw joint of the knee and of the articulating joint of the ankle. In the eventuality of any component wearing out the individual part can be removed and replaced. Hence, the principal requirements of a prosthesis suitable for developing countries are met.

REFERENCES

1. R. S. Hertenberg and J. Denavit, *Kinematics Synthesis of Linkage*, McGraw-Hill (1964).
2. L. W. Lamoreux, Measurements in the study of human walking, *Bulletin of Prosthetic Research*, 3-84 (1971).
3. K. K. Chaudhry, S. K. Guha, and S. K. Verma, Modular concept in a multifunctional above knee prosthesis, *Engg. Med.*, **10**, 155-158 (1981).
4. K. K. Chaudhry, S. K. Guha, and S. K. Verma, An improved above knee prosthesis with functional versatility, *J. Orth. Pros. Int.*, **10**, 157-162 (1982).

Automedica, 1989, Vol. 11, pp. 247–261
Reprints available directly from the publisher only
Photocopying permitted by license only

AN INTERACTIVE SYSTEM OF ELECTRONIC STIMULATORS AND GAIT ORTHOSIS FOR WALKING IN THE SPINAL CORD INJURED

CHANDLER A. PHILLIPS

Department of Biomedical Engineering, Wright State University, Dayton, Ohio 45435

(Received September 1988)

By utilizing commercially available electronic muscle stimulators (E.S) for functional electrical stimulation (F.E.S.) and interfacing them with the reciprocating gait orthoses (G.O.), a new system has been developed and is described herein as an E.S./G.O. Various advantages of the system include: (a) commercially available subsystems from various manufacturers, and (b) subsystems recommended for applications such as gait training. The system itself currently employs six E.S. units worn on a belt, controlled by remote switches and interfacing to electrodes placed over the quadriceps, hamstring and gluteal muscle groups of each leg. Four E.S. units (for quadriceps stimulation) function primarily for stand-up and sit-down. Two other E.S. units (for stimulation of the hip extensors) function primarily for ambulation. The system has been evaluated on a C-7 level quadriplegic individual who sustained a spinal lesion and is six years post injury. Each E.S. unit is powered by a nine volt alkaline transistor battery that provides about 30 stand-up's and sit-down's and approximately 2.5 kilometers of walking before a replacement is necessary. It is emphasized that successful E.S./G.O. walking exercise must be preceded by a physical conditioning program of active physical therapy. New battery technology (such as lithium batteries) should improve the useful lifespan of the system. New electrode technology (such as T.G.'s) should improve patient compliance with the system.

Keywords: Electrical stimulation, orthosis, rehabilitation, spinal cord injury

INTRODUCTION

During the past ten years, significant effort has been directed toward the application of neural prostheses for standing and walking in the spinal cord injured individual. Such neural prostheses require an intact neuromuscular system below the spinal cord lesion so that the patient can respond to the electrical stimulation. Central command has been lost, however, and the majority of these patients remain permanently paralyzed below the level of injury.

Different groups have utilized functional electrical stimulation (F.E.S.) for standing and walking through a variety of lower extremity neural (F.E.S.) prostheses [1–6]. In 1982, computer-controlled standing and walking with F.E.S. was introduced by the Wright State University group. This group utilized closed-loop control [5, 7] and their approach has recently been reviewed [8]. Closed-loop control requires that data from sensors modify the output of a controller to help coordinate walking. Sensors are placed on the hips, knees, and/or ankles of the paralyzed individuals to provide positional data for the computer controller [7, 9]. Closed-loop control was developed in animal experiments [10–12], although it was originally proposed on humans [13, 14]. The Wright State approach has also been directed toward functional electrical rehabilitation [15–18].

Various problems have limited the application of F.E.S. technology for ambulation of the spinal cord injured subject [19]. For example, there is the high energy cost associated with movement induced by F.E.S., and the result is rapid muscular fatigue [20].

F.E.S. is optimally applied during the "swing phase" of gait and results in forward movement of the individual. However, F.E.S. is also applied during the "stance phase" of gait in order to maintain an upright posture. In the "stance phase", there is no movement, but the tonic (isometric) contraction of the paralyzed muscle contributes significantly to fatigue.

Another serious problem is that the individual is not protected from falling. Postural instability results from stimulation of only a few muscle groups (out of the entire population of paralyzed muscles). Furthermore, electronic components can fail, resulting in partial or total loss of the tonic electrical activity (needed to maintain the patient upright).

The reciprocating gait orthosis (G.O.) has been utilized to resolve these problems. The G.O. is a long-leg brace with a hip and knee joint, a pelvic band and thoracic supports. It is composed of plastic materials and lightweight aluminum [21–23]. The pelvic band and thoracic support offer more postural stability for mid- and high-level paraplegics than is available with other braces.

The cabling system is one significant feature of the G.O. Two cables are connected to the anterior and posterior aspects of the hip joints. A reciprocal action at the hips is produced by shifting the upper body weight (through posterior movement of the shoulders). This results in the forward thrust of the pelvis (i.e., hip extension) followed by a transfer of weight onto the arms (and onto the walker). There is also a slight rotation of the central axis of the body. In order to walk, F.E.S. of contralateral thigh muscles produces contralateral hip extension (of the anteriorly placed leg). Simultaneously, the posteriorly-placed ipsilateral leg is moved forward (by action of the interconnecting cables). The operation and application of E.S. with this gait orthosis is described elsewhere [24, 25]. The E.S./G.O. (computer-directed F.E.S. combined with a G.O.) walking system is used in our laboratory as the more energy efficient system [26, 27] when compared to free standing F.E.S. or braces-alone walking [20].

A major direction of our laboratory activity during the past year has been to make this technology available to the medical community for use by the general spinal cord injured population. The limitations of our prior approach [24, 25] have been:

1) A heavy, separate portable power pack (consisting of four Kodavision 8-volt batteries and eight lead-acetate 2-volt batteries). This power pack weighs approximately 8 pounds and is carried on the back of the individual.

2) A heavier, more bulky F.E.S. electronic package when the sit-down/stand-up capability was combined with the basic walking system.

3) A specialized, custom-built electronic package and power pack.

The purpose of this paper is to report on an innovative system which overcomes these limitations. By utilizing commercially available electrical muscle stimulators (E.S.) as the F.E.S. component (in combination with the G.O.), a new system has been developed and termed an E.S./G.O. The advantages of this system are:

1) The power pack is reduced to a single 9-volt alkaline battery contained within each E.S. unit.

2) A total of six E.S. units (each weighing less than 6 ounces, with battery) provide complete stand-up/sit-down and walking functions.

3) Each component of the E.S./G.O. system is:
a) Commerically available from various manufacturers;
b) Recommended for applications such as "gait training";
c) Physician-prescriptable technology so that it is available as a prescription item from any licensed physician.

A second purpose of this paper is to evaluate the battery capability when the system is applied to a quadriplegic subject.

MATERIALS AND METHODS

A spinal cord injured person participated in this study. G.B. is a 28-year-old, male, C-7 level quadriplegic who was injured 6 years ago. He sustained a spinal injury and is Frankel Class C. A somewhat modified system has recently been described for a paraplegic person [28].

A. *Materials*

1. *Electrical muscle stimulator* The unit employed in this study was the E.M.S.-8100 muscle stimulator (NTRON Corp., 3833 Redwood Highway, Post Office Box 7000, San Rafael, CA 94912). The technical specifications are given in table 1. Six E.S. units were employed, hereafter designated as A, A′, B, B′, C and D. A, A′, B and B′ were worn in the front on a belt (see Figure 1) and E.S. units C and D were worn on the back of the belt.

Table 1 Technical specifications of EMS-8100

Power:	0–80 milliamps continuously adjustable (into a 1000 ohm load) for each channel
Rate:	20–100 pulses per second (Hz)
Width:	60–800 microseconds, continously adjustable
Rise:	0–10 seconds
Fall:	0–5 seconds
Duty cycle:	On time 0–30 seconds; Off time 0–30 seconds; both continuously adjustable
Synchronous/reciprocal modes:	Channels operate in a simultaneous or alternating fashion, per set duty cycle
Power source:	9 Volt alkaline battery (supplied)
Waveform:	Symmetrical rectangular pulsatile AC
Manual control option:	Remote jack allows for manual control by hand or foot switches (optional)
Size:	2 5/8 × 3 5/8 × $1\frac{1}{4}''$ (6 × 9.5 × 3.2 cm)
Weight:	5 3/4 oz (160 grams) with battery
Warranty:	5 years parts, 1 year labor (excludes case, cables, battery and electrodes)

The initial operating parameters were adjusted for one set of values for E.S. untis A, A′, and B, B′ and another set of values for E.S. units C and D (see Table 2). These operating parameters were subsequently adjusted during the course of the study for optimal individual performance.

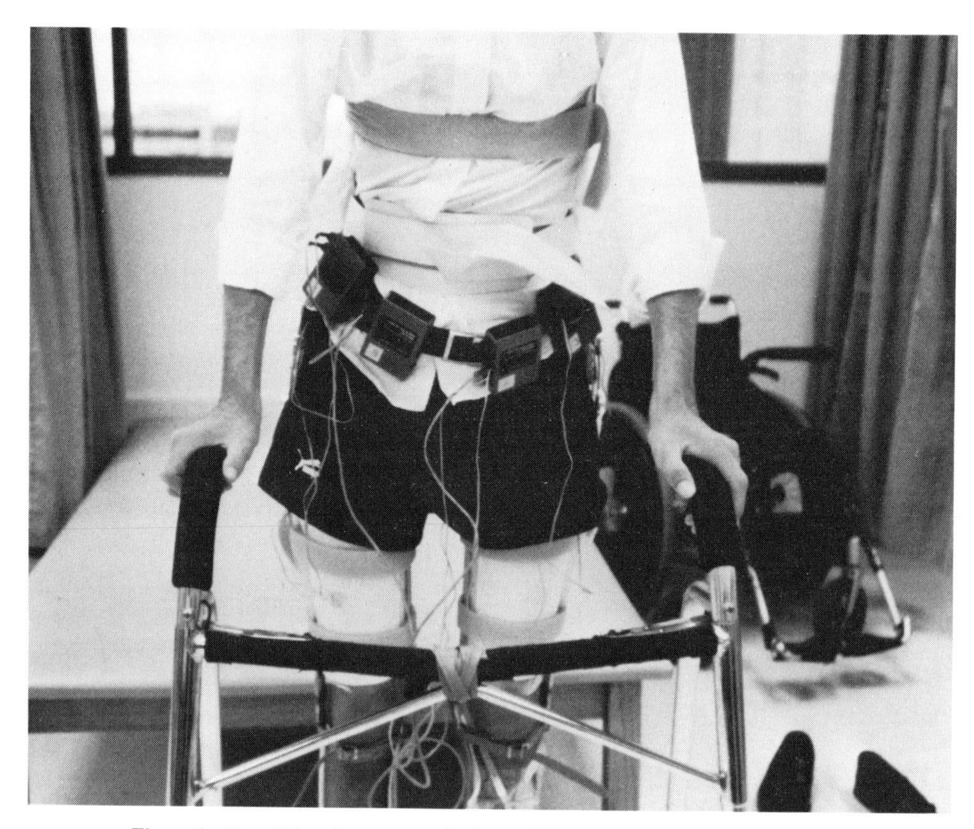

Figure 1 Four E.S. units worn on the front of the belt of a quadriplegic subject.

Table 2 Initial operating parameters

Parameter	E.S. units A/A′/B/B′	E.S. units C/D
Rise time	1.0–2.0 seconds	0 seconds (instantaneous)
Rate (frequency)	100 Hz	70–85 Hz
On-time	30 seconds	30 seconds
Off time	0 seconds	0 seconds
Fall time	5 seconds	0 seconds (instantaneous)
Pulse width	800 μseconds	800 μseconds
Amplitude (thumbwheel reading)	7.0–8.0	7.0–8.0

2. *Switches* The commercially available remote control switch (NTRON part no. 473D) was part of the system. This is a slide type switch, normally "off" which will slide to and hold in the "on" state. Pressing down on the switch slide, while in the "on" position, activates a spring return to return the slide to the "off" position. The switch connects to a legnth of cable that is terminated by a subminiature phone plug. This plug inserts into the manual override jack on the side of the E.M.S.-8100. When the switch is "off", the contacts on the phone plug are open and no

stimulation current is delivered from the E.M.S.-8100 channels (two per unit) to the electrodes. When the switch is "on", the contacts on the phone plug are shorted and stimulation current is delivered via both channels of the E.S. unit to the electrodes.

The quadriplegic subject had some movement of the thumb, index and middle finger of the left hand, but no thumb or finger movement of the right hand. In this case, one slide switch and two push button switches were mounted near the left handle of a conventional walker. Two push button switches were mounted in-line and adjacent to each other on the medial side aspect near the walker handle. These were connected by ribbon cable to a subminiature cable connector (male) which interconnected with a subminiature cable connector (female). The latter connector separated as two cables each terminating with a subminiature phone plug that then connected to the manual override jacks of E.S. units C and D.

One remote slide switch was mounted near the left handle and oriented 45 degrees obliquely between the medial side aspect and the superior aspect (see Figure 2). The remote slide switch cable was then divided into four cables connnected in parallel. Each cable was then terminated with a subminiature phone plug which connected to the manual override jacks of E.S. units A, A', and B, B'.

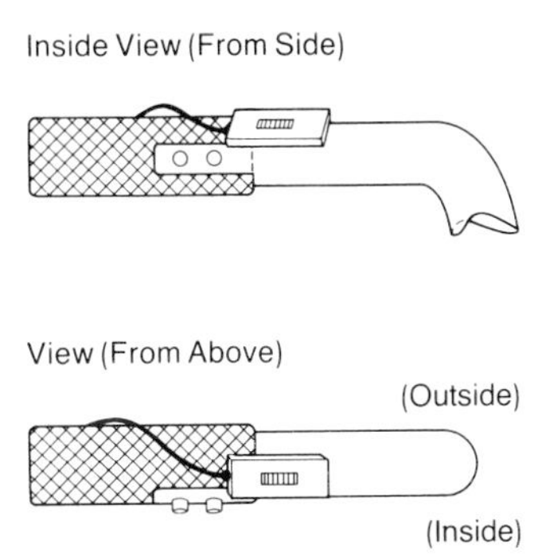

Figure 2 Configuration of the remote control switches located near the handle of a conventional walker (as used by quadriplegic subject).

In operation, the anterior and posterior push-button switches were actuated with the index and middle fingers while the slide swtich was actuated with the thumb. The left palm and remaining fingers assisted with positioning and gripping of the walker handle.

3. *Batteries* Nine-volt alkaline batteries were used since each E.S. unit was designed to operate with nine-volt batteries. 7.2 volt rechargeable Ni-Cad batteries were tried, but did not provide sufficient voltage amplitude to meet the particular current demands of the application.

Battery capability was tested in a continuing series of experiments in which the battery voltage was measured at the terminals before and after each experimental session. The activities of that session were also recorded. When the battery could no longer provide sufficient voltage output to meet the current demands of the application, the results were calculated. The total type of activity performed was divided by the total battery voltage drop as an indication of battery capability. This type of analysis is reasonably valid since for nine volt alkaline batteries, the fall in voltage at the terminals is proportional to the decrease in current supplying capability [29].

4. *Electrodes* These studies utilized ten 2″ × 4″ carbon rubber electrodes (Medtronic® Model 3793, Neuro division, 6951 Central Ave., N.E., P.O. Box 1250, Minneapolis, MN 55440), and four 1½″ × 1½″ carbon rubber electrodes (Medtronic® Model 3795).

A Transcutaneous Transducer Garment, T.G. (Bio-Stimu Trend Corp., 14851 N.W. 27th Avenue, Opa Locka, Florida 33054) has also been successfully tested and utilized in our walking protocols, but has not been fully tested for stand-up and sit-down type activities.

As shown in Figure 3, four 2″ × 4″ carbon rubber electrodes are placed over each quadriceps muscle group (for a total of eight electrodes). Liqui-Cor electrode gel is first applied and both adhesive patches and adhesive tape are utilized to secure the electrodes.

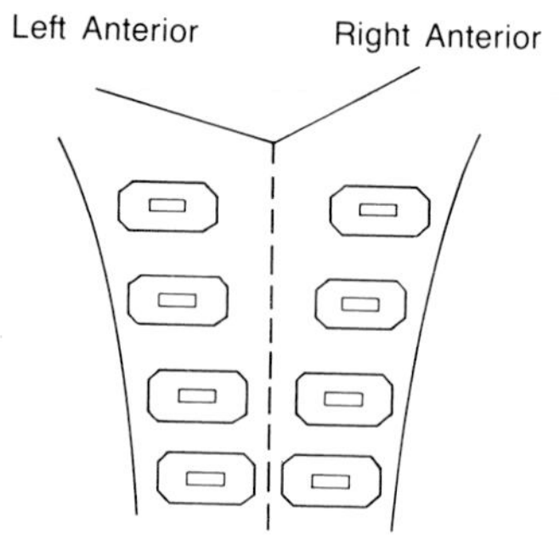

Figure 3 Electrode configuration over the anterior aspect of upper half of the legs (quadriceps muscles).

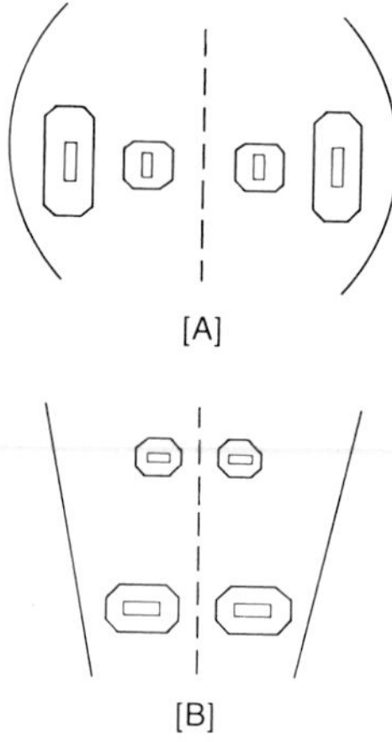

Figure 4 Electrode configuration over the hip extensor muscles: (a) Gluteal muscle group; (b) Hamstring muscle group.

As shown in Figures 4a and 4b, one 2″ × 4″ carbon rubber electrode and one 1½″ × 1½″ carbon rubber electrode are placed over each gluteal muscle group and each hamstring muscle group (for a total of eight electrodes, four of each type). Electrode gel and adhesive are used as before.

Left Leg Subsystem

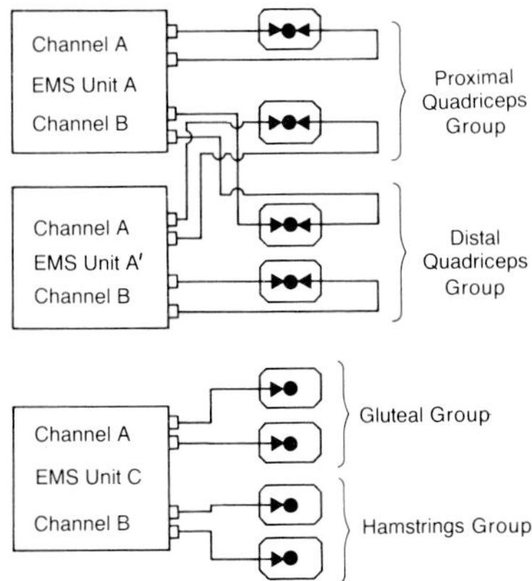

Figure 5 Interfacing of the E.S. units with the electrode configuration over the various muscle groups.

Referring to Figure 5, E.S. units A and A′ are connected to the right quadriceps, E.S. units B and B′ are connected to the left quadriceps. E.S. unit C is connected to the right hip extensors and E.S. unit D is connected to the left hip extensors.

B. *Methods*

The E.S./G.O. system interacts with the subject to perform a variety of tasks including standing-up, walking, turning and sitting-down. These procedures are reviewed.

1. *Stand-up/sit-down procedure* These are treated as two entirely different procedures.
a) Procedure for Stand-Up (Figure 6)

1) Subject with forelegs bent slightly backward at knees. Subject leaning forward at hips (shoulders over knees).

2) Seat elevated so that hip flexed less than ninety degrees.

3) Activate E.S. units A, A′ and B, B′ in the Constant-On Mode.

4) Subject rises as though in a "Forward/Upward Dive" configuration. (On balls of feet.)

5) When erect, subject inclines backward (on heels of feet) to lock knees.

6) Activate E.S. units C and D in Continous Mode for Hip Lock—hold only until locked.

7) Deactivate E.S. units C and D.

8) Deactivate E.S. units A, A′ and B, B′—allow time to "ramp down."

9) Proceed to "Walk" Mode.

b) Procedure for Sit-Down:

1) Subject standing upright about 4″–6″ forward of sitting platform.

2) Unlock hip locks with each hand separately (other hand on walker).

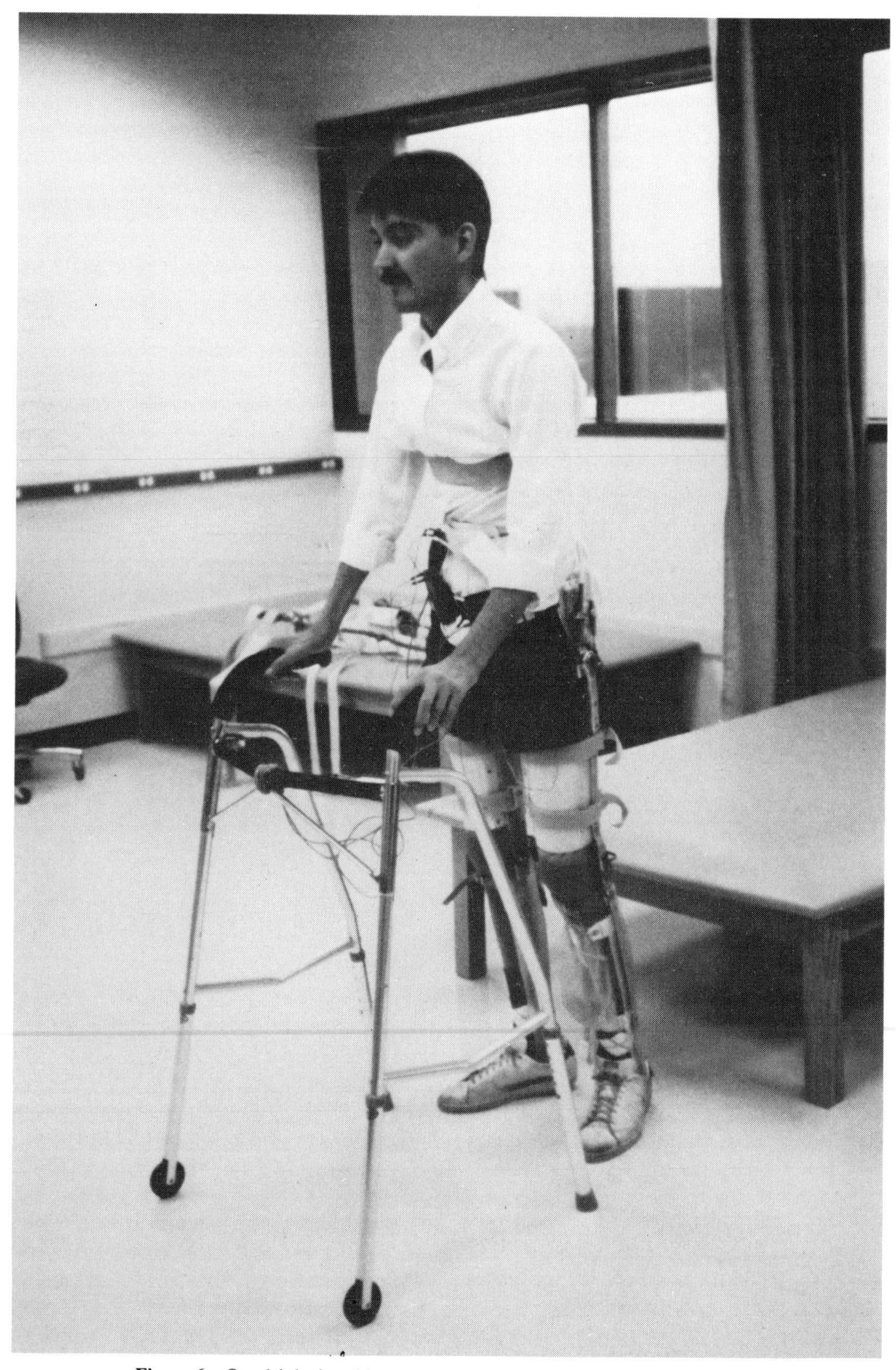

Figure 6 Quadriplegic subject standing erect and outfitted with E.S./G.O.

3) Activate E.S. units C and D continously or intermittently as necessary to maintain upright balance.

4) Activate E.S. units A, A′ and B, B′ in Continous Mode—allow time to "ramp up."

5) Release Knee Locks.

6) Deactivate E.S. units A, A′ and B, B′.

7) Proceed to "sit down" as stimulation "ramps down."

2. *Walking procedure* Let us begin (rather arbitrarily) with the right foot somewhat forward of the left foot and weight equally distributed on both feet. We now wish to take the left step *forward.*

a) Weight is shifted so that all weight bearing is on the right foot.

b) E.S. unit C is activated resulting in right hip extension.

c) Through the reciprocal connection of the cross-connected hip cables, there is contralateral (left hip) flexion.

d) The left leg swings forward and simultaneously E.S. unit C is deactivated.

3) Weight is redistributed equally on both feet with the left foot now somewhat forward of the right foot.

In order to continue forward walking, we can now repeat the procedure by shifting weight to the left foot and activating E.S. unit D as appropriate. By alternating the shifting of weight and activation of the appropriate unit, the person walks forward (Figure 7).

Let us now return to the situation of the right foot somewhat forward of the left foot and the weight equally distributed on both feet. We now wish to take a right step *backward.*

a) Weight is shifted so that all weight bearing is on the left foot.

b) E.S. unit C is activated resulting in right hip extension.

c) The right leg now swings backward and simultaneously E.S. unit C is deactivated.

d) Weight is redistributed equally on both feet with the right foot now somewhat behind the left foot.

In order to continue backward walking, we can now repeat the procedure by shifting weight to the right foot and activating E.S. unit D as appropriate. By alternating the shifting of weight and activation of the appropriate unit, the person walks backward.

3. *Turning procedure* The same turning procedure is employed whether the person is making a 90 degree or 180 degree turn.

a) No F.E.S. is necessary.

b) Pattern follows a "military maneuver" style.

c) RIGHTWARD TURN is on *heel* of right foot and *ball* of left foot.

d) LEFTWARD TURN is on *heel* of left foot and *ball* of right foot.

e) Pattern proceeds as sections of an arc—in between which the walker is repositioned (Figure 8).

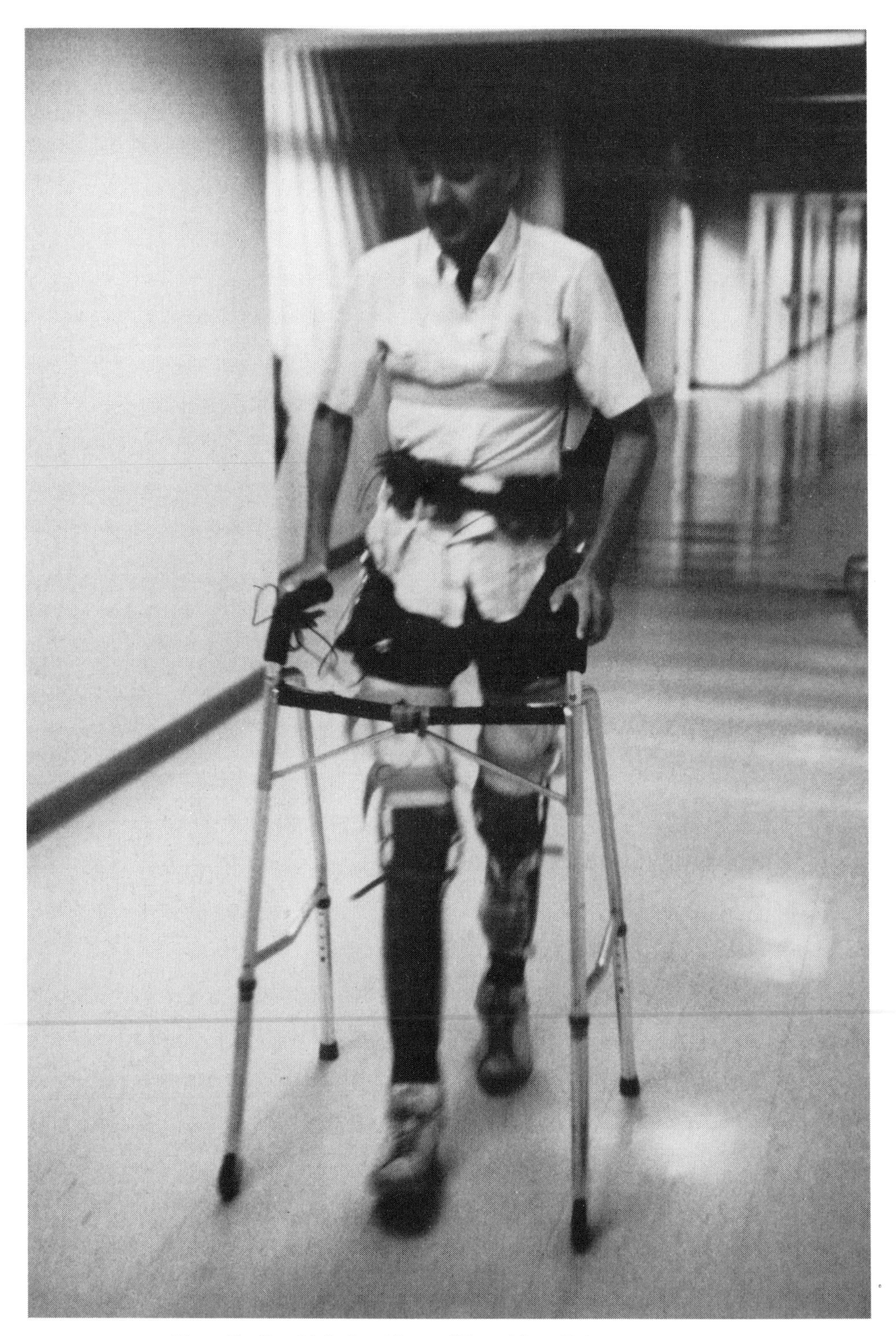

Figure 7 Quadriplegic subject walking while utilizing the E.S./G.O.

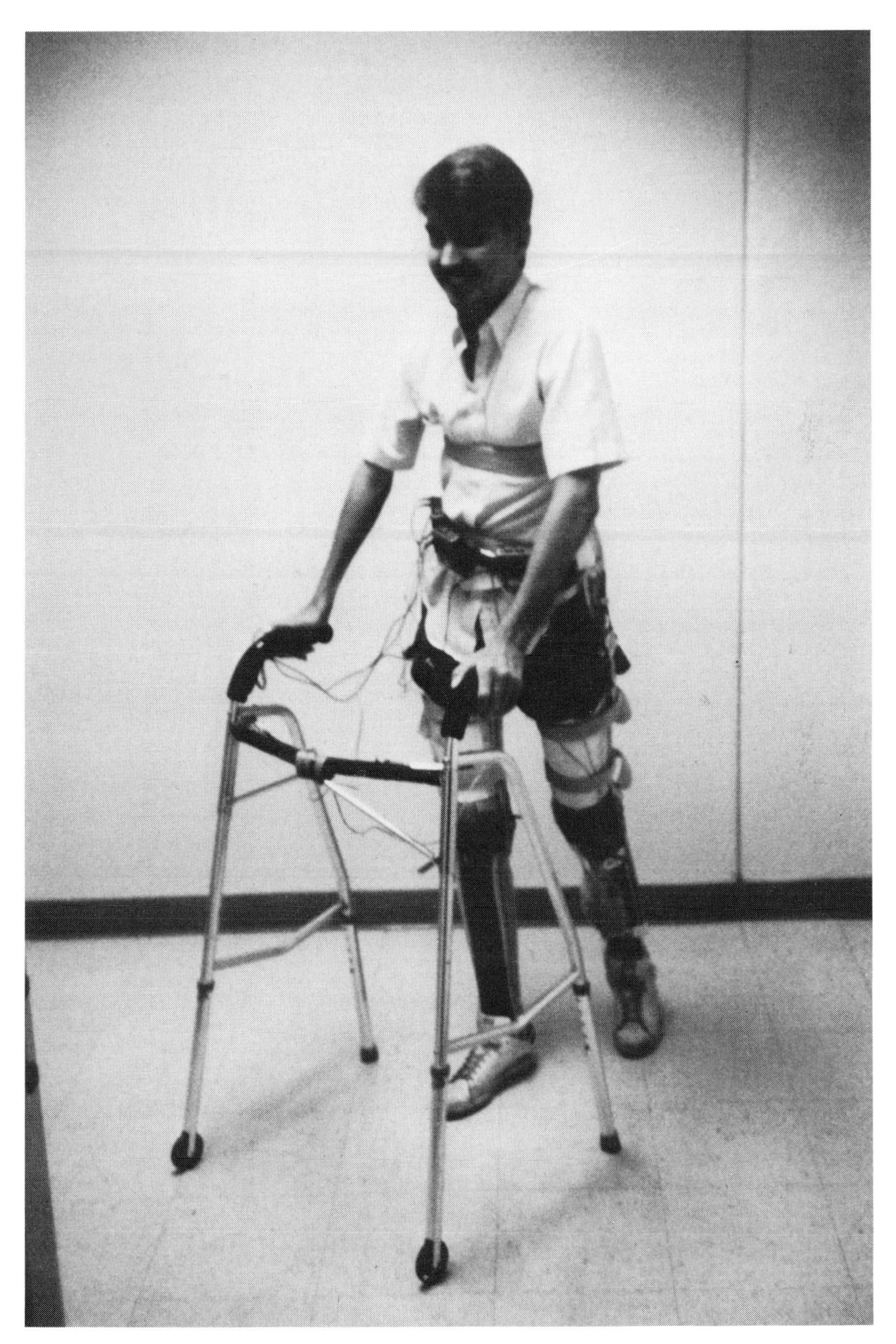

Figure 8 Quadriplegic subject turning while equipped with the E.S./G.O.

RESULTS

A. *Stand-Up and Sit-Down*

The subject performed forty acceptable stand-ups and sit-downs in the E.S./G.O. system. Ten unsuccessful stand-ups were encountered, primarily due to incorrect body position: e.g., feet not behind knees, hip angle greater than 90 degrees, etc. Inadequate quadriceps strength conditioning was a limiting factor for this quadriplegic subject. After two or three stand-ups and sit-downs, the E.S. units A, A′ and B, B′ had to be increased to the maximum amplitude setting of 9.0. At this point, all four units were providing maximal stimulation. Quadriceps fatigue then occurred after another two or threee stand-ups and sit-downs.

B. *Walking*

The quadriplegic individual routinely covered 0.8 kilometer per experimental session at velocities ranging from 1.2 k.p.h. to 2.0 k.p.h. Muscle fatigue was usually the limiting factor for this quadriplegic subject. The subject altered the initial E.S. parameter settings (before the end of the walking episode) such that the frequency (rate) setting was increased to 100 Hz (maximum) and the amplitude setting was increased to 9.0 (the maximum).

C. *Battery performance*

Battery performance was calculated as amount of a specific activity per volt since with alkaline batteries the output voltage at the terminals is proportional to the current delivery capability of the battery.

With respect to the batteries of E.S. units A and B, there was approximately thirty stand-up/sit-downs per volt drop at the battery terminals. With respect to the batteries of E.S. units C and D, there was approximately 2.5 kilometer of walking per volt drop at the battery terminals.

These figures also approximate the useful lifespan of the respective batteries since after about a 1 volt drop of the output terminal voltage, there is no longer sufficient current delivery capability for satisfactory performance of the E.S./G.O. system.

DISCUSSION

The E.S./G.O. described above provides walking exercise that is actually only one phase in a comprehensive program of active physical therapy (A.P.T.). In A.P.T., the functionally electrically stimulated muscles act upon their enivronment to produce the exercise effort. This may be contrasted to conventional physical therapy (no F.E.S.) in which the environment acts upon the muscle to produce the exercise effort [30]. Phillips [17] recently defined A.P.T. to consist of four modalities capable of producing functional electrical rehabilitation: (i) an isokenetic leg trainer, (ii) a stationary bicycle ergometer, (iii) an outdoor exercise bicycle, and (iv) an F.E.S.-orthosis for amublation exercise.

There is a definite requirement for the first two exercise modalities prior to proceeding to walking exercise with the E.S./G.O. The isokinetic leg trainer (I.L.T.) stimulates the quadriceps muscle groups of each leg. Various weights are connected to the foreleg via an ankle harness and repetitive isovelocity contractions are produced at the knee. The I.L.T. has previously been described in detail [31]. Numerous groups have confirmed that this type of exercise increases the strength of the quadriceps muscle group [15, 32, 33].

The F.E.S. bicycle ergometer (F.E.S./B.E) provides six channels of stimulation, one each to the quadriceps, hamstring and gluteal muscle groups of each leg. The paralyzed legs are stimulated to pedal at 50 revolutions per minute against progressive loads adjusted via a friction belt attached to a flywheel. The F.E.S./B.E. has been described in detail elsewhere [34]. Various investigators have reported physical conditioning (training) effects including increased pedaling endurance [35], improved resting blood pressure levels [30], increases in resting heart rate and cardiac output [36] and finally increases in maximal workloads and minute ventilation [37].

As with any treatment modality, the patient must medically qualify for the technology being prescribed. Specifically, there are medical criteria that must be satisfied for prescription of active physical therapy including walking exercise with the E.S./G.O. The initial evaluation of the patient candidate includes history, physical examination, lab tests, report summary and an F.E.S. prescription (level of clearance) as previously described by Phillips [17].

In the Wright State program, the I.L.T. is currently considered to be the entry level F.E.S. exercise modality. Progression to the F.E.S./B.E. requires adequate flexion and medial-lateral stability at the knee, as well as minimal dorsiflexion at the ankle (Table 1 of Phillips [17]). Progression to the E.S./G.O. and walking exercise then requires muscular, cardiovascular and respiratory system function sufficient for 30 min of continuous F.E.S. bicycling at minimal (zero k.p.m.) load. Other medical criteria for walking exercise with the E.S./G.O. are summarized in Table I of Phillips [17].

The useful lifespan of the E.S./G.O. on one set of 9.0 volt alkaline batteries is 30 stand-ups/sit-downs and 2.5 kilometers of walking. New battery technology, such as lithium batteries, should achieve much higher current densities for comparable package sizes. These batteries are now regularly utilized in commercial photographic equipment. They commonly are rated at 1200 milliampere hour. However, the battery connectors are often specific to the camera equipment for which they are intended. These would have to be modified to interface with the E.S. units. Some lithium battery package sizes are larger than the conventional nine volt alkaline cell, but others are of similar size. Lead acetate batteries and gel cells (as used in our previous system) are definitely too heavy and bulky to conveniently satisfy everyday user requirements.

Patient compliance may be a problem in a system where so many electrodes (total of fourteen) need to be applied to the surface of the body. The application of these electrodes is a somewhat tedious and cumbersome task. Furthermore, good positioning over appropriate motor units is important for optimal performance of the system. New stimulation delivery technology (such as the Transcutaneous Transducer Garment, T.G.) allows stimulation delivery at multiple sites by means of a customized garment (similar to leotards). The actual electrical conducting surfaces are interwoven into the fabric of this garment. The patient now simply dons the garment and could then interface with the E.S. units via a single universal connector. Such a connector should be made available by the manufacturers (Bio-Stimu Trend, Opa-locka, FL) for a new version of the E.S./G.O. that is soon to be tested in our laboratories. The results of this new system, defined as an E.S./T.G./G.O. will be the subject of a subsequent report.

REFERENCES

1. G. S. Brindley, G. E Polkey and D. N. Rushton, Electrical splinting of the knee in paraplegia, *Paraplegia*, **16,** 428 (1978).
2. J. Holle, H. Gruber, M. Frey, L. Kern, H. Stohn and H. Thoma, Functional electrical stimulation in paraplegics, *Orthopaedics*, **7,** 1145–1160 (1984).
3. A. Kralj, T. Bajd and R. Turk, Electrical stimulation providing functional use of paraplegic patient muscles, *Med. Prog. Technol.*, **7,** 3–15 (1980).
4. E. B. Marsolais and R. Kobetic, Functional walking in paralyzed patients by means of electrical stimulation, *Clin. Orthop. Rel. Res.*, **175,** 30–36 (1983).
5. J. S. Petrofsky, C. A. Phillips and H. H. Heaton, Feedback control system for walking in man, *Comput. Biol. med.*, **14,** 135–149 (1984).
6. L. Vodovnik, T. Bajd, A. Kralj, F. Gracanin and P. Strojnik, Functional electrical stimulation for control of locomotor systems, *CRC Crit. Rev. Bioengr.*, **6,** 63–131 (1981).
7. J. S. Petrofsky and C. A. Phillips, Computer-controlled walking in the neurological paralyzed individual, *J. Neuro. Ortho. Surg.*, **4,** 153–164 (1983).
8. J. S. Petrofsky, and C. A. Phillips, Closed-loop control of movement of skeletal muscle, *CRC Crit. Rev. Biomed. Eng.*, **6,** 35–94 (1985).
9. J. S. Petrofsky and C. A. Phillips, Electrically-controlled movement of muscle: A potential aid to muscle paralysis, in *Spinal Cord Injury Medical Engineering* (Eds. D. N. Ghista and H. L. Frankel), Charles C. Thomas, Pubs. Springfield, Illinois (1986), pp. 393–437.
10. J. S. Petrofsky and C. A. Phillips, Constant velocity contractions in skeletal muscle of the cat, *Med. Biol. Eng. Comp.*, **17,** 583–592 (1979).
11. C. A. Phillips and J. S. Petrofsky, Velocity of contraction of skeletal muscle as a function of electrical activation and fiber composition, *J. Biomech.*, **13,** 549–558 (1980).
12. C. A. Phillips and J. S. Petrofsky, The passive element force-velocity relationship: Influence on maximal contractile element velocity, *J. Biomech.*, **14,** 399–403 (1981).
13. U. Stanic and A. Trnkoczy, Closed-loop positioning of hemiplegics patient's joint by means of functional electrical stimulation, *IEEE Trans. Biomed. Eng.*, **BME-21,** 365–370 (1974).
14. L. Vodovnik, W. J. Crochetiere and J. B. Reswick, Control of a skeletal joint by electrical stimulation of antagonists, *Med. Biol. Eng.*, **5,** 97–109 (1967).
15. J. S. Petrofsky and C. A. Phillips, Active physical therapy—a modern approach for rehabilitation of the disabled, *J. Neuro. Ortho. Surg.*, **4,** 165–173 (1983).
16. J. S. Petrofsky and C. A. Phillips, The use of functional electrical stimulation for rehabilitation of spinal cord injured patients, *CNS Trauma J.*, **1,** 57–74 (1984).
17. C. A. Phillips, The medical criteria for active physical therapy: physician guidelines for patient participation in a program of functional electrical rehabilitation, *Amer. J. Phys. Med.*, **66,** 269–286 (1987).
18. C. A. Phillips and J. S. Petrofsky, Computer-controlled movement of paralyzed muscle: the medical perspective, in *Proc.* 1*st Vienna Inter. Workshop on Func. Electrostim*, October 19–22 (1983), Paper 5.2.
19. J. Gruner, Considerations in designing acceptable neuromuscular stimulation systems for restoring function in paralyzed limbs, *CNS Trauma J.*, **3,** 37–47 (1986).
20. E. B. Marsolais and B. G. Edwards, Energy costs of walking and standing with functional neuromuscular stimulation and long leg braces, *Arch. Phys. Med. Rehab.*, **69,** 243–249 (1988).
21. R. Douglas, P. Larson, R. D'Aubrosia and R. McCall, LSU reciprocating gait orthosis, *Orthopaedics*, **6,** 834–839 (1983).
22. R. E. McCall, R. Douglas and R. Nicholas, Surgical treatment in patients with myelodyplasia before using the LSU reciprocating system, *Orthopaedics*, **6,** 843–848 (1983).
23. D. Yngve, R. Douglas and J. M. Roberts, The reciprocation gait orthosis and myelomeningocele, *J. Ped. Ortho.*, **4,** 304–310 (1984).
24. J. S. Petrofsky, C. A. Phillips, R. Douglas and P. Larson, A computer-controlled walking system: The combination of an orthosis with functional electrical stimulation, *J. Clin. Engrg.*, **11,** 121–133 (1986).
25. J. S. Petrofsky, C. A. Phillips, P. Larson and R. Douglas, Computer-synthesized walking—an application of orthosis and functional electrical stimulation (FES), *J. Neuro. Ortho. Med. Surg.*, **6,** 219–230 (1985).
26. D. M. Hendershot, M. L. Moore and C. A. Phillips, Cardiopulmonary conditioning when walking with and without FES in the paralyzed, *Fed. Proc.*, **46,** 680 (1987).
27. D. M. Hendershot and C. A. Phillips, Improvement of efficiency in a quadriplegic individual using an FES-RGO system, in *Proc. IEEE/EMBS Tenth Internat. Conf.* **10,** 1577–1578 (1988).

28. C. A. Phillips, Physician directed ambulation for rehabilitation of spinal cord injury. *Surgical Rounds for Orthopaedics* (in press).
29. *Enercell Battery Guidebook*, Master Publishing Inc. Dallas, Texas (1985).
30. C. A. Phillips, J. S. Petrofsky, D. M. Hendershot and D. Stafford, Functional electrical exercise: a comprehensive approach for physical conditioning of the spinal cord injured individual, *Orthopaedics*, **7**, 1112–1123 (1984).
31. J. S. Petrofsky, H. H. Heaton and C. A. Phillips, Leg exerciser for training of paralyzed muscle by closed-loop control, *Med. Biol. Eng. and Comp.*, **22**, 298–303 (1984).
32. S. R. Collins, R. M. Glaser and N. D. Feinberg, Progressive training of paralyzed leg muscles with electrical stimulation, *Fed. proc.*, Abstract no. 4262 (1984).
33. C. J. Robinson, J. M. Bolam, M. Chinoy *et al.*, Response to surface electrical stimulation of the quadriceps in individuals with spinal cord injury, *Proc. RESNA*, **6**, 282–284 (1986).
34. J. S. Petrofsky, C. A. Phillips, H. H. Heaton and R. M. Glaser, Bicycle ergometer for paralyzed muscle, *J. Clin. Eng.*, **9**, 13–19 (1984).
35. J. S. Petrofsky, C. A. Phillips and D. Hendershot, The cardiorespiratory stresses which occur during dynamic exercise in paraplegics and quadriplegics, *J. Neuro. Ortho. Surg.*, **6**, 252–258 (1985).
36. D. Danopulos, P. Kezdi, E. L. Stanley *et al.*, Changes in cardiovascular dynamics after a twelve week active bicycle rehabilitation in young tetraplegics, *J. Neuro. Ortho. Surg.*, **7**, 179–184 (1986).
37. M. L. Moore, D. M. Hendershot and C. A. Phillips, Cardiopulmonary conditioning by F.E.S-induced bicycle ergometry, *Fed. Proc.*, **46**, 679 (1987).

Automedica, 1989, Vol. 11, pp. 263–275
Reprints available directly from the publisher only
Photocopying permitted by license only

LOCOMOTION OF PARAPLEGIC PATIENTS BY FUNCTIONAL NEURO-STIMULATION

J. HOLLE, H. THOMA, M. FREY, H. KERN, W. MAYR, G. SCHWANDA and H. STÖHR

University of Vienna

(Received September 1988)

The first clinical application of implantable FES devices in 4 patients are demonstrated and the experiences during a stimulation period of up to 5 years are demonstrated. With the round-about stimulation system no nerve damages were recognized. There were several complications such as technical implant failure, electrode perforation, dislocation of electrodes and infection. With the aid of crutches two totally paraplegic patients were able to walk with a 4-point gait for about 100 meters and two other patients walked with a swinging through gait.

Keywords: FES of limbs, locomotion therapy, paraplegic rehabilitation.

INTRODUCTION

Standing up and walking, a major desire for paraplegic patients, can be achieved with the aid of orthosis. Since these devices are bulky, uncomfortable and time-consuming to adjust, they mostly are not accepted by disabled persons. Functional Electro-Stimulation (FES) by means of surface electrodes, which has been introduced into clinical use by Kralj [1], Jaeger [2] and Petrofsky [3], represents an alternative concept. Standing up and walking with walkers has been realized with percutaneous, intramuscular electrodes by Marsolais [4]. Surface electrodes and transcutaneous, intramuscular electrodes seem to be scientific steps to substitute locomotion in paraplegic patients with implantable nerve stimulation devices. Brindley [5] was the first to describe the clinical application of implantable, functional electro-stimulation systems to mobilize paraplegic patients.

In 1984 we first described our own experiences with an implantable multi-channel stimulation device to substitute locomotion in paraplegic patients [6, 7]. A 6 year's period of research and development was necessary to prepare these first clinical experiments. Several limiting factors, e.g., electrically induced fatigue of the nerve-muscle system, tissue reaction at the site of the electrode-nerve-connection and various others, prevented a wide clinical application of FES to susbtitute locomotion. A new patented stimulation strategy called "round-about electro-stimulation" with less fatigue reaction of the nerve-muscle system and with the main advantage of fiber selectivity has been developed in our department [8, 9].

In several animal experiments and in our own clinical experience in electrophrenic respiration therapy the round-about stimulation has proved to be most effective [10–12]. New electrode designs fixed to the epineureum of the selected nerves using microsurgical techniques, have been tested in various animal studies and demonstrated little or almost no nerve damage at all [13]. To test the

Author's name and address: Univ. Prof. Dr. Juergen Holle, IInd Surgical University-Clinic, Spitalgasse No. 23, A-1090 Vienna, Austria.

durability of our stainless steel electrodes, extensive *in-vitro* tests were carried out with constant current devices by varying polarity, duration, amplitude, frequency and mode of pulses [14–17]. In order to test the stress imposed towards the electrodes under *in-vivo* conditions, implantable, free-channel, constant voltage generators were developed and implanted into sheep. The result of all these experiments to construct an ideal stimulation electrode was a helix lead of stainless steel consisting of 12 stranded wires (diameter: 50 my) isolated with silicone rubber, the ends of the lead being unisolated; they are curved in a loop of 1 mm in diameter.

In 1978 the first design of an implantable, 8-channel nerve stimulator unit to stimulate two nerves by means of the round-about stimulation technique was developed. This device as well as recent devices consist of an extracorporal control unit with two inductive transmitting coils to the implants. The implant consists of the following components: receiver coil, electronic supply generator and pulse generator. The connections lead to the electrodes (Figure 1).

It seems very important to point out that all the necessary information concerning stimulation parameters are tansmitted before each stimulation pulse. This is essential for FES in locomotions with movements, that have to be adapted at any time. Decoupling capacitors in all output lines to the nerves prevent DC-curents. The size of the implant is comparable to that of a heart pacemaker.

The external control unit is battery powered and able to supply two implants via two transmission coils. Control is warranted by means of a CMOS microprocessor. According to the principle of the round-about electrode, all the electrode combinations have to be stimulated with an adequate current. Incorrect electrode combinations can be deleted. Once programmed, all the data remain stored in a CMOS-RAM. Usually two values of each combination for the adjustment of the minimum and maximum of muscle force are programmed. In accordance with our special stimulation strategy (round-about stimulation), 4 electrodes are fixed to the epineureum of each nerve (Figure 2). Therefore, 50 combinations per nerve are possible by changing the switch-mode and polarity.

CLINICAL APPLICATION OF THE VIENNESE IMPLANTABLE ROUND-ABOUT STIMULATION SYSTEM

Table 1 presents the basic data for our four patients. In October 1982 the system was implanted into two female patients with a posttraumatic lesion of the thoracic spinal cord at the level of Th-9 to Th-10. One year later the system was implanted in another female posttraumatic case and into a male patient following tumour-resection of the spinal cord at the level of Th-5 to Th-8. All of them were complete paraplegics. In all four patients FES was meant to stabilize the hip- and knee-joint in stretched position in order to stand up and walk in a four point gait using lower arm crutches. Functional electro-stimulation of the quadriceps femoris muscle and the infragluteal nerve stimulating the gluteus maximus muscle was meant to function like an internal orthosis.

SURGERY

In the gluteal region, just above the foramen infrapiriformae, a transverse incision was made. To reach the foramen without cutting, the muscle fibers of the gluteus

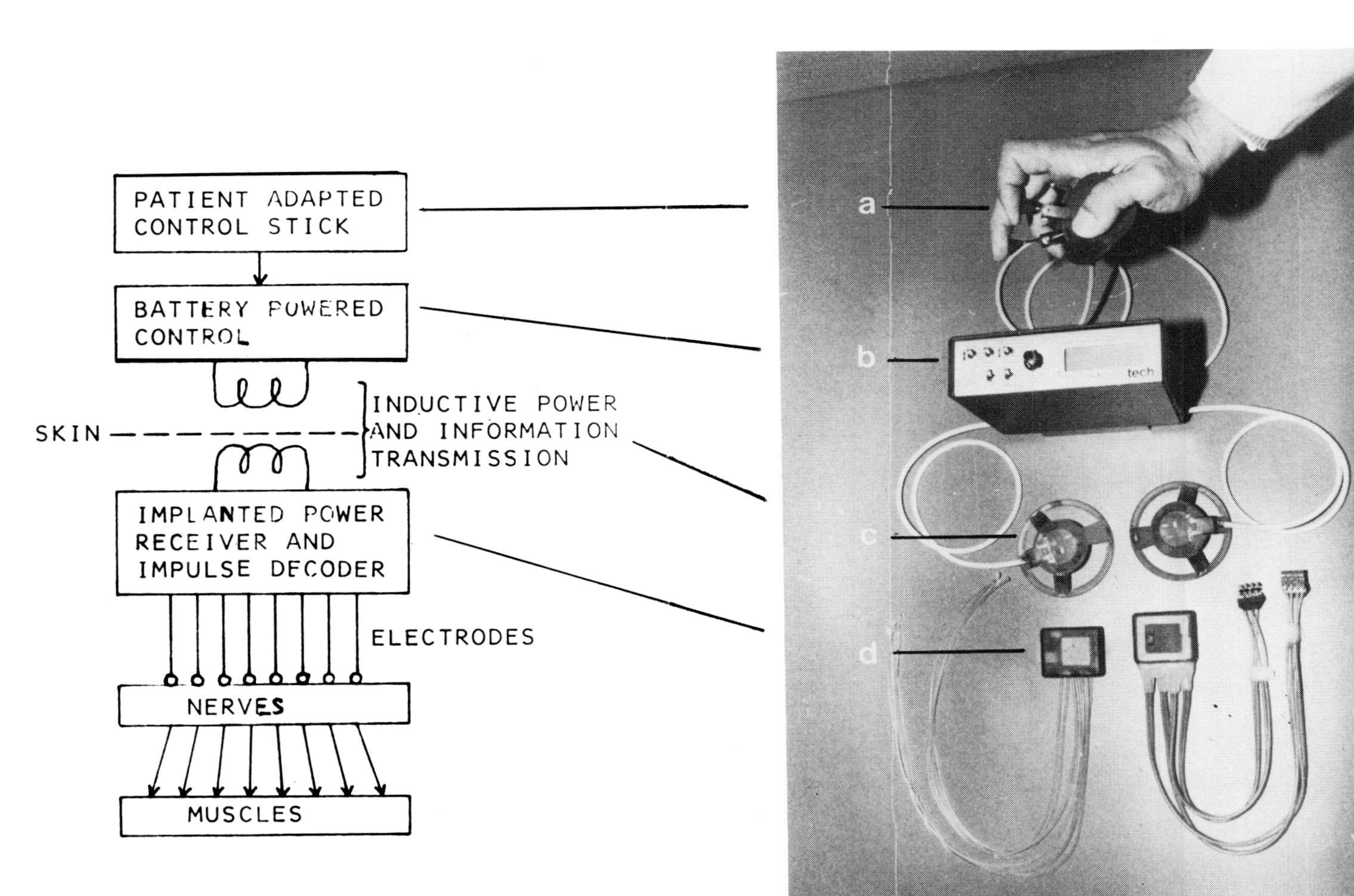

Figure 1 8-Channel nerve stimulator.

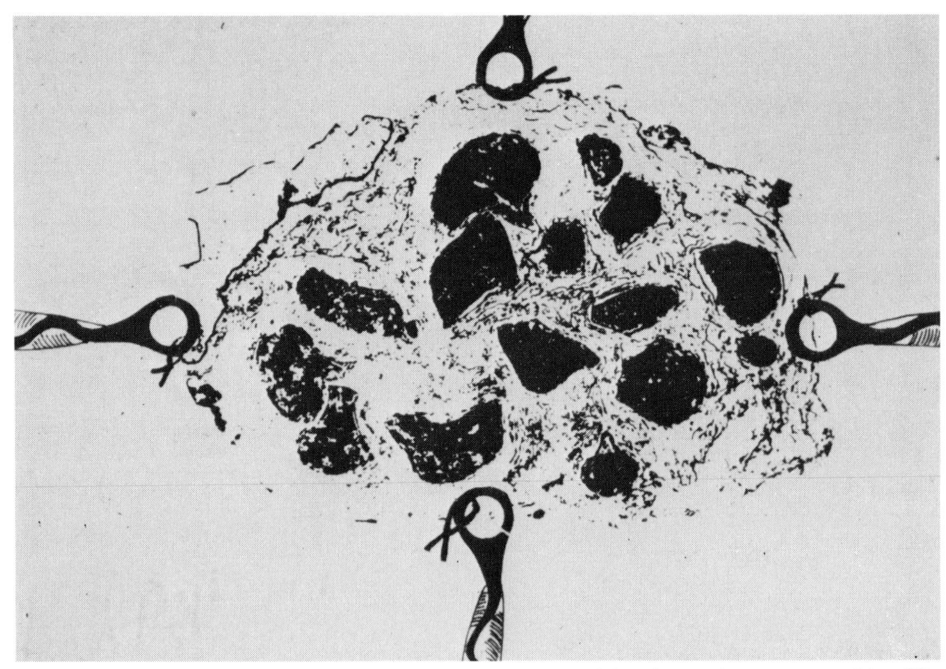

Figure 2 Four electrodes attached to the epineurium of one peripheral nerve in a square position. The electrodes are sutured to the epineurium with 8/O prolene.

maximum were separated. Using an electrostimulating device, branches of the infragluteal nerves innervating the various parts of the gluteus maximus muscle responsible for dorsal flexion of the hip joint, were located. Four electrodes were fixed to the epineureum of the nerve branches with 8/O sutures. Fixation of the four electrodes to the epineureum of each nerve in a square position, the electrodes being 1 cm apart from each other, was performed. Additionally, the electrodes were fixed to the muscle fascia with a distance of 5 cm from the electrode tips. Another vertical incision in the groin was made and the femoral nerve was located. Again the nerve branches to the quadriceps femoris, which produce the maximum extension of the knee joint, were selected. To the epineureum of the nerve four electrodes were connected with 8/O sutures and all electrodes from the groin and gluteal region were pulled through subcutaneously into the lower abdominal wall of each side. The implants were placed into a subcutaneous pocket of the abdominal wall. The postoperative period of all patients was uneventful and the function of the stimulation system was tested on the 10th postoperative day. The patients were discharged from the hospital on the 15th postoperative day.

TRAINING AND REHABILITATION

The process of muscle training and rehabilitation, summarized in Table 2, demonstrates that only muscle training was performed during the first four months after the operation. The patients were admitted to a rehabilitation center five

Table 1 Process of life time and events in 6 patients, 4 patients equipped with leg pacemakers

PATIENT INITIALS	LEFT/RIGHT	LEG PACEMAKER NO. OF CHANNELS	OCTOBER	NOVEMBER	DECEMBER	JANUARY	FEBRUARY	MARCH	APRIL	MAY	JUNE	JULY	AUGUST	SEPTEMBER	OCTOBER	NOVEMBER	DECEMBER	JANUARY	FEBRUARY	MARCH	APRIL	MAY	JUNE	JULY	AUGUST	SEPTEMBER	OCTOBER	NOVEMBER	DECEMBER	JANUARY	FEBRUARY	MARCH	APRIL	MAY	JUNE	JULY	AUGUST	SEPTEMBER	OCTOBER	NOVEMBER	DECEMBER	JANUARY
			1982			1983												1984												1985												
B. W.	L	8	1													2 5																								6		
	R	8	1													5																							8 →			
J. R.	L	8	1												2 5																						8 →					
	R	8	1																			4 5															8 →					
M. S.	L	8														1																						8 →				
	R	8														1																						8 →				
R. K.	L	8														1					3 6																	7				
	R	8														1																						7				

Solid line ——— first implant
broken line --- second (replaced) implant
Events:
1. Implantation
2. Failure (implant)
3. Infection
4. Dislocation (electrodes)
5. Implant change
6. Explantation
7. Patient died
8. Reduced function

Table 2 An example of the coordination of 16 combination numbers with the real switching mode of the 4 electrodes

Combination	0	1	2	3	4	5	6	7	8	9	A	B	C	D	E	F
Electrode 1			+	−	+	−			+	−	−	+			−	+
Electrode 2	−	+			−	+			−	+	−	+	−	+		
Electrode 3			−	+			−	+	−	+			+	−	+	−
Electrode 4	+	−					+	−			+	−	+	−	−	+

months after the operation to learn to stand up and to walk. The patients were able to walk with 2 crutches 9 months after the operation—first in a swinging-through gait, later in a four-point gait.

The isometric muscle force of the musculus quadriceps was measured in sitting position with 90° bent knee joints using a strain gauged load cell. A special training was offered to the patients in order to strengthen the atrophied and weak muscles. The most important point was to stimulate the musculus quadriceps and gluteus maximus on both sides using the round-about stimulating method. This training was performed four times a day. Depending on the muscle force, weights (1–3 kg) were applied to both ankles during the exercises.

Increasing muscle force first succeeded in attempts to stand in parallel bars. At this stage the patients started coordination programmes including standing in hyperlordosis as well as balance exercises in a standing position with as little use of the arms as possible. For home training they were supplied with special stance parallel bars. To stand up and walk was the main goal of the 4 weeks training in the rehabilitation center. This could be achieved only with the help of intensive individual care by two physiotherapists as well as the infrastructure of a rehabilitation center.

RESULTS

Muscle force increased in all patients to 3–4 times during the postoperative training period. All four patients were able to leave their wheelchair and sit back into it without any help one year after the operation (Figure 3). Three patients

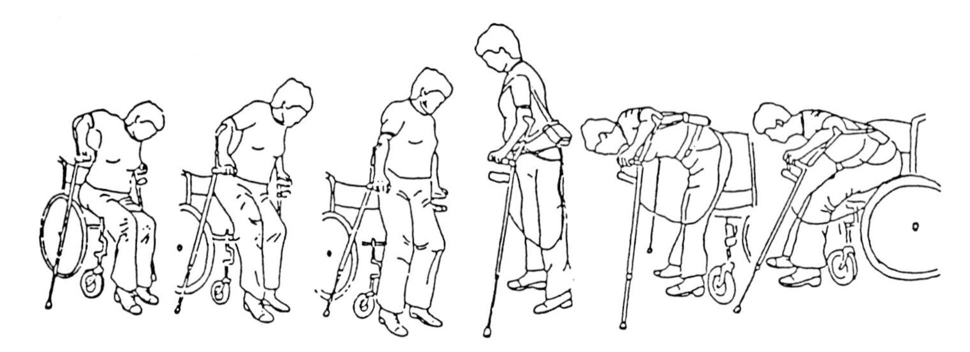

Figure 3 Standing up (left) and sitting down (right) with an implanted 16-channel FNS device.

learned to walk for about 100 meters in a swinging-through gait technique. Performing the swinging-through gait technique, all 4 muscle groups are stimulated simultaneously; with the help of lower arm crutches this kind of walking is much simpler to learn than the 4-point gait technique, which only two patients were able to perform adequately. The trunk muscles play an important role in walking with the 4-point gait (Figure 4). The trunk on one side is stabilized by the erector spinae muscle while hip- and knee-joint are stabilized by electrostimulation. By contracting the latissimus dorsi muscle on the other side, the opposite hip can be elevated; thus the non-stimulated leg will be moved forwards. This leg now will be stabilized by electrostimulation, the erector spinae muscles on this side stabilizes the trunk and the opposite hip then will be elevated by the opposite latissimus dorsi muscle. In that way the opposite leg is brought forward and while moving the lower arm crutches alternatively forward a 4-point gait can be realized successfully. Two of our four patients were able to walk with this technique for about 100 meters. Several complications had to be dealt with in the postoperative course. Dislocation of one electrode occurred in one case. The electrode could be readapted to the nerve without major problems. Moreover, in one patient the implant of one electrode had to be removed on one side because of infection. Because of failure

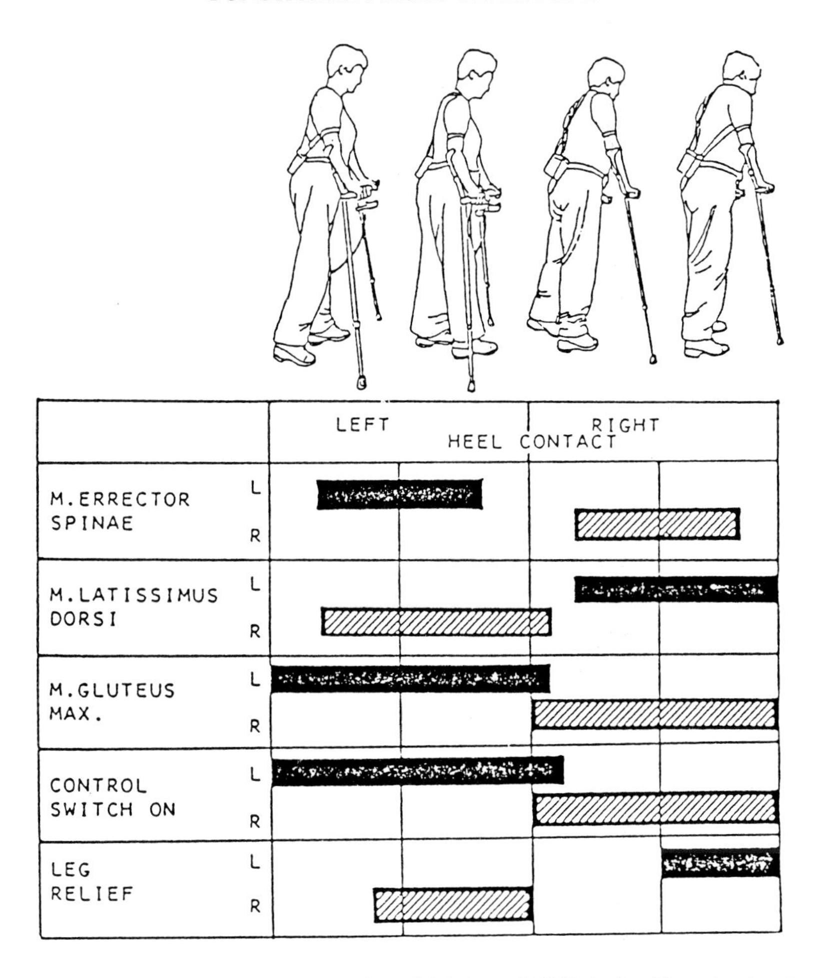

Figure 4 The mode of walking with an implanted 16-channel FNS device. Note that because of the absence of flexion of knee and hip joint the patient has to use his m.latissimus dorsi and errector spinae for leg relief.

and perforation through the skin the implants had to be exchanged. Revision was necessary. One patient died two years after the implantation because of a cerebral bleeding. So it was possible to examine the stimulated nerves and muscles histologically. In table 2 an example of the coordination of 16 different electrode combinations with the switching mode of the 4 electrodes is shown. In Figure 5 the amplitude gain of patient R.K. two years after the implantation is demonstrated. The amplitude gain is high in 6 electrode combinations, but the absolute value is low (6 Nm). The stimulation current of the efficient combination is within the normal range of 1–1.5 mA. Six combinations are low or without any effect. The amplitude range was between 0–6.4 mA, pulse duration 0.5 msec and frequency 30 Hz; number of pulses: 4; stimulated muscles: right quadriceps femoris; angle of the knee-joint 90°. Pulse duration gain of one patient is shown in Figure 6. Main

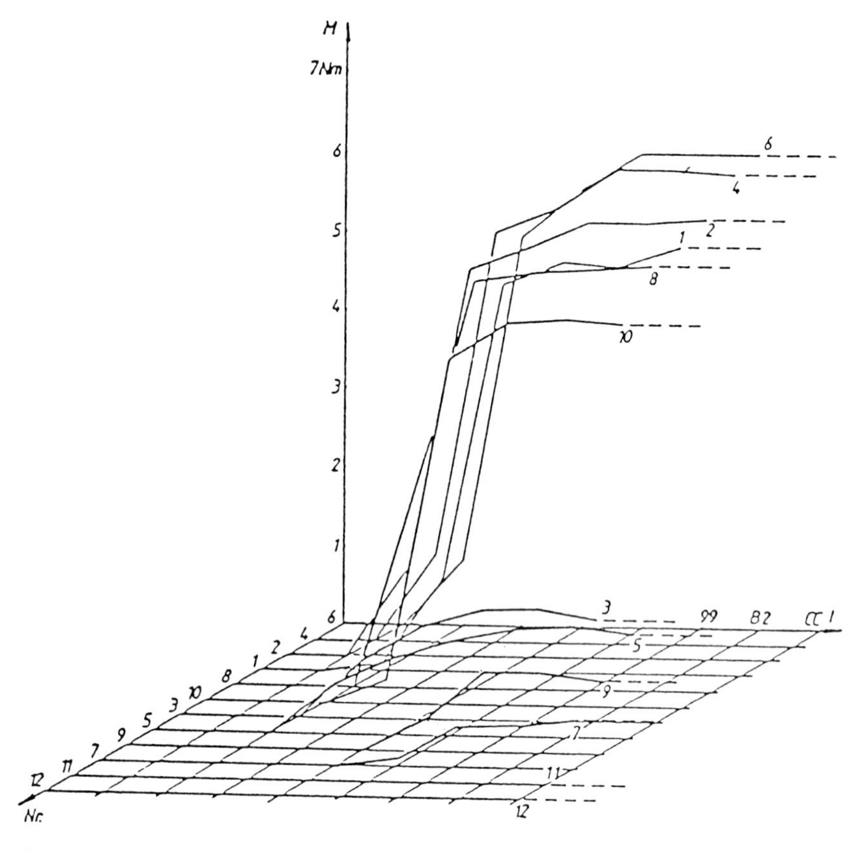

Figure 5 Amplitude gain, Date 14.5.1985, patient R.K., implant 5, amlitude range (OO..CC), (0..6.4 mA), pulse duration 0.5 ms, frequency 30 Hz, number of pulses 4, stimulated muscle: right quadriceps femoris, angle of knee joint 90°.

force response is in the range of 0.1–0.2 msec. Another gain characteristic is shown in Figure 7. The other parameters are on a fixed medium level in this frequency gain graph. The combination 7 f.e. with the pulse duration of 0.5 msec and a constant current of 2.8 mA induces the high torque of 65–70 Nm at an angle of 30° and 60°. In a 90° position only half of this value could be reached.

The changes in functional response of the left femoris nerve due to different electrode combinations during 1 month are demonstrated in Table 3. This table shows that there still exists a change in functional response to stimulation with the same electrode combinations two years after the implantation. Up to now there is no explanation why one electrode combination functions with weak muscle contraction on one day and with a strong muscle contraction on the next. In Tables 4 and 5 we can recognize that electrode combinations used for single muscle contraction change in their function when used in combination between the femoral and infragluteal nerve electrodes. Measures of the function of the electrodes change when compared between walking and resting state on both sides. This phenomenon is difficult to explain.

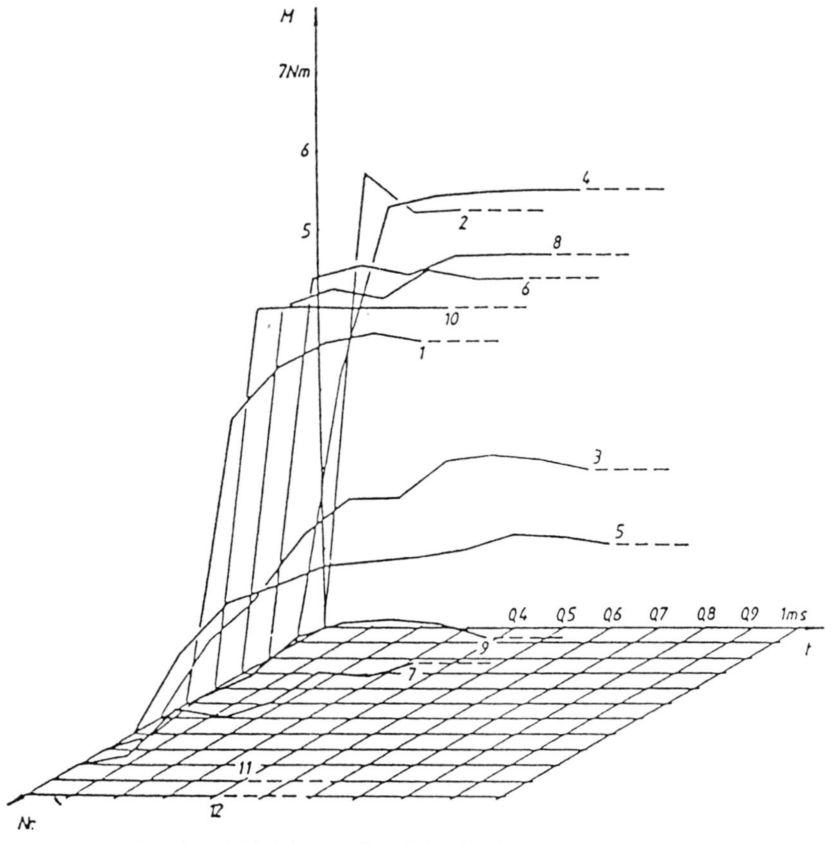

Figure 6 Impulse gain, date 14.5.1985, patient R.K., implant 5, impulse duration range 0–1 ms, amplitude 8 mA, frequency 30 Hz, number of pulses 4, stimulated muscle: right quadriceps femoris, angle of knee joint 90°.

By observing the stimulation pulses a failure of the implant can be checked. Skin electrodes are positioned close to the electrode-nerve connection region for this purpose. No spontaneous dislocation of electrode was observed during the observation time of 11 patient years. Although all 4 electrodes were disconnected from one nerve in one case due to trauma, no lesion of the nerve was evident. The electrodes were replaced easily and the function of the nerve restored. During the entire observation time no nerve-muscle group lost its functional response to electrostimulation demonstrating that the used electrode design usually is atraumatic to the nerve structures.

Since one patient died $1\frac{1}{2}$ years after the implantation, there was a chance to investigate the stimulated nerve structures and the nerve-electrode connections histologically. No changes in the nerve sectors could be found—neither proximal, at the side of the electrode combination, nor distal of this point in the investigated nerves. Connective tissue had developed between the tips of the electrodes and the nerve structures. The stimulated muscle structures did not differ histologically or histochemically from the non-stimulated contralateral ones.

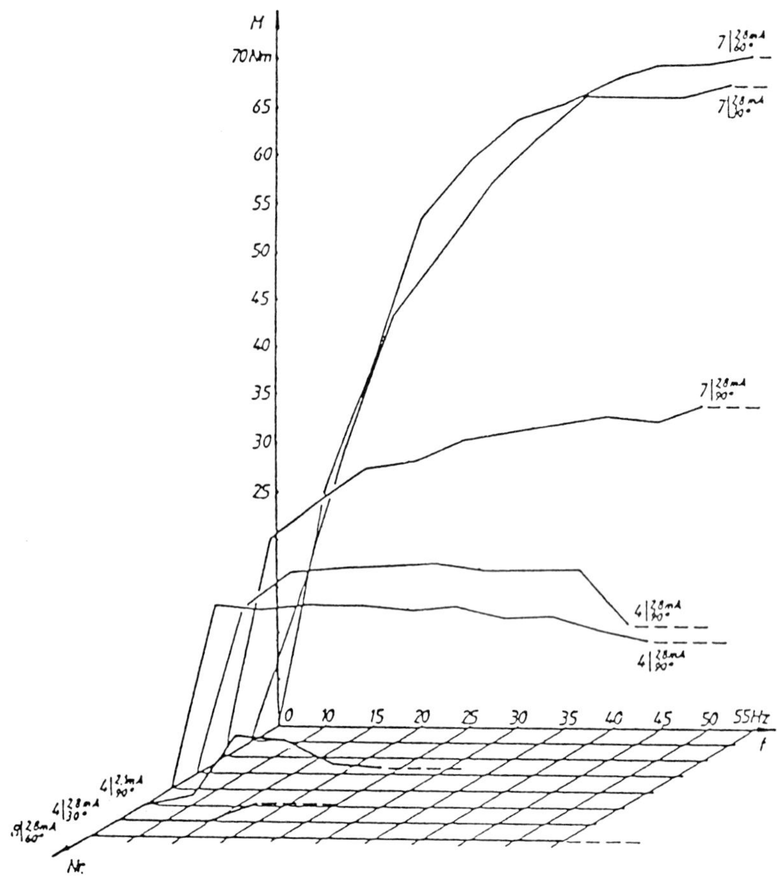

Figure 7 Frequency gain, date 31.5.1985, patient M.S., implant 6, frequency range 0–55 Hz, pulse duration 0.5 ms, duration of measuring cycle 500 ms, stimulated muscle left quadriceps femoris, angle of knee joint 30°/60°/90°.

DISCUSSION

Our experience with FES in four paraplegic patients during a time interval of 6 years demonstrates the ability of our stimulation devices in long standing application. Using the round-about electrode stimulation method with 4 electrodes attached to the epineureum of one nerve we can achieve smooth and strong motion of the selected muscle groups without limiting reactions in the nerve-electrode connection site. Thus, electrically induced fatigue can be reduced and a maximum of 3 mA peak current for sufficient stimulation is necessary because of the close connections of the electrodes with the nerve fibers. In contrast to direct muscle stimulation the electrodes can be connected with the nerve fibers in regions with relatively poor motion. The electrodes show high stability against low DC currents and there is little connective tissue reaction around the electrodes due to the special design. The electrodes can be disconnected without any nerve damage by pulling

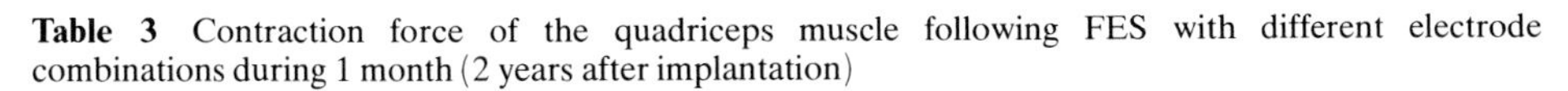

Table 3 Contraction force of the quadriceps muscle following FES with different electrode combinations during 1 month (2 years after implantation)

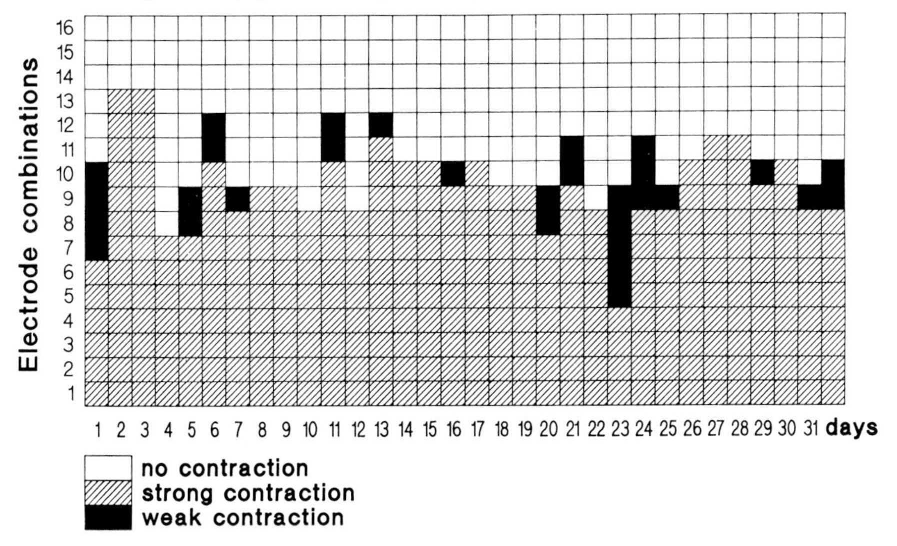

Table 4 Interaction of different electrode combinations during FES of the right gluteus maximus and quadriceps muscles in the rest and walking state

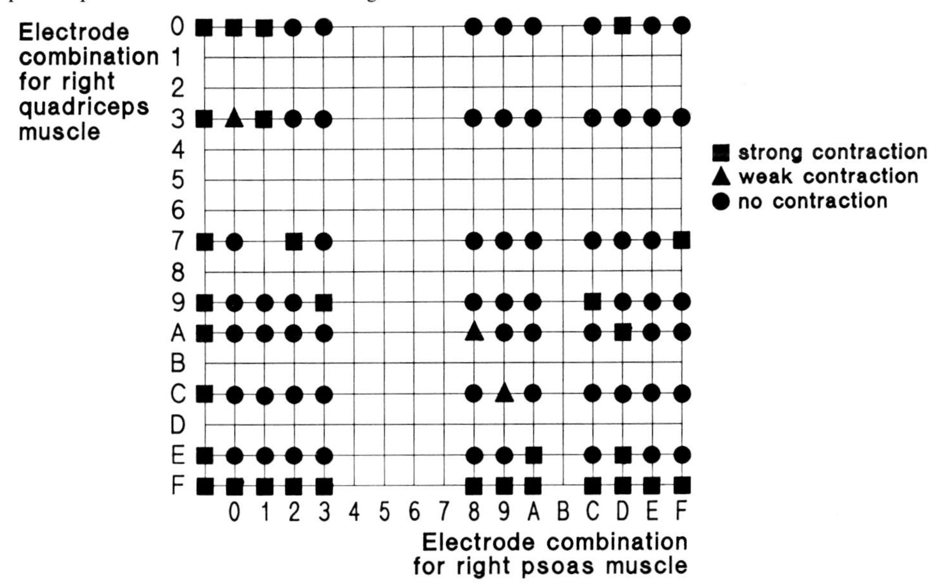

out. They can be replaced without severe difficulties using microsurgical techniques. The presented postoperative complications (infection, electrode perforation and failure of implant function) may be reduced in future by means of our new electrode- and implant-developments. To substitute acceptable locomotion in paraplegic patients special features and possibilities have to be developed in future. A 4-point gait with the aid of lower arm crutches could only be realised in one patient.

Table 5 Interaction of different electrode combinations during FES of the left gluteus maximus and quadriceps muscles in the rest and walking state

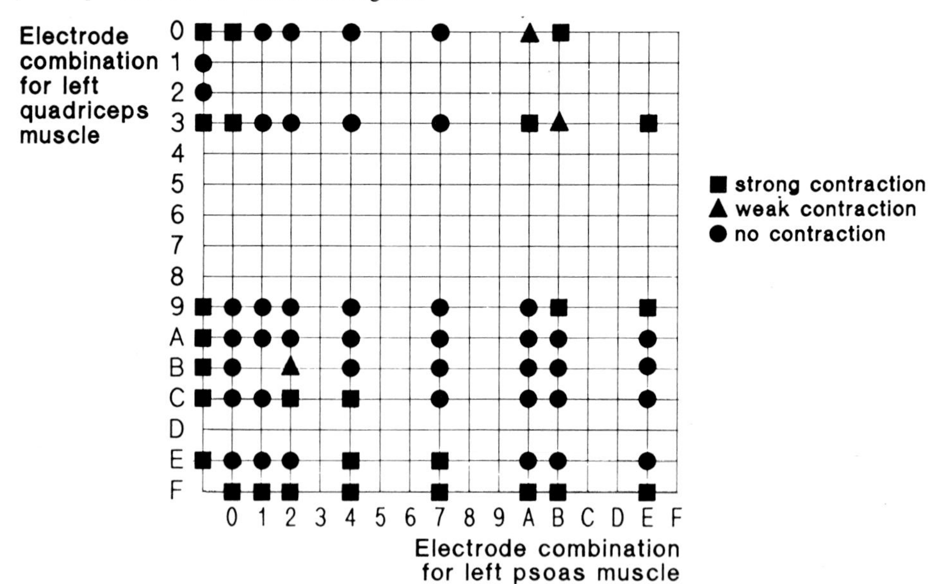

To be accepted by most of the paraplegic patients the everyday practical use of the substituted locomotion needs to be far more comfortable for handicapped persons to be preferred to the wheel chair. Although the technical and medical possibilities of our applied systems are advanced, several new developments have to take place. In order to decrease the patient's fatigue more muscles have to be stimulated and more electrodes have to be added to our system. Nerve groups responsible for bending of the knee- and hip-joint have to be acquired. Another technology to develop better adaptable electrodes is already realized. More than 3 years of stability of the electrode leads and the implants will have to be achieved. Since the clinical results have demonstrated no definite technical or biological limitation for the application of multi-chnnel FES systems for locomotion of paraplegic patients, we are encouraged to continue research in this field.

REFERENCES

1. A. Kralj, T. Bajd, R. Turk, M. Stefancic, H. Benko and J. Sega, Experiences with FES enabled standing in complete paraplegic patients, *Yugoslav Committee for Electronics and Automation: Advances in External Control of Human Extremities* (1981).
2. R. J. Jaeger, Electrical stimulation for standing in paraplegia, *IEEE Frontiers of Engineering and Computing in Health Care*, 655 (1983).
3. J. S. Petrofsky and C. A. Phillips, Computer controlled walking in the paralyzed individual, *J. Neuro. and Ortho. Sug.*, **4,** 153–164 (1983).
4. E. B. Marsolais, R. Kobetic, H. Chizeck and J. Mansour, Improved synthetic walking in the paraplegic patient using implanted electrodes, in *Proceedings 2nd International Conference on Rehabilitation Engineering*, Otttawa, Canada, (1984), p. 439.
5. G. S. Brindley, C. E. Polkey and D. N. Rushton, Electrical splinting of the knee in paraplegia, *Paraplegia*, **16,** 428–435 (1978/79).
6. J. Holle, M. Frey, H. Gruber, H. Kern, H. Stöhr and H. Thoma, Functional Electrostimulation of Paraplegics, *Orthopaedics*, **7,** 1145–1155 (1984).

7. H. Thoma, M. Frey, H. Gruber, J. Holle, H. Kern, E. Reiner, G. Schwanda and H. Stöhr, First implantation of a 16-channel electric stimulation device in human, *ASAIO Transactions*, **XXIX,** 301–306 (1983).
8. J. Holle, E. Moritz, H. Thoma and A. Lischka, Die Karussellstimulation. Eine neue Methode zur elektro-phrenischen Langzeitbeatmung, *Wien. Klin. Wschr.*, **86/1,** 23–27 (1974).
9. H. Stöhr, W. Hammerschmid, I. Hochmair and H. Thoma, A new system for fatigue-free nerve stimulation, *ESAO Proceedings*, **5,** Zürich, 112–115 (1978).
10. H. J. Gerner and P. Kluger, High quadriplegia accompanied by neurogenic respiratory insufficiency. Potentialities, limits and outlook, in *Proceeding 2nd Vienna International Workshop on Functional Electrostimulation* (1986), 157–161.
11. H. Kern, T. Bochdansky, M. Frey, J. Holle, G. Schwanda, H. Stöhr and H. Thoma, Funktionelle Elektrostimulation querschnittgelähmter Patienten. 1 Jahr praktische Anwendung, Erfolge der Patienten und Erfahrungen, *z.f. Orthopädie und ihre Grenzgebiete*, **123,** 1–12 (1985).
12. W. Mayr, M. Frey, J. Holle, H. Kerne, G. Schwanda, H. Stöhr and H. Thoma, Fahrrad als Trainingsgerät für Paraplegiker mit implantierten Neuralstimulatoren, *Schriftreihe Österr. Computer Gesellschaft*, **24,** 320–324 (1984).
13. D. Rosenkranz, G. Fenzl, W. Lack, W. Lipp, U. Losert and H. Thoma, Histologic examinations of nerves after application of low DC by electrodes, in *Proceedings 1st Vienna International Workshop on Functional Electrostimulation*, (1983), 1.9.
14. G. Schwanda, W. Mayr, H. Stöhr and H. Thoma, Concepts in quality assurance for implantable neuromuscular stimulators, in *Proceedings ISAO*, Chicago (1985).
15. H. Stöhr, M. Frey, U. Loseert, D. Rosenkranz and H. Thoma, Reaction of the electrode nerve connection on lowest DC currents and high peak voltages, in *Proceedings of ISAO Paris* (1981), pp. 264–266.
16. H. Stöhr, M. Frey, J. Holle, H. Kern, W. Mayr, G. Schwanda and H. Thoma, Six years progress in technological developments of implantable nerve stimulators, in *Proceedings 2nd Vienna Muscle Symposium*, (1985), pp. 185–189.
17. H. Thoma, M. Frey, J. Holle, U. Losert, D. Rosenkranz and H. Stöhr, Experiments on the electrode nerve connection, in *Proceedings 7th International Symposium on External Control of Human Extremities*, Dubrovnik (1981), pp. 121–135.

NCTEPO
LIBRARY
131 ST JAMES RD, GLASGOW G4 0LS